# Social odours in mammals

# Social odours
# in mammals

## Volume 2

Edited by

## RICHARD E. BROWN

*Department of Psychology, Dalhousie University,*
*Nova Scotia, Canada*

and

## DAVID W. MACDONALD

*Department of Zoology, University of Oxford, England*

CLARENDON PRESS · OXFORD
1985

Oxford University Press, Walton Street, Oxford OX2 6DP
London New York Toronto
Delhi Bombay Calcutta Madras Karachi
Kuala Lumpur Singapore Hong Kong Tokyo
Nairobi Dar es Salaam Cape Town
Melbourne Auckland
and associated companies in
Beirut Berlin Ibadan Mexico City Nicosia

Oxford is a trade mark of Oxford University Press

Published in the United States
by Oxford University Press, New York

© The several contributors listed on p. xi

**British Library Cataloguing in Publication Data**

Social odours in mammals.
Vol. 2.
  1. Mammals—Behaviour          2. Smell
  I. Brown, Richard E.          II. Macdonald, David W.
  599.01'826          QL 739.3
  ISBN 0-19-857617-X

**Library of Congress Cataloging in Publication Data**

Main entry under title:

Social odours in mammals.

  Bibliography: v. 2., p.
  Includes index.
  1. Mammals—Behavior.    2. Odours.    3. Smell.
4. Social behavior in animals.    I. Brown, Richard E.
II. Macdonald, David W. (David Whyte)
QL734.3.S6    1985    599'.059    83-17290
ISBN 0-19-857617-X (v. 2)

Typeset by Joshua Associates, Oxford
Printed in Great Britain by
St Edmundsbury Press, Bury St Edmunds, Suffolk.

# Preface

Generations of countryfolk, naturalists, and huntsmen have wondered at the acute sense of smell which opens a dimension to the lives of many other mammals, but from which we are barred. Some have doubtless cursed, rather than marvelled, at the sense which drives the once reliable dog besotted for a far-off bitch, or which thwarts the stalker at the whim of a breeze. It does not take a training in science to recognize our nose-blindness, or to hunger for a knowledge of what Kenneth Grahame, in *The wind in the willows*, called the mysterious fairy calls from out of the void, for which 'we have only the word smell, to include the whole range of delicate thrills which murmur in the nose of the animal night and day, summoning, warning, inciting, repelling'.

The realization that our own species is dismally ill-equipped to notice, far less to decipher, this profusion of mammalian odours, seems not to have deterred, but rather to have spurred biologists to excesses of curiosity. The result has been, over the past decade or so, a rapid expansion of information on these scents which function in mammalian communication, and which we shall call social odours. However, information on these social odours, their nature, sources, deployment, and functions, is widely scattered in the scientific literature and has become, in two senses, unusually fragmented. The first reason for this fragmentation is that the strength and the weakness of studies of olfactory communication is that they can be launched from diverse perspectives, involving the expertise of biologist or chemist, ecologist or histologist, field naturalist, endocrinologist or experimental psychologist. There is thus the opportunity for the fruitful combination of interdisciplinary forces in the common quest, or for their artifical division along the barriers of jargon. One practical consequence of these varied inputs is that it is not always obvious where to look in order to read about either recent advances or established background.

The second sense in which information on social odours is fragmented is a combined result of the tendency for biologists to study intensively only a few species, and the fact that members of thousands of mammalian species are each endowed with several sources of social odours. In consequence, a survey of current publications concerning social odours often gives an imbalanced impression of the prevalence of olfactory communication amongst mammals, disguising the fact that existing studies have tackled only the tip of the iceberg. Furthermore, the same focus on only a few species fails to direct the reader to the enormous variety of relevant snippets of information on a much wider array of mammals.

The growth of theoretically exciting ideas concerning social odours, the scattered publications and expertise, the feeling that a systematic review of literature on the whole Class could provide a realistic perspective and a useful reference work all prompted us to produce this two volume book. The structure of the book very much reflects our aims, as elaborated in our introductory chapter (pp. 1-18). Briefly, the two volumes are organized taxonomically, each chapter dealing with one or more orders. We hope it has an encyclopaedic quality which will be enduring useful, in that a vast store of literature is summarized in the text and, particular, in the summary tables. The possibilities for reference use are enhanced by comprehensive species and author indexes. At the same time, there is also an emphasis on reviewing modern ideas and even, in some chapters, otherwise unpublished data.

One of the most exciting fields of modern biology is the study of social behaviour; and the role of social odours in mammalian societies could hardly be more forcefully emphasized than in the chapters which follow. Indeed, we would argue that odours are a fundamental part of mammalian sociality and adaptation. While the secretions of the mammary glands may legitimately have pride of place, the scent-producing glands of the skin do not rank far behind in the list of qualities that are essentially mammalian.

We gratefully acknowledge all those who have helped in the production of these two volumes. In particular, at Dalhousie and Oxford, we were aided in proof-reading by Geoffrey Carr, Stefen Natynczuk, Jenny Ryan and Lilyan White, and in indexing by Gillian Kerby and Colin Pringle, together with Peter King and Dierdre Harvey of the Kellogg Health Sciences Library at Dalhousie.

*Halifax and*                                                      R.E.B.
*Oxford*                                                           D.W.M.
August, 1984

# Contents

VOLUME 1

# Plates

# Contributors

Diana J. Bell,
School of Biological Sciences,
University of East Anglia,
Norwich, UK.

Richard E. Brown,
Department of Psychology,
Dalhousie University,
Halifax, Nova Scotia, Canada.

Richard L. Doty,
Smell and Taste Research Center,
University of Pennsylvania Hospital,
Philadelphia, USA.

Gisela Epple,
Monell Chemical Senses Center,
Philadelphia, USA.

L. M. Gosling,
Coypu Research Laboratory,
Ministry of Agriculture, Fisheries and Food,
Norwich, UK.

David W. Macdonald,
Department of Zoology,
University of Oxford, UK.

Patricia D. Moehlman,
School of Forestry and Environmental Studies,
Yale University,
New Haven, Conn., USA.

Amos B. Smith III
Department of Chemistry,
University of Pennsylvania,
Philadelphia, USA.

# 12 The rabbits and hares: order Lagomorpha

DIANA J. BELL

Despite the considerable volume of research into the scent-signalling system of the European rabbit, *Oryctolagus cuniculus* L., our knowledge of those scents produced by other lagomorph species remains fairly sparse. The few species for which information about the distribution and morphology of scent glands has appeared are listed in Table 12.1. It should be emphasized that the absence of a reference does not necessarily imply that the gland is absent from that species.

Details about the location and gross structure of the chin, anal, inguinal, and Harderian glands found on *O. cuniculus* are presented below together with points of particular interest about the histochemistry and odour of the glands and/or their secretions. Further information about the morphology, histochemistry/enzymology and control of these glands may be found in those

Table 12.1. Occurrence of scent glands in lagomorphs

| | Submandibular cutaneous (chin) | Cheek | Inguinal | Anal | Harderian |
|---|---|---|---|---|---|
| **Family Leporidae** | | | | | |
| *Oryctolagus cuniculus* | ♂ > ♀ <br> 6, 10 | | ♂ > ♀ <br> 8 | ♂ > ♀ <br> 7, 10 | ♂ > ♀ <br> 9 |
| *Sylvilagus floridanus* | + <br> 5, 11 | | ♂ = ♀ <br> 1 | + <br> 10 | |
| *Sylvilagus aquaticus* | + <br> 11 | | | + <br> 10 | |
| *Lepus europaeus* | ♂ > ♀ <br> 6 | | ♀ > ♂ <br> 8, 12 | ♂ = ♀ <br> 7, 10 | ♂ > ♀ <br> 9 |
| *Peolagus marjorita* | | | ♂ , ♀ <br> 4 | | |
| **Family Ochotonidae** | | | | | |
| *Ochotona princeps* | | ♂ , ♀ <br> 2, 3, 13, 14 | | | |

+ present, sex unspecified; ♂ , ♀ in ♂ and ♀; ♂ > ♀ larger in ♂ than ♀; ♀ > ♂ larger in ♀ than ♂; ♂ = ♀ same size in ♂ and ♀.

References cited
1. Elchlepp (1952).
2. Harvey and Rosenberg (1960).
3. Kawamichi (1976).
4. Kingdon (1974).
5. Parris (1969).
6. Mykytowycz (1965).
7. Mykytowycz (1966a).
8. Mykytowycz (1966b).
9. Mykytowycz (1966c).
10. Mykytowycz (1968).
11. Mykytowycz (1970).
12. Mykytowycz (1974).
13. Sharp (1973).
14. Svendsen (1979).

references listed in Table 12.2. Similar details are provided for the cheek glands found on North American pikas, *Ochotona princeps.*

Anal gland secretion is released via the faeces, while chin and cheek gland secretions are deposited by characteristic chin- and cheek-rubbing actions (Plate 12.1). Secretions from glands around the eye, chin, and cheek may also be rubbed over the feet and body surface during self-grooming (Schneider 1977). No specialized behaviour has been associated with the release of inguinal secretion which collects in pouches of skin and may also form a yellowish encrustation around the genitalia and upper thigh region. The functional significance of these lagomorph scent glands receives detailed consideration elsewhere (Bell 1980;

Table 12. 2. Sources of further information about the structure, chemistry, and control of scent glands in *O. cuniculus*

| Gland | Gross structure | Ultrastructure/histochemistry/ /histoenzymology | Control |
|---|---|---|---|
| Inguinal | 1, 2, 3, 4, 5, 6 | 2, 4, 7, 8, 9, 10, 11, 47 | 3, 4, 7, 10, 12 |
| Anal | 3, 6, 13, 14, 15 | 10, 11, 16, 17, 18, 47 | 3, 10, 13, 19 |
| Chin | 7, 20, 21, 22 | 7, 20, 21, 22, 23, 24, 25, 26, 27, 47 | 7, 12, 19, 22, 47, 48 |
| Harderian | 28, 29, 30, 31, 32 | 28, 30, 31, 32, 33, 34, 35, 36, 37, 38, 39, 40, 41, 42, 43, 44, 45, 46 | |

References cited
1. Holtz and Foote (1978).
2. Montagna (1950).
3. Coujard (1947).
4. Mykytowycz (1966*b*).
5. Andersson (1980).
6. Schaffer (1940).
7. Heath (1974).
8. Kühnel and Wrobel (1969).
9. Marois and Salesses (1968).
10. Salesses and Marois (1969).
11. Goodrich and Mykytowycz (1972).
12. Wales and Ebling (1971).
13. Mykytowycz (1966*a*).
14. Hesterman and Mykytowycz (1968).
15. Mykytowycz and Dudzinski (1966).
16. Marois and Salesses (1967).
17. Galli and Parmeggiani (1968).
18. Goodrich *et al.* (1978).
19. Mykytowycz (1968).
20. Kurosumi *et al.* (1961).
21. Lyne *et al.* (1964).
22. Mykytowycz (1965).
23. Mineda (1973).
24. Oikawa *et al.* (1976).
25. Bondi *et al.* (1978).
26. Imai *et al.* (1978).
27. Materazzi *et al.* (1976).
28. Björkman *et al.* (1960).
29. Mykytowycz (1966*c*).
30. Kühnel (1971*a*).
31. Jost *et al.* (1974).
32. Rock (1977).
33. Kühnel (1966).
34. Kühnel (1971*b*).
35. Kühnel and Wrobel (1966*a*).
36. Kühnel and Wrobel (1966*b*).
37. Kühnel and Wrobel (1968).
38. Schneir and Hayes (1951).
39. Radominska-Pyrek *et al.* (1979).
40. Blank, Kasama, and Snyder (1972).
41. Kasama, Rainey, and Snyder (1973).
42. Rock and Snyder (1975*a*).
43. Rock and Snyder (1975*b*).
44. Rock, Fitzgerald, Rainey, and Snyder (1976).
45. Rock, Fitzgerald, and Snyder (1977).
46. Rock, Fitzgerald, and Snyder (1978).
47. Strauss and Ebling (1970).
48. Mykytowycz (1962).

Mykytowycz 1974), but for purposes of completeness a summarized discussion is also included here.

The second part of this chapter presents further evidence for a signalling function for the other source of lagomorph scent—urine—and considers the possible role that these odours may play during the process of mate selection in these species.

## Anal glands

These are saddle-shaped organs consisting of two masses of brown lobes lying on either side of the rectum. Each mass is surrounded by a layer of connective tissue, muscle, and fat and the two are joined dorsally by connective tissue. The secretion passes into the anus via separate pores found at the junction of the rectal mucus membrane and skin epidermis (Mykytowycz 1966a) and is malodorous to man (Hesterman and Mykytowycz 1968).

### Histochemistry

From those references listed in Table 12.2 I have selected a number of findings which may be of particular interest to those concerned with identifying the active constituents of these glandular secretions (see also Mykytowycz 1979).

Marois and Salesses (1967) for example report an abundance of acid (mostly phospholipids) and neutral lipids in the anal-gland secretion of *O. cuniculus* while Salesses and Marois (1969) found numerous neutral polysaccharides in anal glands from females. Gonadectomy caused a reduction in cell size and the appearance of mucopolysaccharides in the anal glands of male rabbits and a contrasting increase in cell size in females. Oestrogen therapy increased the number of PAS granules in glands of castrated males and females; androgen treatment resulted in the disappearance of all PAS granules. Lipid concentrations in the gland also decreased in castrates following hormone treatment.

Hesterman, Goodrich, and Mykytowycz (1976) found that both acid and neutral fractions of male anal glands elicited strong avoidance responses in unrestrained juveniles and hand-held nestlings.

Goodrich, Hesterman, Murray, Mykytowycz, Stanley, and Sugowdz (1978) have since developed an elegant system whereby the heart rate of restrained rabbits can be monitored during exposure to gas chromatographic effluents from headspace volatiles of homogenized anal glands. Using this technique, a number of saturated and mono-unsaturated normal and methyl-branched aldehydes in the $C_9$–$C_{12}$ range have been found to elicit a significant deceleration in heart rate.

## Harderian gland

This is a large ovoid, tubuloalveolar organ situated deep within the orbit of the eye. It consists of two lobes, a large pink lobe and a smaller white one.

*Histochemistry*

Lipids are the major components of both the pink and white lobes (see Table 12.3). The pink portion contains a preponderance of alkyl glycerolipids and is a rich source of those enzymes involved in their biosynthesis (Blank, Kasama, and Snyder 1972; Kasama, Rainey, and Snyder 1973; Rock and Snyder 1974, 1975*a,b*; Rock, Fitzgerald, and Snyder 1977, 1978). A unique subclass of alkyldiacylglycerols has been isolated from this pink lobe, where the 3-position of the glycerol moiety is occupied exclusively by isovaleric acid (Blank *et al.* 1972; Rock and Snyder 1975*a*). It is interesting to note that isovaleric acid has been isolated as the active constituent in the subauricular gland of the North American pronghorn (Müller-Schwarze, Müller-Schwarze, Singer, and Silverstein 1974); it may be worth testing the rabbit's behavioural response to this substance.

The major lipid component of the white lobe is a mixture of 2-(*O*-acyl) hydroxy-fatty acid esters (Rock, Fitzgerald, Rainey, and Snyder 1976). These have not been found in the pink lobe nor reported in the Harderian tissue of other species.

Table 12. 3. Glycogen, protein, and lipid content of Harderian glands from adult female rabbits. (From Jost *et al.* (1974))

|  | Glycogen | Soluble protein | Lipid (mg/g of wet weight) |
|---|---|---|---|
| White lobe | 0.3 | 33.6 | 167 |
| Pink lobe | 0.3 | 24.8 | 415 |

## Submandibular cutaneous (chin) glands

In the European rabbit these glands consist of two deeply embedded lateral groups of lobes and a third centrally located group just below the skin. In males the lobes in each group may be coalesced into a single mass and the fur beneath the chin matted with a yellow encrustation of the secretion. These glands are less well developed in the European hare, where they occur as numerous small discrete lobes each with its own secretory duct (Mykytowycz 1965).

Two main types of secretory cells are present—see Lyne, Molyneux, Mykytowycz, and Parakkal (1964); Oikawa, Higuchi, Mori, and Okano (1976) and Imai, Mineda, Oikawa, and Okano (1978). Lyne *et al.* (1964) describe the chin glands as 'modified sweat glands of the apocrine type . . . because they develop from hair follicles and because some secretory cells appear to lose part of their protoplasmic substance during secretion', while Kurosumi, Yamagishi, and Sekine (1961) describe the stages in apocrine cell secretion in their EM study of these glands. Heath (1974), on the other hand, argues that the chin glands, like the brown inguinal glands, 'release' their secretions by exocytosis rather than by any apocrine mechanism and indeed rejects previous evidence for the latter as 'artefactual'.

*Odour*

Freshly expressed secretion is odourless to man but develops a highly persistent 'cat-urine' odour over time.

*Histochemistry*

See under Inguinal glands below.


**Inguinal glands**

These are paired organs lying on either side of the penis/vulva. Each organ consists of two distinct glands: a white, pyriform sebaceous gland (approx. 3–5 mm in diameter, 10–15 mm in length in adult male *O. cuniculus*) and a brown, elongated, tubular sweat-producing gland. The mixture of sebum and sweat produced by these glands empties into pouches of hairless skin at the base of the genital opening. It may also be referred to as the *perineal* (not inguinal) gland, and is occasionally mistaken for the neighbouring preputial gland (Holtz and Foote 1978).

*Histochemistry*

Heath (1974) found sex differences in the cytoplasmic structure of the brown inguinal glands, but no differences in glands from oestrous, pregnant, and pseudo-pregnant does. The results of thin-layer chromatographical analysis of lipid extracts from homogenized inguinal and chin glands indicate the presence of hydrocarbons, non-glyceryl esters, free fatty acids, and cholesterol in both glands (Goodrich and Mykytowycz 1972). The other major components of extracts from the chin glands and sebaceous portions of the inguinals were shown by chromatography to be triglycerides, diglycerides, and monoglycerides, although their exact structure has yet to be determined.


**Cheek glands (pika)**

Two main ducts discharge into separate pores in two slightly protruding, highly pigmented papillae situated approximately 5 mm ventrocaudal to the base of the ear. The pelage hairs overlying this area are broken, short (5mm length), and denser than those found elsewhere on the body. The gland is composed of a tight coil of epithelial tubes (secretory portion) embedded in a cell mass ('B gland'). The 'B gland' is larger in oestrous females and sexually mature males than in pregnant females or immature bucks (Harvey and Rosenberg 1960).

Svendsen (1979) observed cheek-rubbing by pikas of all age and sex classes in a free-living population and found no seasonal variation in the frequency of cheek-marking behaviour. The clumped distribution of scent-marks within male territories reflected their greater concentration in those areas frequented by overlapping females rather than the distribution around territory perimeters reported by Barash (1973).

*Odour*

Harvey and Rosenberg (1960) note that this secretion is odourless to humans, but Kawamichi (personal communication) commented that pika cheek-gland secretion emits a urinous odour identical to that of the sample of rabbit chin-gland secretion I gave him to smell.

### Nasal and mammary gland scents

The naked S-shaped pads of richly innervated, vascularized skin found between the nostrils of many lagomorphs have been nominated as additional 'scenting' glands (Degerbøl 1949). Recent histological studies by Westlin, Jeppsson, and Meurling (1982), however, found no specialized sensory or glandular structures in the prominent nasal pads found on the European hare (see also Flux 1970). Experiments reported by Schley (1977) suggest that the teat-searching movements of nestling rabbits may be orientated by olfactory signals from the mammary glands of lactating does. Those youngsters rendered anosmic failed to find teats to suckle.

### Scent glands and social behaviour

In a further useful comparative note on the overall odour emanating from these creatures, J. Flux (personal communication) reports that the two species of African hares he studied in Kenya, *L. capensis* and *L. crawshayi*, differed quite noticeably in smell, the former having 'the same acrid/musky smell as *L. europaeus*' while the latter 'has a curiously sweet smell'.

The references cited for *O. cuniculus* in Tables 12.1 and 12.2 describe how the size and secretory activity of all four glands varies in relation to sex, age, social, and reproductive status. The sexual dimorphism, for example, appears at the onset of sexual maturity and the highest levels of glandular activity coincide with the breeding season. The results of endocrinological manipulations provide further evidence for the androgenic control of these glands, with castration during early life inhibiting their growth in males but stimulating their development in females, while testosterone treatment results in an increase in the size and secretory activity of the glands in both males and females (Coujard 1947); Heath (1974); Mykytowycz, 1965, 1968; Strauss and Ebling 1970; Wales and Ebling 1971).

The scent glands are similarly largest and most active in those socially dominant individuals also responsible for most of the scent-marking activities recorded within the breeding group (see Mykytowycz and Dudzinski 1966; Mykytowycz 1970; Black-Cleworth and Verberne 1975; Bell 1977, 1980, 1981).

Using 'organoleptic panel' techniques, Hesterman and Mykytowycz (1968) found that humans detect differences in the intensity of the odour of anal-gland secretions taken from rabbits of varying age, sex, and reproductive status. Odour intensity was positively correlated with age when samples from a series of 4 28-week-old rabbits were assessed, while anal-gland odours from castrated bucks were rated as less intense than those of intact males, but smellier than those taken from adult does. Again, these glandular secretions were reported to be most pungent during the breeding season.

Further tests suggested that the rabbit can vary the amount of anal-gland secretion applied to voided pellets, since those pellets released by dominant males and females exposed to conspecific odours (so-called 'marking pellets') were judged to be more odorous than the 'non-marking' pellets deposited in the absence of such stimulation.

An analysis of the 'site' and 'behavioural context' of those defecation episodes observed in a free-living study population of *O. cuniculus* on an offshore Welsh island, showed that 'latrines', 'pawscrapes', burrow entrances, ant hills, and other prominent features were popular targets for pellet deposition (Bell 1977, 1980). In general, our own studies of latrine, or 'dung pile', usage (ref. cited) support the earlier work of Mykytowycz and Gambale (1969) and Mykytowycz and Hesterman (1970) in Australia, i.e. a typical latrine visit lasts only 2–3 minutes and includes initial investigation and subsequent deposition of own scent as faeces, urine, and/or chin-gland secretions. Mykytowycz and Gambale (1969) found that males visited latrines both more frequently and for longer than females, while visits by dominant bucks tended to be shorter than those of subordinates. The total time devoted to scent-marking was nevertheless similar for males of differing status.

The frequency of defecation at latrine sites also varies in relation to season and climate, i.e. peak numbers of latrines coincide with peak population density and there is a significant increase in the incidence of pellet deposition after a period of rainfall. Not all latrines are visited on a daily basis and certain dung piles receive more faecal pellets than others.

There appear to be very few references to site-specific defecation by other lagomorph species (see Table 12.4). The European hare certainly appears to maintain dung piles but Flux (personal communication) suggests that, as with *L. capensis*, *L. crawshayi*, and *L. timidus*, these should be regarded as pellets that have simply accumulated at frequently used stopping places rather than special defecation sites. It might be worth pointing out that one cannot infer a less important signalling function for secretion-coated faeces simply because they are distributed in a 'diffuse' rather than site-specific fashion. Indeed, any future investigations of the spatial distribution of excreta in other lagomorph

Table 12.4. Object/site marking with faecal pellets by lagomorph species

| Species | Observations | Reference |
|---|---|---|
| *L. timidus* | defecates over strange objects | Flux (1970) |
| *L. timidus* *L. europaeus* *L. capensis* *L. crawshayi* | latrine and pellet accumulations at regular 'stopping places' along runs | Flux (personal communication) |
| *Ochotona princeps* | typically defecate at one place in individual home range | Eisenberg and Kleiman (1978) |
| *Oryctolagus cuniculus* | defecate over prominent environmental features, e.g. ant hills, also at latrines, warren entrances, pawscrapes and along trails | Bell (1977, 1980) |
| | 'latrine' maintenance studied by direct observation and indirect survey | Mykytowycz and Gambale (1969) |
| | response to latrines by penned wild rabbits | Mykytowycz and Hesterman (1970) |

NB These may also be important sites or targets for the deposition of other *invisible* glandular secretions or urine messages.

species must clearly include direct observations of the excretory act (accompanying displays, social context, etc.) and subsequent responses to those excreta. Bell (1977, 1980), for example, found that dominant males frequently visited a latrine after chasing a conspecific.

Similar 'target' and 'behavioural context' analysis of 80 chin-marking bouts recorded in the above-mentioned island study population also found that rabbits deposited chin-gland secretion over almost any surface in their home-range area, including conspecifics, with burrow entrances and 'pawscrapes' emerging as focal centres for the activity (Bell 1977, 1980). These acts were similarly performed in a wide range of solitary and social behaviour contexts and again most were attributed to the dominant males (Fig. 12.1). Marsden and Holler (1964) report that chinning was performed solely by dominant males in the swamp rabbit *S. aquaticus*, 'in the presence of a female, or another male' or during solitary territorial patrols apparently analogous to those observed in *O. cuniculus* (Bell 1977) (see Plate 12.1).

Recent studies of the effects of domestication on the behaviour of *O. cuniculus* (Kraft 1979) suggest that males of domesticated strains chin-mark more frequently than wild rabbits, with wild rabbits chin-marking only during the breeding season (cf. Mykytowycz 1965; Bell 1977).

Histological comparisons of Harderian glands found them to be more strongly developed in wild than domesticated rabbits (Mykytowycz 1966c). The secre-

Fig. 12.1. (a) A 'mutual identification' function for chin-gland scents is supported by the stereotyped nose-to-chin posture seen during pair encounters. (See Bell 1980. (Inset) The active chin glands of the dominant males produce a yellowish encrustation on the chin fur. (b) Chin-rubbing protruding vegetation. (c) Defecation at a 'latrine' site. (d) 'Pawscraping' and subsequent scent-marking with faeces, urine, and chin-gland secretion, represents a form of threat display when performed by dominant males. (e) 'Urine-squirt' scent-marking at warren entrance.

tions from these Harder's glands are thought to serve a protective function in the rabbit, contributing to tear production by the lachrymal system (Jost, Kuehnel, and Schimassek 1974) and facilitating movement of the third eyelid (Rock 1977). The possibility that these secretions have evolved secondary signalling significance in the rabbit as they have in the gerbil (Thiessen 1977) remains to be explored. Caged individuals certainly devote appreciable time to mutual sniffing/licking around the eyes but Mykytowycz (1966c) notes that rabbits do not react to dissected lachrymal or Harderian glands in the same way they do to anal, inguinal, or submandibular tissue.

Mykytowycz (1968) suggests that these scent glands 'may convey information about age, sex, reproductive stage, and group membership. They may also warn of danger and they definitely serve to define the "territory" of the rabbit.' He suggests that the two externally secreting anal and chin glands function primarily for territorial purposes, while inguinal-gland secretions may be important for individual identification and as sexual attractants (see also Wales and Ebling 1971).

Mykytowycz (1973, 1975) has emphasized the 'confidence-enhancing' effects of being surrounded by familiar odours rather than any 'repellent' effects that territorial marking may exert on intruding conspecifics. When 'strange' rabbits were paired together in a 'neutral' pen in the presence of own/sexual partner's chin-gland secretion, those individuals surrounded by familiar odours dominated the behaviour of their 'partner'.

Mykytowycz derives further support for a territorial function for the anal and chin glands from a comparative study of these glands in lagomorph species with quite different social systems—namely the European hare, the cottontail, and the swamp rabbit. He found that despite a larger body size both the chin and anal glands of the solitary, far-ranging European hare are far smaller than those of highly gregarious, territorial *O. cuniculus*, being approximately one-quarter and one-tenth their size respectively. Also, unlike the rabbit, the anal gland of the hare shows no sexual dimorphism. Similarly, the chin and anal glands of the territorial swamp rabbit are larger than those of the non-territorial cottontail (Mykytowycz 1968; Marsden and Holler 1964).

One must be cautious when attempting to determine the functional significance of scent glands from such cross-species comparisons when current knowledge of both social systems and scent glands of these species is so fragmentary (see Tables 12.1 and 12.5). Apparent differences in aspects of social organization of these various species could merely reflect the differing conditions under which they were studied, e.g. whether they were enclosed or free-ranging populations, their density, the timing of behavioural observations, the provision/non-provision of foodstuffs (see e.g. Marsden and Holler 1964). It is also important to note that although chin-marking may be absent in the cottontail, dominant males were seen (*ibid.*) to rub targets '*with the corner of the eye*' during interactions with females and subordinate males.

Table 12.5 Social organization in different lagomorph species

| Species | Territorial* | Breeding group formation | 'Dominance' hierarchies in breeding groups ♂ | ♀ | References |
|---|---|---|---|---|---|
| *Oryctolagus cuniculus* (European wild rabbit) E & FR | √ | √ ♂ 1–5 ♂ 1–6 | √ | √ | 1, 2, 3, 4, 5, 6, 7, 8, 9, 10, 11, 12, 13 |
| *Sylvilagus floridanus* (cottontail rabbit) E | ✗ but ♀ defend nest sites | 'partial breeding group formation' | √ | √ | 14 |
| FR | ✗ | — | — | — | 15, 16, also 17 |
| *Sylvilagus aquaticus* (swamp rabbit) E | √ | √ ♂ 1–2 ♀ 2–3 | √ | ✗ | 14 |
| *Lepus timidus* (mountain hare) FR | ✗ | ✗ | | | 17, 18 |
| *Lepus americanus macfarlani* (snowshoe hare) FR | ✗ | — | — | — | 19, 20 |
| *Lepus californicus* (black-tailed jack rabbit) FR | ✗ | ✗ | | | 21 |
| *Ochontona princeps* (North American pika) FR | √ ♂ and ♀ defend individual territories | ✗ | | | 22, 23 |

E = enclosed study population; FR = free-ranging study population; √ = yes; ✗ = no;
— = no data given.
*Territorial = 'shows active defence of an area'.

References cited
1. Southern (1948).
2. Mykytowycz (1958).
3. Mykytowycz (1959).
4. Mykytowycz (1960).
5. Myers and Poole (1959).
6. Myers and Poole (1961).
7. Lockley (1961).
8. Myers and Schneider (1964).
9. Dunsmore (1974).
10. Bell (1977).
11. Parer (1977).
12. Gibb *et al* (1978).
13. Daly (1979).
14. Marsden and Holler (1975).
15. Jurewicz *et al.* (1981).
16. Dixon *et al.* (1981).
17. Flux (1970).
18. Lemnell and Lindlöf (1981).
19. Grange (1932).
20. Adams (1959).
21. Lechleitner (1958).
22. Broadbrooks (1965).
23. Barash (1973).

The sexual attractant function ascribed to the inguinals is primarily based on their larger size in *L. europaeus* compared to *O. cuniculus* together with their converse sexual dimorphism in the hare where they are far larger in the female (Mykytowycz 1966*b*, 1968). Male hares certainly appear to track oestrous females by scent (Flux, personal communication) but the exact source of the attractant has yet to be investigated.

The idea that inguinal secretions also convey individual identity derives mainly from experiments testing mothers' response to progeny smeared with inguinal or anal-gland secretions from strange females (Mykytowycz and Goodrich 1974, p. 129). Under natural conditions, adult does will severely attack and kill kittens born to other females and, in these tests, smearing with inguinal secretions appeared to mask the identity of progeny which were similarly attacked and severely bitten by their mothers. It is, however, worth noting that those kittens anointed with anal-gland secretions were also attacked significantly more than untreated controls. Analogous tests with either urine or chin-gland secretions have not been reported but there is mounting evidence to suggest that these scents may also signal individual identity (see, e.g. Bell 1977; Schalken 1976; Goodrich and Mykytowycz 1972; Mykytowycz 1975; Mykytowycz, Hesterman, Gambale, and Dudzinski 1976). Indeed, I have argued elsewhere (Bell 1980) that at present there is little evidence to support the notion of discrete signalling functions for these several glands in *O. cuniculus*, and discussed the possible advantages of an alternative system whereby basically similar information is broadcast by scent products with differing physicochemical properties.

**Urine**

The release of urine by lagomorphs may often be both target-directed and accompanied by stereotyped movements or postures (Table 12.6). In *O. cuniculus* three forms of urine-depositing behaviour may be distinguished on the basis of the volume evacuated and the target at which it is directed; namely normal squat urination, urine-spraying, and urine-squirting (Fig. 12.1). The last two are regarded as forms of scent-marking behaviour where small volumes of urine are ejected at conspecifics and prominent inanimate environmental features respectively (see Bell 1980). Most of the target-spraying observed in a free-living population was performed by dominant males engaged in aggressive or courtship interactions. In males, the seminal vesicle may function as a storage organ for urine evacuated in this way (see Flux 1970); Bern and Krichesky 1943).

Rabbits also show a characteristic sequence of prolonged olfactory/gustatory investigation and subsequent deposition of own scents over the urine voided by other conspecifics (Bell 1976, 1977; Black-Cleworth and Verberne 1975).

The results of a series of two-choice discrimination tests with domesticated *O. cuniculus* suggest that urine may carry information about the individual identity and sexual and social status of the depositing animal. When presented

Table 12.6. Target-spraying/squirting of urine by lagomorphs

| Species | Target | Reference |
|---------|--------|-----------|
| *Oryctolagus cuniculus* | conspecific adults, juveniles, kittens; warren entrances; anthills; paw-scrapes; breeding 'stops' | Southern (1948) Lockley (1961) Mykytowycz and Rowley (1958) Bell (1976) Heath (1972) |
| *Sylvilagus floridanus* *Sylvilagus aquaticus* | of bucks by ♀'s and does by ♂'s during sexual chases | Marsden and Holler (1964) Marsden and Conaway (1963) |
| *Lepus californicus* | of bucks by ♀'s and does by ♂'s during sexual chases | Lechleitner (1958) |
| *Lepus europaeus* | conspecifics | Boback (1954) |
| *Lepus timidus scoticus* | conspecifics, e.g. during sexual chases | Flux (1970) |
| *Ochotona princeps* | haypiles | Kilham (1958) |

with a choice between 'own' urine and that of another male, adult bucks spend longer sniffing and scent-marking over the 'strange' male sample. Similarly, overall responsiveness to conspecific urine measured in terms of sniffing and chin-marking times, declines with increasing age of urine samples (Bell 1976, 1977; Schalken 1976).

Sexually mature males also spent longer investigating urine from low- (subordinate) rather than high- (dominant) frequency chin-marking males, while females showed greater attraction to the latter (Bell 1981).

Experiments 1 and 2 below tested the hypothesis that males would also 'prefer' oestrous female urine to that of intact, adult males in a similar two-choice context. Male response to oestrous versus pregnant female urine was compared in Experiment 3, and the preferences of sexually immature and mature females confronted with adult male and oestrous female urine were examined in Experiments 4 and 5.

### Experimental procedure

Two strains of domesticated rabbits were used: Dutch-belted (Experiments 1 and 2) and Flemish giants (Experiments 3-5). The Dutch were bred in the Zoology Department at University College, Cardiff, weaned at 4-5 weeks, caged individually at 12 weeks and maintained on an *ad-lib*. diet of SG1 Pilsbury pellets and water and a 12/12 h light/dark cycle. The Flemish giants were bred in the Physiology Department at University College, Cardiff, and reared according to a similar regime.

Urine samples were collected in bowls left overnight beneath home-cages and equal volumes pooled from two donors for each experiment.

The simultaneous two-choice discrimination tests were conducted in either an open arena ($0.8 \times 1.8 \times 0.6$ m)—Experiments 1 and 2, or the subject's home-cage—Experiments 3–5.

*Experiment 1*

The six male subjects and four urine donors ($2\male$, $2\female$) were approximately 59 weeks old. Samples (1.5 ml) of male and female urine were dried on to glass plates ($10.0 \times 15.0$ cm) in hot-air currents and positioned on opposite sides of the test arena. Each subject received four 5-min choice tests at intervals of 24, 72, and 24 h respectively. Behavioural responses were recorded using at 12-channel event recorder. 'Preference' for either odorant was determined from time spent:

(a)  with snout $\leqslant 1$ cm from odorant = total contact (TC);

(b)  TC − time 'scent-marking' odorant = total sniff/licking;

(c)  chin-marking odorant;

(d)  mean visit duration, i.e. TC/No. of separate visits during trial;

(e)  duration of first visit, i.e. time from first approach to $\leqslant 1$ cm—to body movement away.

The positioning of samples was fully counterbalanced and the arena thoroughly cleaned and urine samples renewed between subjects. Results are presented in Table 12.7.

Table 12.7. Male response to air-dried samples of male and female urine. (Test duration = 1200 s)

| Preference parameter | Mean times (in s) ± SE | | $N$ | $p^*$ | PR† |
|---|---|---|---|---|---|
| | $\female$ | $\male$ | | | |
| Total contact | 147.38 ± 38.80 | 110.79 ± 25.37 | 6 | 0.047 | 5/1 |
| Total Sn/Li | 118.67 ± 21.62 | 91.25 ± 19.33 | 6 | 0.047 | 5/1 |
| Chin-marking | 28.71 ± 8.51 | 19.54 ± 9.11 | 6 | NS | 5/1 |
| Average contact | 7.41 ± 0.59 | 3.03 ± 0.54 | 6 | 0.02 | 5/1 |
| First visit | 138.38 ± 26.16 | 99.04 ± 26.03 | 6 | 0.015 | 6/0 |

*Probabilities derived from Wilcoxon's matched-pairs signed-ranks test (one-tailed).

†PR = preference ratio = $\dfrac{\text{number of subjects spending longer at } \female \text{ urine}}{\text{number of subjects spending longer at } \male \text{ urine}}$.

*Experiment 2*

Four 25-week-old rabbits (2♂: 2♀) provided the urine which was presented on glass plates in fresh liquid form to 11 males (30-weeks (6) or 100-weeks (5)-old). Experimental procedure was otherwise identical to that of Experiment 1 (Table 12.8).

Table 12.8. Male response to fresh liquid samples of male and female urine.
  (Test duration = 1200 s)

| Preference parameter | Mean times (in s) ± SE | | N | p* | PR[†] |
|---|---|---|---|---|---|
| | ♀ | ♂ | | | |
| Total contact | 148.14 ± 28.52 | 123.22 ± 30.64 | 11 | 0.04 | 9/2 |
| Total Sn/Li | 125.30 ± 22.86 | 107.86 ± 26.37 | 11 | NS | 9/2 |
| Chin-marking | 22.86 ± 6.73 | 15.36 ± 4.78 | 8 | 0.019 | 7/1 |
| Average contact | 5.53 ± 0.82 | 4.44 ± 0.99 | 11 | 0.041 | 9/2 |

  *, † As Table 12.7.

*Experiment 3*

Urine donors were pairs of oestrous (5–6-month-old) and 28-day pregnant (20–28 month-old) females. Ten sexually intact (5–17-month-old) males each received two urine-choice tests on consecutive days where 2-ml samples were presented in two glass specimen tubes (5.0 × 2.5 cm) suspended 20 cm apart beneath cage floors. Time spent sniffing within 2 cm of each odorant was recorded by accumulative stopwatch (Table 12.9). Side of presentation was reversed in consecutive trials.

*Experiments 4 and 5*

Ten sexually mature (5–21-month) and 12 immature (9–10-week-old) females received 2 × 3-min and 1 × 3-min exposure(s) respectively to oestrous female and intact male urine in similar home-cage texts (Table 12.9).

*Discussion*

These results suggest that rabbits of both sexes are able to discriminate male from female urine. Sexually intact males spent longer investigating fresh urine from oestrous females than that from adult male conspecifics. Female urine was still preferred when urine samples had been air-dried. The simultaneous choice-paradigm has a number of disadvantages (see Doty (1975) and Stevens (1975) for further critiques). Firstly, odorant (A) may be 'investigated' longer than another (B) because the animal finds that:

  (i)   (A) is positively attractive and/or (B) is positively aversive,
  (ii)  (B) carries less information than (A), or

Table 12.9. Response to conspecific urine by male and female rabbits

| Experiment | Subjects | | $N$ | Test duration (s) | Mean investigation time ± SD (s) | | $p^*$ | PR[†] |
|---|---|---|---|---|---|---|---|---|
| | | | | | A | B | | |
| 3 | mature ♂ | 10 | | 360 | pregnant ♀ 31.70 ± 20.64 | oestrous ♀ 62.95 ± 43.97 | < 0.02 | 8/2 |
| 4 | mature ♀ | 10 | | 360 | mature ♂ 28.45 ± 16.40 | oestrous ♀ 15.63 ± 13.89 | < 0.05 | 7/1 |
| 5 | immature ♀ | 12 | | 180 | 21.33 ± 16.27 | 7.25 ± 5.12 | < 0.01 | 10/0 |

*Probabilities derived from Wilcoxon's matched-pairs signed-ranks test (two tailed)

†PR = preference ratio = $\dfrac{\text{number of subjects spending longer at urine A}}{\text{number of subjects spending longer at urine B}}$

(iii) (B) carries important information which is more readily available than that carried by (A). For these reasons phrases such as 'preference for' and 'greater attraction towards' should be read in inverted commas throughout this paper. Secondly, the messages conveyed by (A) and (B) may be quite different but take similar time to 'process', i.e. an inability to distinguish (A) from (B) cannot be inferred from similar investigation scores.

The male 'preferences' expressed here, for example, might be explained in terms of aversion towards urinary components released by male conspecifics (Bell 1981) and/or the presence of attractants in the urine of oestrous females. The greater attraction shown towards oestrous over pregnant female urine in Experiment 3 may be interpreted as additional evidence for the latter. These problems may be resolved by further research measuring alternative behavioural responses in the 'receiver', e.g. degree of sexual arousal.

It is interesting to note that in European, cottontail, and swamp rabbits both oestrous and pseudopregnant females become 'strongly sexually attractive' to bucks at intervals of 6–7 days (Myers and Poole 1961, 1962; Marsden and Holler 1964; Heath 1972; Marsden and Conaway 1963). This coincides with a similar 6–7-day cyclicity in certain behavioural activities and physiological processes in non-pregnant does, e.g. blood oestrogen levels and vaginal pH (see May and Simpson 1975). It appears that odour stimuli often initiate this male arousal, as bucks 'become excited by' and follow the scent-trails left by oestrous does (Myers and Poole 1962; Bell 1977). Indeed, the sexual pursuit following an approach/scent-marking (urine-squirting, chin-marking, etc.) display by a female, might be regarded as 'active solicitation' of the buck by the female.

Immature and mature females spent longer investigating urine from intact males when presented with samples from each sex. Their further discrimination between urine from dominant and subordinate males, i.e. greater attraction towards the former, was mentioned earlier. These preferences may be attributed to the release of androgen-dependent male-aversive/female-attractant constituents by adult bucks similar to those reported in mice (Jones and Nowell 1973a,b, 1974) and rats (Krames, Carr, and Bergman 1969; Gawienowski, de Nicola, and Stacewicz-Sapuntzakis 1976).

Those urine donors of high-ranking social status also emerged as behaviourally dominant in paired round-robin tests (Bell 1981), and showed the highest rates of chin-marking and plasma testostorone levels and the largest testes to body-weight ratio.

In view of reported correlations between (a) testis weight and sperm number per ejaculation and (b) concentration of blood androgens and the level of gelatinous substance in semen important for sperm transport (Kihlström and Degerman 1963) one might speculate that those androgen-dependent constituents present in male urine may effectively provide a mechanism for female assessment of male fertility reproductive potential.

Indeed the 3–7-day sex cycle reported in sexually intact bucks may cause day-to-day fluctuations in male fertility. This involves a synchronized cyclicity in behavioural sex drive (Degerman and Kihlström 1961), number and motility of sperm released, and semen volume (Kihlström and Degerman 1963; Doggett 1956; Hafez 1970). Again, females may be able to detect these daily fluctuations in male fertility by urine inspection.

Males frequently spray urine over females during courtship while the overt 'solicitation' and also active rejection of particular bucks by both wild and domestic females may be regarded as male selection by females.

A genetic correlation between testis weight and ovulation rate has been found in both mice and sheep (Land 1973, 1974; Land, Carr, and Lee 1979). In mice, for example, selective breeding for high and low testis weight over five generations produced a mean divergence of 112 mg between high and low lines from an initial average testis weight of 191 mg while the ovulation rate of females of the lines changed in the same direction as that of selection (Islam, Hill, and Land 1976).

Selecting males with large testes, and hence high sperm production, could increase the inclusive fitness of a female by increasing the reproductive potential of both male and female offspring as well as by maximizing the probability that her own eggs will be fertilized.

## References

Adams, L. (1959). An analysis of a population of snowshoe hares in northwestern Montana. *Ecol. Monogr.* 29, 142–70.

Andersson, M. (1980). Studies of the reproductive biology of the wild rabbit *Oryctolagus cuniculus* in southern Sweden. Ph.D. thesis, University of Lund, Sweden.

Barash, D. P. (1973). Territorial and foraging behaviour of pika (*Ochotona princeps*) in Montana. *Am. Midl. Nat.* 89, 202–7.

Bell, D. J. (1976). Pheromonal communication in rabbits. Paper presented at Second ECRO Congress, Reading, September 1976. *Chemorecep. Abstr.* 5, No. 1.

— (1977). Aspects of the social behaviour of wild and domesticated rabbits, *Oryctolagus cuniculus* L. Unpublished Ph.D. thesis, University of Wales.

— (1980). Social olfaction in lagomorphs. *Symp. zool. Soc. Lond.* 45, 141–63.

— (1981). Chemical communication in the European rabbit: urine and social status. In *Proc. World Lagomorph Conf.*, Guelph, 1979 (ed. C. MacInnes and K. Myers). pp. 271–9.

Bern, H. L. and Krichesky, B. (1943). Anatomic and histologic studies of the sex accessories of the male rabbit. *Univ. Calif. Publ. Zool.* 47, 175–96.

Björkman, N., Nicander, L., and Schantz, B. (1960). On the histology and ultrastructure of the Harderian gland in rabbits. *Z. Zellforsch.* 52, 93–104.

Black-Cleworth, P. and Verberne, G. (1975). Scent-marking, dominance and flehmen behaviour in domestic rabbits in an artificial laboratory territory. *Chem. Senses Flavor* 1, 465–94.

Blank, M. L., Kasama, K., and Snyder, F. (1972). Isolation and identification of an alkyldiacyl glycerol containing isovaleric acid. *J. lipid Res.* **13**, 390–5.

Boback, A. (1954). Zur Frage des 'Harnspritzens' beim Feldhasen, *Lepus europaeus* Pallas, 1778. *Säugetierk. Mitt.* **2**, 78–9.

Bondi, A. M., Materazzi, G., Maraldi, N. M., and Baldoni, E. (1978). Fine structure of the rabbit submandibular gland during embryonic development. *Anat. Anz.* **144**, 179–94.

Broadbrooks, H. E. (1965). Ecology and distribution of the pika of Washington and Alaska. *Am. Midl. Nat.* **73**, 299–335.

Coujard, R. (1947). Etude des glandes odorantes du lapin et de leur influencement par hormones sexuelles. *Revue Can. Biol.* **6**, 3–14.

Daly, J. C. (1979). The ecological genetics of the rabbit in Australia. Unpublished Ph.D. thesis. Australian National University, Canberra.

Degerbøl, M. (1949). Gnavere (Rodentia). In *Vort lands dyreliv*, Vol. 1 (ed. F. W. Braestrup, G. Thorson, and E. Wesenberg-Lund) pp. 51–82. Glydendal, København.

Degerman, G. and Kihlström, J. E. (1961). Brief cyclic variations in some sexual functions of the male rabbit. *Acta physiol. scand.* **51**, 108.

Dixon, K. R., Chapman, J. A., Rongstad, O., and Wilde, D. (1981). A comparison of home range size in *Sylvilagus floridanus* and *S. bachmani*. In *Proc. World Lagomorph Conf.*, Guelph, 1979 (ed. C. MacInnes and K. Myers), pp. 541–8.

Doggett, V. C. (1956). Periodicity in the fecundity of male rabbits. *Am. J. Physiol.* **187**, 445–50.

Doty, R. L. (1975). Determination of odor preferences in rodents: a methodological review. In *Methods in olfactory research* (ed. D. G. Moulton, A. Turk, and J. Johnston). Academic Press, London.

Dunsmore, J. D. (1974). The rabbit in subalpine south-eastern Australia. I. Population structure and productivity. *Aust. wildl. Res.* **1**, 1–16.

Eisenberg, J. F. and Kleiman, D. G. (1978). Communication in lagomorphs and rodents. In *How animals communicate* (ed. T. A. Sebeok) pp. 634–54. Indiana University Press, Bloomington.

Elchlepp, J. G. (1952). The urogenital organs of the cottontail rabbit (*Sylvilagus floridanus*). *J. Morph.* **91**, 169–98.

Flux, J. E. C. (1970). Life history of the mountain hare (*Lepus timidus scoticus*) in north-east Scotland. *J. Zool., Lond.* **160**, 75–123.

Galli, G. and Parmeggiani, E. (1968). Histologic and histochemical studies on the anal glands of *Lepus cuniculus*. *Quad. Anat. Prat.* **24**, 317–26.

Gawienowski, A. M., de Nicola, D. B., and Stacewicz-Sapuntzakis, M. (1976). Androgen dependence of a marking pheromone in rat urine. *Horm. Behav.* **7**, 401–5.

Gibb, J. A., Ward, C. P., and Ward, G. D. (1978). Natural control of a population of rabbits, *Oryctolagus cuniculus* (L.), for ten years in the Kourarau enclosure. *DSIR Bull.* 223.

Goodrich, B. S., Hesterman, E. R., Murray, K. E., Mykytowycz, R., Stanley, G., and Sugowdz, G. (1978). Identification of behaviorally significant volatile compounds in the anal gland of the rabbit *Oryctolagus cuniculus* L. *J. chem. Ecol.* **4**, 581–94.

— and Mykytowycz, R. (1972). Individual and sex differences in the chemical composition of pheromone-like substances from the skin glands of the rabbit *Oryctolagus cuniculus* (L.). *J. Mammal.* **53**, 540–8.

Grange, W. B. (1932). Observations on the snowshoe hare, *Lepus americanus phaenotus* Allen. *J. Mammal.* **13**, 1–19.

Hafez, E. S. E. (1970). Rabbits. In *Reproduction and breeding techniques for laboratory animals* (ed. E. S. E. Hafez). Lea & Febiger, Philadelphia.

Harvey, E. B. and Rosenberg, L. E. (1960). An apocrine gland complex of the pika. *J. Mammal.* **41**, 213–19.

Heath, E. (1972). Sexual and related territorial behaviour in the laboratory rabbit (*Oryctolagus cuniculus*). *Lab. Anim. Sci.* **22**, 684–91.

— (1974). Cytologic observations on the secretory cells of the chin-submandibular gland and brown inguinal gland in the rabbit. *Cell Tissue Res.* **154**, 399–408.

Hesterman, E. R., Goodrich, B. S., and Mykytowycz, R. (1976). Behavioural and cardiac responses of the rabbit *Oryctolagus cuniculus* to chemical fractions from anal gland. *J. chem. Ecol.* **2**, 25–37.

— and Mykytowycz, R. (1968). Some observations on the odours of anal gland secretions from the rabbit *Oryctolagus cuniculus* (L.). *CSIRO Wildl. Res.* **13**, 71–81.

Holtz, W. and Foote, R. (1978). The anatomy of the reproductive system in male Dutch rabbits (*Oryctolagus cuniculus* L.) with special emphasis on the accessory sex glands. *J. Morph.* **158**, 1–20.

Imai, M., Mineda, T., Oikawa, M., and Okano, T. (1978). Some affinities and differences between the submandibular glands of the primates, tree-shrew (*Tupaia glis*), insectivora and some other kinds of animals. *Okajimas Folia anat. jap.* **55**, 209–28.

Islam, A. B. M. M., Hill, W. G., and Land, R. B. (1976). Ovulation rate of lines of mice selected for testis weight. *Genet. Res., Camb.* **27**, 23–32.

Jones, K. and Nowell, N. (1973a). Aversive and aggression-promoting properties of urine from dominant and subordinate male mice. *Anim. Learn. Behav.* **1**, 207–10.

— — (1973b). The coagulating glands as a source of aversive and aggression-inhibiting pheromone(s) in the albino mouse. *Physiol. Behav.* **11**, 455–62.

— — (1974). A comparison of the aversive and female attractant properties of urine from dominant and subordinate male mice. *Anim. Learn. Behav.* **2**, 141–4.

Jost, U., Kuehnel, W., and Schimassek, H. (1974). A morphological and biochemical analysis of the Harderian gland in the rabbit. *Cytobiologie* **8**, 440–56.

Jurewicz, R. l., Cary, J. R., and Rongstad, O. J. (1981). Spatial relationships of breeding female cottontail rabbits in south-western Wisconsin. In *Proc. World Lagomorph Conf.*, Guelph, 1979 (ed. C. MacInnes and K. Myers), pp. 295–309.

Kasama, K., Rainey, W. R., and Snyder, F. (1973). Chemical identification and enzymatic synthesis of a newly discovered lipid class hydrozyalkyl glycerols. *Archs Biochem. Biophys.* **154**, 648–58.

Kawamichi, T. (1976). Hay territory and dominance rank of pikas (*Ochotona princeps*). *J. Mammal.* **57**, 133–48.

Kihlström, J. E. and Degerman, G. (1963). Hormonally regulated cyclic variations in the sexual functions of the male rabbit. *Ark. Zool.* **15**, 357–8.

Kilham, L. (1958). Territorial behaviour in pikas. *J. Mammal.* **39**, 307.

Kingdon, J. (1974). *East African mammals*, Vol. 2B. *An atlas of evolution in Africa (hares and rodents)*. Academic Press, London.

Kraft, R. (1979). Vergleichende Verhaltensstudien an Wild- und Haus-kaninchen. II. Quantitative Beobachtungen zum Sozialverhalten. *Z. Tierzüchtg. Züchts-Biol.* **95**, 165–79.

Krames, L., Carr, W. J., and Bergman, B. (1969). A pheromone associated with social dominance among male rats. *Psychol. Sci.* **16**, 11–12.

Kühnel, W. (1966). Enzymhistochemische Untersuchungen an der Harderschen Drüse des Kaninchens. *Histochemie* **7**, 230–44.

— (1971a). Structure and cytochemistry of the Harderian gland in rabbits. *Z. Zellforsch. Mikrosk. Anat.* **119**, 384–404.

— (1971b). Fine structure and possibilities of metabolism of the Harderian gland in the rabbit. *Verh. anat. Ges.* **66**, 79–82.

— and Wrobel, K. H. (1966a). Die Histotopik von Aldolase und Alkohol-Dehydrogenase in der Harderschen Drüse des Kaninchens. *Histochemie* **7**, 245–50.

— — (1966b). Über die histochemisch faßbare Aktivität der β-D-Glucuronidase und der β-D-Galaktosidase in der Harderschen Drüse des Kaninchens. *Albrecht v. Graefes Arch. klin. exp. Ophthal.* **171**, 173–83.

— — (1968). Histochemische Studien und der Harderschen Drüse des Kaninchens. *Gegenbaurs morph. Jahrb.* **111**, 493–500.

— — (1969). Zür morphologie der braunen Inguinaldrüse des Kaninchens. *Z. Zellforsch.* **93**, 505–15.

Kurosumi, K., Yamagishi, M., and Sekine, M. (1961). Mitochondrial deformation and apocrine secretory mechanism in the rabbit submandibular organ as revealed by electron microscopy. *Z. Zellforsch. microsk. Anat.* **55**, 297–312.

Land, R. B. (1973). The expression of female, sex-limited characters in the male. *Nature, Lond.* **241**, 208–9.

— (1974). Physiological studies and genetic selection for sheep fertility. *Anim. Breed. Abstr.* **42**, 155–8.

— Carr, W. R., and Lee, G. J. (1979). A consideration of physiological criteria of reproductive merit in sheep. Paper presented at the Symposium on Selection Experiments in Animals, Harrogate, 1979.

Lechleitner, R. R. (1958). Certain aspects of behavior of the black-tailed jack rabbit. *Am. Midl. Nat.* **60**, 145–55.

Lemnell, P. A. and Lindlöf, B. (1981). Diurnal and seasonal activity pattern in the mountain hare. In *Proc. World Lagomorph Conf.*, Guelph, 1979, (ed. C. MacInnes and K. Myers) pp. 349–56.

Lockley, R. M. (1961). Social structure and stress in the rabbit warren. *J. anim. Ecol.* **30**, 385–423.

Lyne, A. G., Molyneux, G. S., Mykytowycz, R., and Parakkal, P. F. (1964). The development, structure and function of the submandibular cutaneous (chin) gland in the rabbit. *Aust. J. Zool.* **12**, 340–8.

Marois, M. and Salesses, A. (1967). Contribution to histochemical studies of the anal gland in rabbits. *C. r. Ass. Anat.* **138**, 850–3.

— — (1968). Contribution to the histochemical study of the rabbit's inguinal gland. *Ann. Histochim.* **13**, 97–104.

Marsden, H. and Conaway, C. H. (1963). Behaviour and the reproductive cycle in the cottontail. *J. wildl. Mngt* **27**, 161–70.

— and Holler, N. R. (1964). Social behaviour in confined populations of the cottontail and swamp rabbit. *Wildl. Monogr.* **13**, 1–39.

Materazzi, G., Bondi, A. M., Baldoni, E., and Maraldi, N. M. (1976). The ultrastructure of the rabbit submandibular gland. *Anat. Anz.* **140**, 154–61.

May, D. and Simpson, K. B. (1975). Reproduction in the rabbit. *Anim. Breed. Abstr.* **43**, 253–61.

Mineda, T. (1973). Electron microscopic investigations on the submandibular gland in the rabbit. *Acta anat. nippon.* **48**, 39.

Montagna, W. (1950). The brown inguinal glands of the rabbit. *Am. J. Anat.* **87**, 213–37.

Müller-Schwarze, D., Müller-Schwarze, C., Singer, A. G., and Silverstein, R. M. (1974). Mammalian pheromone: identification of active component in the subauricular scent of the male pronghorn. *Science, NY* **183**, 860–2.

Myers, K. and Poole, W. E. (1959). A study of the biology of the wild rabbit, *Oryctolagus cuniculus* (L.) in confined populations. I. The effects of density on home range and the formation of breeding groups. *CSIRO wildl. Res.* **4**, 14–26.

—— —— (1961). A study of the biology of the wild rabbit, *Oryctolagus cuniculus* (L) in confined populations. II. The effects of season and population increase on behaviour. *CSIRO Wildl. Res.* **6**, 1–41.

—— —— (1962). A study of the biology of the wild rabbit *Oryctolagus cuniculus* (L.) in confined populations. III. Reproduction. *Aust. J. Zool.* **10**, 225–67.

—— and Schneider, E. C. (1964). Observations on reproduction, mortality and behaviour in a small, free-living population of wild rabbits. *CSIRO Wildl. Res.* **9**, 138–43.

Mykytowycz, R. (1958). Social behaviour of an experimental colony of wild rabbits, *Oryctolagus cuniculus* (L.). I. Establishment of the colony. *CSIRO Wildl. Res.* **3**, 7–25.

—— (1959). Social behaviour of an experimental colony of wild rabbits, *Oryctolagus cuniculus* (L.). II. First breeding season. *CSIRO Wildl. Res.* **4**, 1–13.

—— (1960). Social behaviour of an experimental colony of wild rabbits, *Oryctolagus cuniculus* (L.). III. Second breeding season. *CSIRO Wildl. Res.* **5**, 1–20.

—— (1962). Territorial function of chin gland secretion in the rabbit, *Oryctolagus cuniculus* (L.). *Nature, Lond.* **193**, 799.

—— (1965). Further observations on the territorial function and histology of the submandibular cutaneous (chin) glands in the rabbit, *Oryctolagus cuniculus* (L.). *Anim. Behav.* **13**, 400–12.

—— (1966*a*). Observations on odoriferous and other glands in the Australian wild rabbit, *Oryctolagus cuniculus* (L.), and the hare, *Lepus europaeus*. I. The anal gland. *CSIRO Wildl. Res.* **11**, 11–29.

—— (1966*b*). Observations on odoriferous and other glands in the Australian wild rabbit, *Oryctolagus cuniculus* (L.), and the hare, *Lepus europaeus*. II. The inguinal glands. *CSIRO Wildl. Res.* **11**, 49–64.

—— (1966*c*). Observations on odoriferous and other glands in the Australian wild rabbit, *Oryctolagus cuniculus* (L.), and the hare, *Lepus europaeus*. III. Harder's, lachrymal and submandibular glands. *CSIRO Wildl. Res.* **11**, 65–90.

—— (1968). Territorial marking by rabbits. *Scient. Am.* **218**, 116–26.

—— (1970). The role of skin glands in mammalian communication. In *Advances in chemoreception. I. Communication by chemical senses* (ed. J. W. Johnson, D. G. Moulton, and A. Turk) pp. 327–60. Appleton-Century-Crofts, New York.

—— (1973). Reproduction of mammals in relation to environmental odours. *J. Reprod. Fert.* Suppl. **19**, 431–44.

—— (1974). Odour in the spacing behaviour of mammals. In *Pheromones* (ed. M. C. Birch) pp. 327–43. North-Holland, Amsterdam.

—— (1975). Activation of territorial behaviour in the rabbit *Oryctolagus cuniculus* by stimulation with its own chin gland secretion. In *Olfaction and taste,*

Vol. 5 (ed. D. A. Denton and J. P. Coughlan) pp. 425–32. Academic Press, New York.

— (1979). Some difficulties in the study of the function and composition of semiochemicals in mammals, particularly wild rabbits, *Oryctolagus cuniculus* L. In *Chemical ecology: odour communication in animals* (ed. F. J. Ritter). Elsevier/North-Holland, Amsterdam.

— and Dudzinski, M. L. (1966). A study of the weight of odoriferous and other glands in relation to social status and degree of sexual activity in the wild rabbit, *Oryctolagus cuniculus* (L.). *CSIRO Wildl. Res.* **11**, 31–47.

— and Gambale, S. (1969). The distribution of dung-hills and the behaviour of free-living rabbits, *Oryctolagus cuniculus* (L.) on them. *Forma funct.* **1**, 333–49.

— and Goodrich, B. S. (1974). Skin glands as organs of communication in mammals. *J. invest. Derm.* **62**, 124–31.

— and Hesterman, E. R. (1970). The behaviour of captive wild rabbits, *Oryctolagus cuniculus* (L.) in response to strange dung-hills. *Forma funct.* **2**, 1–12.

— — Gambale, S., and Dudzinski, M. L. (1976). A comparison of the effectiveness of the odours of rabbits *Oryctolagus cuniculus* in enhancing territorial confidence. *J. chem. Ecol.* **2**, 13–24.

— and Rowley, I. (1958). Continuous observations of the activity of the wild rabbit, *Oryctolagus cuniculus* (L.), during 24-hour periods. *CSIRO Wildl. Res.* **3**, 26–31.

Oikawa, M., Higuchi, K., Mori, T., and Okano, T. (1976). Embryological and histochemical investigations on the submandibular glands in the rabbit fetuses. I. Histogenesis and carbohydrates in the submandibular glands of the fetuses. *Aichi-Gakuin J. dent. Sci.* **13**, 1–16.

Parer, I. (1977). The population ecology of the wild rabbit *Oryctolagus cuniculus* (L.) in a Mediterranean-type climate in New South Wales. *Austr. Wildl. Res.* **4**, 171–205.

Parris, W. R. (1969). Histological survey of the chin glands of the cottontail rabbit *Sylvilagus floridanus*. *Trans. Mo. Acad. Sci.* **3**, 91.

Radominska-Pyrek, A., Dabrowiecki, Z., and Horrocks, L. A. (1979). Synthesis and content of ether-linked glycero-phospho-lipids in the Harderian gland of rabbits. *Biochim. biophys. Acta* **574**, 248–57.

Rock, C. O. (1977). Harderian gland. In *Lipid metabolism in mammals,* Vol II (ed. F. Snyder) pp. 311–21. Plenum Press, New York.

— Fitzgerald, V., Rainey, W. J. Jr, and Snyder, F. (1976). Mass spectral identification of 2-(0-acyl)hydroxy fatty-acid esters in the white portion of the rabbit Harderian gland. *Chem. Phys. Lipids* **17**, 207–12.

— — and Snyder, F. (1977). Properties of dihydroxyacetone phosphate acyltransferase in the Harderian gland. *J. biol. chem.* **252**, 6363–6.

— — — (1978). Coupling of the biosynthesis of fatty acids and fatty alcohols. *Archs Biochem. Biophys.* **186**, 77–83.

— and Snyder, F. (1974). Biosynthesis of 1-alkyl-sn-glycero-3-phosphate via ATP 1-alkyl-sn glycerol phospho transferase. *J. biol. Chem.* **249**, 5382–7.

— — (1975*a*). A short chain acyl coenzyme A 1-alkyl-2-acyl-sn glycerol acyl transferase from a microsomal fraction of the rabbit Harderian gland. *Biochim. biophys. Acta* **388**, 226–30.

— — (1975*b*). Metabolic interrelations of hydroxy-substituted ether-linked glycerolipids in the pink portion of the rabbit harderian gland. *Archs Biochem. Biophys.* **171**, 631–6.

Salesses, A. and Marois, M. (1969). Contribution a l'étude du controle hormonale de deux glandes d'origine ectodermique: sensibilité des glandes anales et inguinales du lapin a la testosterone et a l'oestradiol. *C. r. Ass. Anat.* **143**, 1458–67.

Schaffer, J. (1940). *Die Hautdrüsenorgane der Säugetiere.* Urban & Schwarzenberg, Berlin.

Schalken, A. P. M. (1976). Three types of pheromones in the domestic rabbit *Oryctolagus cuniculus* (L.). *Chem. Senses Flavor* **2**, 139–55.

Schley, P. (1977). The destruction of sense of smell and its influence on suckling behaviour in young animals. *Berl. Münch. teirärztl. Wschr.* **90**, 382–5.

Schneider, E. (1977). Beobachtungen zum Körperflegeverhalten des Feldhasen (*Lepus europaeus*, Pallus). *Beitr. Naturk. Niedersachs.* **30**, 7–19.

Schneir, E. S. and Hayes, E. R. (1951). The histochemistry of the harderian gland of the rabbit. *J. natn. Cancer Inst.* **12**, 257–8.

Sharp, P. L. (1973). Behavior of the pika (*O. princeps*) in the Kananaskis region of Alberta. Unpublished M.Sc. thesis, University of Alberta, Edmonton.

Southern, H. N. (1948). Sexual and aggressive behaviour in the wild rabbit. *Behaviour* **1**, 173–94.

Stevens, D. A. (1975). Laboratory methods for obtaining olfactory discrimination in rodents. In *Methods in olfactory research* (ed. D. G. Moulton, A. Turk, and J. Johnston). Academic Press, London.

Strauss, J. S. and Ebling, F. J. (1970). Control and function of skin glands in mammals. *Mem. Soc. Endocr.* **18**, 341–71.

Svendsen, G. E. (1979). Territoriality and behavior in a population of pikas (*O. princeps*). *J. Mammal.* **60**, 324–30.

Thiessen, D. D. (1977). Thermoenergetics and the evolution of pheromone communication. In *Progress in psychobiology and physiological psychology* (ed. E. M. Sprague and A. N. Epstein). Academic Press, New York.

Wales, N. A. M. and Ebling, F. J. (1971). The control of the apocrine glands of the rabbit by steroid hormones. *J. Endocr.* **51**, 763–70.

Westlin, L. M., Jeppsson, B., and Meurling, P. (1982). The nasal pad of the European hare *Lepus europaeus*—a histologic and scanning electron microscopic study. *Säugetierk. Mitt.* **30**, 221–6.

# 13 The odd-toed ungulates: order Perrisodactyla

PATRICIA D. MOEHLMAN

The order Perrisodactyla is composed of three living families: Tapiridae (tapirs —four species), Rhinocerotidae (rhinos—five species), and Equidae (horses, asses, and zebras—six species) (Table 13.1). This order is a relatively small one today, but included a greater number of species in the geological past (Kingdon 1979). The six living genera each include at least one or more species designated as endangered or vulnerable by the International Union for the Conservation of Nature (IUCN). Perrisodactyls have a non-ruminant digestive system and hence can feed opportunistically on lower-quality forage than ruminant artiodactyls of equivalent body size by eating greater quantities per unit time. However, as an order their reproductive biology is conservative and the resistance of their populations to heavy mortality seems generally to be low.

## Tapiridae

Family Tapiridae consists of one genus with four living species: *Tapirus terrestris* (Brazilian tapir); *T. pinchaque* (mountain tapir); *T. bairdii* (Central American tapir); and *T. indicus* (Malayan tapir). Fossils of primitive tapirs (Hydrachyidae) are known from the Eocene and present day forms have retained many primitive characteristics (Colbert 1969). In fact they are often classed as one of the oldest living mammals. Typically both the American and Asiatic species are found in mesic forests and savannas in association with swamps, streams, and/or rivers. All tapir species appear to be primarily browsers and semi-aquatic in their habits. However, due to their low density (most are on the endangered list), densely vegetated habitat, and shy behaviour, few data are available on free-ranging individuals in their natural habitat.

Data are not available on the occurrence of scent glands or the vomeronasal organ of tapirs. Von Sontag (1974) noted that dark circumorbital spots on *Tapirus terrestris* appeared to be glandular. Adult individuals repeatedly rubbed this region against conspecifics and against urine-marking spots in their compound.

Tapirs appear to be largely asocial, with the only stable group being mother and offspring. Studies in captivity of *T. terrestris* (Hunsaker and Hahn 1965; von Sontag 1974) indicate that the Brazilian tapir may be territorial. Hunsaker and Hahn recorded aggressive behaviour by both males and females when their 'territories' were encroached upon by other tapirs. However, these observations were made on five tapirs enclosed within a 0.2 ha compound where proximity may have prompted aggression. In captivity, Brazilian tapirs regularly urinate, defecate, and bathe (wallow) in artificial pools. Both males and females 'squirt-

Table 13.1. Perrisodactyl scent glands and olfactory communication

| | Circumorbital | Pedal | Preputial | Circumanal | Circumoral | Perineal | Vomeronasal organ | Flehmen | Dung piles | Dung scraping-pawing | Squirt urination | Wallow and mud rubbing | Territorial and/or overlapping home-range | References |
|---|---|---|---|---|---|---|---|---|---|---|---|---|---|---|
| **Family Tapiridae** | | | | | | | | | | | | | | |
| *Tapirus terrestris* (Brazilian tapir) | ? | | | | | | | + | + | | ♂,♀ | + | T? | 1, 2, 5 |
| *Tapirus pinchaque* (mountain tapir) | | | | | | | | + | + | | ♂,♀ | ? | | 3 |
| *Tapirus bairdii* (Central American tapir) | | | | | | | | + | + | | ♂,♀ | + | | 4 |
| *Tapirus indicus* (Malayan tapir) | | | | | | | | + | + | | + | + | OH | 2, 5 |
| **Family Rhinocerotidae** | | | | | | | | | | | | | | |
| *Dicerorhinus sumatrensis* (Sumatran rhino) | | | | | | | | + | +rare | +rare | ♂,♀ | + | OH | 6, 7 |
| *Rhinoceros sondaicus* (Javan rhino) | | + | | | | | | + | +rare | +rare | ♂,♀ | + | OH | 8 |
| *Rhinoceros unicornis* (Indian rhino) | | + | | | | | | + | + | +rare | ♀,♂ | + | T/OH | 9 |
| *Diceros bicornis* (black rhino) | | | | | | | | + | + | + | ♂,♀ | + | T/OH | 10–12 |

| | | | | | | | | | | Tδ | + | T | | |
|---|---|---|---|---|---|---|---|---|---|---|---|---|---|---|
| *Ceratotherium simum* (white rhino) | + | + | | | | | | + | | | | + | T | 13 |
| **Family Equidae** | | | | | | | | | | | | | | |
| *Equus caballus* (horse) | + | + | | + | + | + | + | + | + | + | + | | T/OH | 14–18, 25 |
| *Equus burchelli* (Burchell's zebra) | | | | | + | + | + | + | + | | | | OH | 19, 25 |
| *Equus zebra* (mountain zebra) | | | | | + | + | + | + | + | | | | OH | 20, 25 |
| *Equus grevyi* (Grevy's zebra) | | | | | + | + | + | + | + | | | | T | 21, 25 |
| *Equus asinus* (African ass) | | | | | + | + | + | + | + | | | | T/OH | 22–25 |
| *Equus hemionus* (Asian ass) | | | | | | + | | | + | | | | | 25 |

References
1. von Sontag (1974).
2. Schneider (1936).
3. Schauenberg (1969).
4. Terwilliger (1978).
5. von Richter (1966).
6. Groves and Kurt (1972).
7. Borner (1977).
8. Schenkel and Schenkel-Hulliger (1969a).
9. Laurie (1978).
10. Goddard (1967).
11. Schenkel and Schenkel-Hulliger (1969b).
12. Frame (1980).
13. Owen-Smith (1973, 1975).
14. Feist and McCullough (1975, 1976).
15. Welsh (1975).
16. Rubenstein (1978).
17. Turner et al. (1981).
18. Schaffer (1940).
19. Klingel (1975a).
20. Joubert (1972).
21. Klingel (1969a, b, 1975a).
22. Klingel (1972).
23. Moehlman (1974, 1979).
24. Woodward (1979).
25. Estes (1972).

urinate' (a backwards stream or burst of urine), with the adult male most frequently marking particular spots within the compound. Head-rubbing with the putative 'scent gland' was also concentrated on these urination marks. They also concentrate their defecations onto dung piles, and scrape with their hind-legs.

The mountain tapir, *Tapirus pinchaque*, lives at high altitudes (2500–6000 m) and is particularly elusive (Schauenberg 1969). However, limited observations indicate that mountain tapirs also squirt-urinate and defecate on to dung piles. During reproductive behaviour in captivity, males are very aggressive and inflict severe lacerations on the female (Bonney and Crotty 1978). These observations may reinforce the conjecture that tapirs are rather solitary.

*Tapirus bairdii*, the Central American tapir, has been observed in a small cohesive group on Barro Colorado Island (Terwilliger 1978; Overall 1980). However, these tapirs are provisioned and the grouping pattern may be an artefact of the situation. The Central American tapir often feeds alone, but social interactions were not limited to mother–offspring or reproductive contexts. Tapirs made quiet contact noises and also touched nose to nose. Smell appeared to be important for locating other individuals and the direction of proboscis movement correlated with movements towards conspecifics. *Tapirus bairdii* defecates most frequently in water and spray-urinates.

The Malayan tapir, *Tapirus indicus*, is essentially solitary but is sometimes found in associations of two to three individuals. A radio-collared male tapir had a home range of 12.75 km$^2$ that overlapped the home-ranges of several other tapirs in the West Malaysian National Park (Williams and Petrides 1980). Captive studies on the Malayan tapir indicate that their defecation and urination behaviour is similar to that of the rest of the genus, i.e. they use dung piles, squirt-urinate, eliminate in water and/or mud pools, and wallow therein.

The accumulation of odour through urination, defecation, and wallowing may be an important means of leaving a scent trail for a genus in which all species typically inhabit moist and dense vegetation and which generally occur in low numbers. Dung piles and squirt-urination rituals may function as territorial markers or may allow other communication between solitary individuals that have overlapping home-ranges (Eisenberg and Kleiman 1972). All species of *Tapirus* do exhibit flehmen which is commonly accepted to indicate an ability to differentiate between reproductive states on the basis of olfactory information in the urine.

### Rhinocerotidae

The family Rhinoceroditidae is composed of four genera and five species, all of which are on the endangered list: *Dicerorhinus sumatrensis* (Sumatran rhino), *Rhinoceros sondaicus* (Javan rhino), *Rhinoceros unicornis* (Indian rhino), *Diceros bicornis* (black rhino), and *Ceratotherium simum* (white rhino). Rhinos, like tapirs, are mainly solitary animals. Both the Sumatran and Javan rhino live in dense forest where visibility is poor. The Sumatran rhino is considered to be the most primitive species belonging to the genus *Dicerorhinus* which is believed to be ancestral to all living Rhinocerotidae (Groves and Kurt 1972). Males are solitary and the only stable group is the mother–offspring unit. Associations of more than two animals are rare (Pollok and Thom 1900; Wroughton and Davidson 1918). Sumatran rhinos are browsers and live in widely over-lapping home-ranges (10–30 km$^2$; van Strien 1974; Groves and Kurt 1972). They inhabit hilly forests near water and range from lowland rain-forest to mountain moss-forest (1000–1500 m). Borner (1977) reports localized defecation areas along trails and near streams in his study area. Out of 316 observed dung heaps 17.4 per cent were the defecation of a single animal, 69.4 per cent were the result of 2–4 defecations, and 13.2 per cent were the result of 5+ defecations. Scraping of dung was rare and was associated with only 5 per cent of defecations ($n$=316). The majority (93 per cent) of defecations appeared to involve no more than the individual stopping, defecating, and walking away and piles of dung were rare.

Both male and female Sumatran rhinos may squirt urine, with the spray being directed backwards for up to 6 m. In one sample of 61 urinations, 95 per cent were ritualized urine squirts (Borner 1977). Sumatran rhinos, like tapirs, defecate in water and mud and mud wallow in mud. By wallowing they may be covering their bodies with scent deliberately and subsequently leaving a trail of body marks the vegetation by twisting saplings, beating shrubs, and scraping and scratching trees.

Olfactory communication amongst the Javan rhino (*Rhinoceros sondaicus*) is similar to that of the Sumatran rhino. They are also primarily browsers, living in overlapping home-ranges, in dense lowland rain-forests. In general Javan rhinos are more gregarious than the Sumatran rhinos, but they are nevertheless mainly solitary. They rarely create dung piles or dung scrape but instead probably use pedal scent glands to lay a scent trail. They squirt urine and urinate in their mud wallows. Schenkel and Schenkel-Hulliger (1969*a*) describe the strong urine smell of rhino wallows, the olfactory impregnation of the rhino's skin, and the transfer of this odour to the soil, vegetation, and air of the 'tunnel-like' paths of a rhino trail.

The Indian rhino (*Rhinoceros unicornis*) inhabits swampy grasslands in India and Nepal. They are predominantly solitary except for mother–offspring groups. Only 9.6 per cent ($n$=4352) of observations involved groups other than solitary individuals or cow-calf pairs (Laurie 1978). Temporary aggregations occur at wallows and feeding grounds. The Indian rhino has been studied by Laurie (1978) who described its social organization and olfactory communication in detail. The Indian rhino displays a particular interest in the urination and defecation sites of other rhino. Laurie describes their ability to use scent to follow and locate each other. Indian rhinos utilize dung piles, defecating onto such sites in 88 per cent of observations ($n$=94). Adult males defecate more frequently than do females on dung piles. Rhinos normally walk away after defecating and scraping is rare (7 per cent of observations). Like the Javan rhino, the Indian rhino has pedal scent glands. Dung piles were found at the borders of riverine forest and grassland, banks of rivers and wallows, paths and especially path junctions, and near man-made roads and ditches. The frequency of fresh faeces on dung piles positively reflected local variations in rhino density. Laurie checked dung piles for ten consecutive days in December 1973 during which they were used 36 times in total. In the same area, over 10 days in July 1974, 24 sites were used 123 times. Laurie translocated dung from the local population and checked for any difference in the rhinos' response to these. There was no difference in reaction to the local and alien dung.

Laurie (1978) recorded 433 instances of urination by Indian rhinos. Both sexes perform squirt-urination. Males spray backwards 3–4 m on vegetation up to a height of about one metre. Females normally squirt-urinate only during oestrus. The majority of urinations observed were in response to observers (27.7 per cent) and other rhinos or ground scents (26.1 per cent). Adult males urinated four times more frequently than females or subadults ($n=433$). In addition to urinating more frequently overall, adult males usually squirt-urinated ($n=217$, 95 per cent). This ritualized urination was usually accompanied by rubbing the head and horn on vegetation, and foot dragging. This behaviour left broken vegetation, furrows, and vegetation-covered urine. Laurie observed the reactions of rhinos to urine, dung, or tracks of known individuals on 103 occasions, of which 61 per cent involved flehmen. Adult and subadult males performed flehmen significantly more frequently than did females. Adult males reacted to female tracks and urine. Other sex and age classes reacted predominantly to the tracks and urine of adult males. Indian rhinos also performed flehmen near and in mud wallows, where they urinated while wallowing (5.5 per cent of $n=433$). It seems probable that they impregnate their skin with the odour of urine while wallowing, as do tapirs and Sumatran and Javan rhinos.

The home-ranges of adult male Indian rhinos overlap but there is some range exclusivity, and in that sense territorially, between dominant males. Individuals of this species generally occupy overlapping home-ranges within which they move and forage alone. Males often locate females by following their tracks. Thus in a dense habitat in which individuals are widely scattered, Indian rhinos' marking behaviour potentially facilitates the exchange of olfactory information between widely spaced animals.

The black rhino (*Diceros bicornis*) is primarily a browser and lives in a variety of habitats ranging from semidesert to montane forest (Joubert and Eloff 1971; Hitchins 1971; Schenkel and Schenkel-Hulliger 1969*b*; Frame 1980; Goddard 1966, 1967). However, the black rhino needs to drink regularly and during the dry season stays within 25 km of water. Both species of African rhinos depend on water for temperature control and they will sweat conspicuously (Kingdon 1979) in order to regulate body temperature. They typically live in overlapping home-ranges although territorial behaviour has been reported by Hitchins (1971) for a population in Zululand. Goddard (1967) found that dung piles were located at random over an individual's home-range. A dung pile might be utilized by several individuals of both sexes but each rhino did not necessarily defecate on every dung pile that it approached. A typical response by a black rhino to a dung pile would be to sniff, wipe with its horn, scrape, defecate, scrape, and walk away. Strange elements in the environment may also elicit defecation behaviour. Goddard investigated aspects of the functional significance of the odours from dung by putting rhino faeces in a net bag and dragging it behind a vehicle in a complicated zig zag trail of up to two miles. Often rhinos followed such trails: (i) 60 per cent followed scent-trails of their own faeces and 50 per cent also defecated on the artificial deposit of their own faeces; (ii) 70 per cent followed scent-trails of dung from animals with which they shared their home-range and 20 per cent also defecated on their home-range companion's dung; and (iii) only 30 per cent followed the scent-trails of dung from animals living several miles distant and also defecated on this artificially laid dung. These interesting data emphasize the importance of scent communication for the 'exchange of information, orienting the movements of individuals, and integrating social and reproductive behavior' (Eisenberg and Kleiman 1972). Black rhino adult males are solitary and the only stable group is mother and offspring. Home-ranges vary in size from 1.33 to 2.6 km². Thus the ability to locate conspecifics through the use of scent-trails may be critical for reproduction. Scent-trails may also enable individuals to avoid one another. In areas of higher density, black rhino in overlapping home-ranges appear to be loosely organized into clans or communities of individuals that are tolerant of each other (Joubert and Eloff 1971; Goddard 1967; Mukinya 1973). Aggregations were temporary but did occur around water and/or wallows. Hitchins (1971), in the one recorded example of territoriality, found three males with subordinant males in their territories (393–465 ha).

Both males and females squirt-urinate. Ritualized urination by females usually occurs when they are in oestrus. Adult males mark bushes, tufts of grass, stumps, and/or stones with 1–4 bursts of urination that spray backwards for distances of 3–4 metres. Head-wiping and horn-beating on bushes sometimes accompanies squirt-urination.

The white rhino (*Ceratotherium simum*) differs from other rhinos and the tapirs in that it is primarily a grazer. In common with the aforementioned perrisodactyls, adult male white rhinos are solitary and the only stable group is mother and offspring. However, white rhinos are most definitely territorial with one dominant territorial male occupying an exclusive home-range of 0.7–2.6 km². The territorial male may tolerate 0–3 subordinate males that occupy the same territory (Owen-Smith 1973).

Owen-Smith (1973, 1975) made a detailed study of the white rhino population in Umfolozi Game Reserve, Zululand. Territorial males performed defecation and urination rituals that resulted in the placement of persistent scent marks in the environment. Females, subordinant adult males, and subadults had a strictly eliminative manner of urinating and defecating.

Territorial males make kicking movements before and after defecation. In all of 69 observations males sniffed, kicked, defecated, kicked, and walked away. Except for two which were deposited on fresh female dung, defecations by territorial males were on dungheaps. On nine occasions the territorial male rubbed his horn in the dung. This seemed to correlate with the presence of the faeces of a strange male on the dung heap. But when Owen-Smith introduced strange dung experimentally into a territory there was no response. There was an average of 30 dung heaps in each territory, within which they were randomly distributed. Owen-Smith's observations of the communicative function of these dung heaps were that they '(i) provided a record of the visits of many different rhinos; (ii) most of the dung heaps within a territory were marked with the dung of the dominant male; (iii) only territorial bulls kicked and scattered the dung which facilitated a more powerful release of scent, increased the surface area exposed, increased the distribution of the territorial male's scent on the dung heap, and left the scent on the bull's feet and thus a path of scent subsequently'.

Only territorial males squirt-urinated regularly and this display typically consisted of sniffing, horn-wiping (in 50 per cent of cases), scraping, and squirt-urination in 3-5 bursts. Squirt-urination was usually performed at territory boundaries. The rate of squirt-urination in the interior of a territory was 2 per hour, while the rate was 10 per hour along boundaries. A flehmen response was usually given after any class of male smelled a female urination site.

White rhinos have a preputial gland that may contribute a special odour to urine (Cave and Aumonier 1964; Cave 1966; Owen-Smith 1973). Owen-Smith speculated that possible communicative functions of urination by white rhino included '(i) advertisement that the territorial male was in residence and might contribute to maintenance of the status quo and (ii) indication, through hormonal levels, of their social and reproductive status'.

**Equidae**

The family Equidae is composed of one genus and six species: *Equus asinus* (African ass), *E. hemionus* (Asian ass), *E. grevyi* (Grevy's zebra), *E. zebra* (mountain zebra), *E. quagga* (common zebra), and *E. caballus* (horse). These species live in a variety of habitats ranging from desert (asses) to tropical grasslands (zebras) to temperate grasslands (horses). Early equids were forest dwellers and may have been similar to tapirs in their solitary habits (Kingdon 1979). Present-day species exhibit social behaviour ranging from that typical of species in arid environments involving male territory, with the only stable group being that of mother and offspring, to the system more characteristic of mesic habitats, which involves stable unimale or multimale harem groups (Klingel 1975*a*, Moehl-man 1974). These two types of social behaviour appear to be the extremes of a continuum of social organization ranging from a system in which territoriality plays an important role and social bonding is limited to the more socially organized and cohesive family groups. Generally, the territorial type of social organization has been described for *Equus asinus* and *E. grevi*, with the stable family group being typical of *E. quagga*, *E. zebra*, and *E. caballus*. However, field research on various populations of feral asses (Moehlman 1979; McCort 1980) and feral horses (Rubenstein 1978) has revealed intraspecies behavioural plasticity and variability of social organization.

*Equus grevyi* lives in arid grasslands and characteristically has a social organization in which there are territorial males and the only stable bonds are between female and offspring (Klingel 1969*a*, 1975*a*). Their territories ($\sim$10 km$^2$) are the largest recorded for herbivores and conspecifics of all age and sex classes are tolerated. However, only the territorial male has access to oestrous females. Territorial males have large dung piles along the boundaries of their areas and these may serve as fixed orientation points for the resident (Klingel 1975*b*). Dung piles are also found near paths, waterholes, and rivers and may be the result of more than one individual's activity (Kingdon 1979).

*Equus asinus* (African ass) has endangered status in its native habitat, Ethiopia. Limited observations by Klingel (1977) indicate that in this arid district (Danakil Desert) the species is territorial and only mother and offspring form stable groups. Thriving populations of feral asses have been studied in North America (Moehlman 1974, 1979; Woodward 1979). Feral asses in Death Valley National Monument (Moehlman 1974) were browsers and displayed social organization typical of equids in an arid habitat, i.e. 10 per cent of adult males held territories adjacent to the major water source, and female and offspring formed the only permanent groups. Males were often solitary (23.9 per cent of $n$=1158 observations) and the general population trend was that of small groups of two to four individuals (57.8 per cent of $n$=1158). Non-territorial individuals live in over-lapping home-ranges of up to 103.6 km$^2$. Large aggregations (8–21) occurred rarely (3.3 per cent of $n$=1158) and were associated with scarce resources, i.e. water and/or oestrous females.

Territorial asses (Moehlman 1974) form dung piles along boundaries and within their territories. Dung piles also occur along trails and in the vicinity of watering areas. These dung piles are not limited to the use of any one animal and all males freely defecated on them. Females of all age classes showed little interest in dung and simply defecated where they stood and walked away (Table 13.2). Adult males often defecated in single deposits ($n$=42). On 38 per cent of these occasions they smelled their defecation, pawed it, flehmened, and/or urinated on it. Adult males also smelled and defecated on dung piles ($n$=87). In 54 per cent of the observations the male simply smelled the dung pile and walked away. The typical display in the remaining observations consisted of

Table 13.2. Age and sex class response to defecations and dung piles

Defecate and walk away
| | |
|---|---|
| Adult females | $n = 17$ |
| Female foals | $n = 3$ |
| Female yearlings | $n = 6$ |

| Adult male defecations | $n = 42$ | % of $n$ |
|---|---|---|
| Defecates, walks away | | 61.9 |
| Defecates, smells | | 21.4 |
| Defecates, smells | | 4.8 |
| Defecates, smells, second male smells | | 2.4 |
| Defecates, smells, paws, smells, paws, urinates | | 2.4 |
| Defecates, threatens second male away | | 2.4 |
| Defecates, urinates | | 2.4 |
| Defecates, on recent defecation, walks away | | 2.4 |

Male foal defecations
| | |
|---|---|
| Defecates and walks away | $n = 2$ |
| Defecates and smells | $n = 1$ |

| Adult male response to dung piles | | $n = 87$ | ( ) = % |
|---|---|---|---|
| Paws dung pile | Smells dung pile | | |
| | 2 (2.3) | | Fights |
| | 4 (4.6) | | Flehmens |
| | 2 (2.3) | | Paws dung pile |
| | 47 (54.0) | | Walks away |
| | 1 (1.1) | | Paws dung pile, flehmens |
| 3 (3.4) | 11 (12.6) | | Defecates |
| 4 (4.6) | 8 (9.2) | | Defecates, smells |
| | 2 (2.3) | | Paws dung pile, defecates, smells |
| 1 (1.1) | 1 (1.1) | | Smells dung pile, defecates |
| 1 (1.1) | | | Smells dung pile, defecates, smells |

smelling the dung, defecating, and walking away. Pawing, flehmen, and fighting were also associated with visits to dung piles.

Adult females made virtually no response to urination, either their own or that of other sex and age classes. Female foals and yearlings occasionally responded to other's urination ($n=2$). Adult males displayed a wide range of responses to their own and other's urinations. Females were observed urinating on 40 occasions, and the subsequent behaviour of males present was recorded (Table 13.3). In 72 per cent of male responses, flehmen was involved. In most of these observations, the male smelled the ground on which the female had urinated, and then flehmened. In 18 per cent of the responses the male urinated on the female's urine and did not flehmen. In 20 per cent of the observations, males were present but made no response to the female's urination. Flehmen posture in feral asses often served as a visual stimulus, as it was followed by the approach of other males that also smelled the urination spot and flehmened. Males assumed flehmen posture while urinating, after smelling the ground

Table 13.3. Age and sex class response to urinations

| Adult female urinates and walks away ($n = 40$) | |
|---|---|
| Male response to female urinations ($n = 61$) | % of $n$ |
|     Male smells ground and flehmens | 55.7 |
|     Male smells ground and flehmens, urinates on it | 8.3 |
|     Male smells ground and flehmens, urinates on it, flehmens | 3.3 |
|     Male urinates on it | 1.6 |
|     Male urinates on it, flehmens | 1.6 |
|     Male smells ground, urinates on it | 1.6 |
|     Male paws ground, urinates on it, flehmens | 1.6 |
|     Male smells ground, shakes head side-to-side | 1.6 |
|     Male copulates female | 1.6 |
|     Male smells female genitals | 1.6 |
|     Male smells ground, flehmens, defecates on it | 1.6 |
|     Male present, no response | 19.7 |

Foal response ($n = 11$)

| | |
|---|---|
| 1 | Female foal urinates |
| 3 | Male foal urinates on it |
| 5 | Male foal smells ground, flehmens |
| 1 | Male foal smells ground, walks away |
| 1 | Male foal flehmens |

Yearling response ($n = 2$)

| | |
|---|---|
| 1 | Female yearling urinates on it |
| 1 | Male yearling smells ground, flehmens |

(presumably an old urination spot), after smelling another male's urine, after mounting and/or copulating with a female, after smelling male defecation, and after smelling a female's genital area. A male foal was observed in the flehmen posture after smelling ejaculate on the ground. Flehmen consists of a raising of the head with the muzzle pointed towards the sky, the upper lip drawn back extremely and puckered, upper teeth and gums exposed, and nostril wrinkled into a longitudinal and closed position (Fig. 13.1). All species of *Equus* exhibit flehmen and have a vomeronasal organ (Jacobson's organ). In equids the openings of the vomeronasal ducts are in the incisive ducts which connect with the nose only. Estes (1972) postulates that equids suck air in through the nostrils and the ventral nasal meatus and fill the vomeronasal organ with air. A build up of air pressure is aided by the closing of the external nares and tilting the head upwards, thus facilitating air diffusion into the vomeronasal ducts. Male asses were observed to inhale deeply before assuming the flehmen posture. Estes concludes that the function of flehmen is the facilitation of introducing odorants into the vomeronasal organ. This organ is considered to be a specialized olfactory receptor for excreted hormones and thus functions in urinalysis. Estes quotes Fraser (1968)

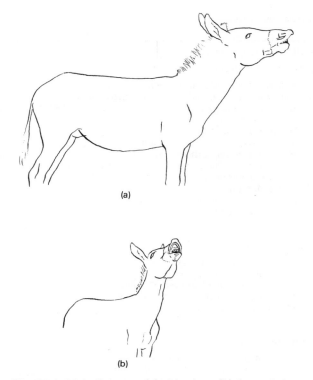

Fig. 13.1. Male flehmen, (a) side view, (b) frontal view.

'Maximal output of urine, with its endocrine products, is a fundamental behavioural character of oestrus—it is a reasonable assumption that oestrous cycle phasing may be recognizable to the male animal by the odour testing of urine'. The existence of the ability to test the reproductive status of the female would enable the male to detect the onset of oestrus and ovulation and would increase the probability of the female's being fertilized.

Feral asses also occur on an island off the coast of Georgia. In this mesic habitat they are grazers and form stable single-male and multi-male harem groups (Moehlman 1979; McCort 1980). The family groups are also territorial and defend areas ranging in size from 0.7 to 5.1 km² (McCort 1980). In this population adult females rarely responded to urine and faeces. Adult males sniffed and then urinated and/or defecated on urine spots and faeces. Dung piles were located throughout territories and were not concentrated on boundaries.

Within stable groups, if a female urinated, males (in order of dominance) sniffed and urinated on the spot (McCort 1980). If the female defecated, the males sniffed and defecated on the faeces. Infrequently, males urinated on faeces and defecated on urine. When the female came into oestrus, males displayed flehmen in response to the female's urination. The urination site of oestrous females was the subject of intense competition between males present. When males excreted, other male members of the group sniffed, urinated, defecated, and exhibited flehmen in response. However, male excretion sites were rarely contested.

McCort (1980) experimentally moved piles of faeces from one territory to another. Without exception, foreign dung piles were fought over by males and females, and were sniffed, defecated on, and flehmened over.

Feral asses will also defecate on 'strange' objects in their environment such as new paint blips on the centre of a road (Moehlman 1974) or horseshoe crabs (McCort 1980).

Dung piles probably serve as communication posts and landmarks. Female urine appears to be important in the detection of hormonal status. When males urinate or defecate on top of female excretions they may (i) establish their presence and (ii) cover and/or dilute the female's scent mark (Trumler 1958).

*Equus quagga* (common or plains zebra) has stable uni- or multimale harems, lives in mesic habitat, and is primarily a grazer (Klingel 1969*b*). They live in overlapping home-ranges and undertake long-distance migrations. Dung deposits can be the object of pushing and shoving, with several males contributing dung (Antonius 1951). Klingel has described males of a harem covering the excreta of the females in their group.

*Equus zebra* (mountain zebra) also lives in stable uni- or multimale harems and is primarily a grazer. Information on defecation, urination, and flehmen behaviour is limited but appears to be similar in occurrence and context to that of plains zebras (Joubert 1972).

*Equus caballus* (feral horses) have been studied extensively in North America (Feist 1971; Feist and McCullough 1975, 1976; Miller and Denniston 1979; Rubenstein 1978; Welsh 1975). Their basic pattern of social organization is also uni- and multimale harem groups that are stable and have overlapping home-ranges. The one exception is Rubenstein's study on the Shakleford Banks where some harem bands were also territorial.

Among Pryor Mountain feral horses (Feist and McCullough 1976) dominant harem males usually responded to excreta of harem members and rival males. In response to urination by adult females ($n = 77$), the harem stallion urinated on the spot in 48 per cent of the observations. The male would approach, smell, urinate, and smell again. Urination was performed in an exaggerated stance with legs extended back, tail lifted higher than normal, and urine released in a short strong spurt. The harem stallions urinated on 20 per cent ($n=24$) of immature, yearling, and foal urinations of both sex classes. In response to defecations by adult females ($n=76$), dominant males defecated on 5.3 per cent and urinated on 34 per cent. Defecations by immatures, yearlings, and foals were seldom responded to (7 per cent of $n=29$). The predominant response of the dominant male to group excreta was urination (92 per cent of $n=76$ responses). Females were never observed to respond to excreta.

Fig. 13.2. Mustang stallions sniffing nostrils at a dung pile.

Order of elimination among males was correlated with dominance, with subordinate male eliminations being covered by that of the dominant male. Most interactions between males involved defecation, usually at dung piles ($n=186$ defecations by stallions with 89.8 per cent at dung piles). Utilization of dung piles was an adult male prerogative. Typical behaviour involved smelling, pawing, defecating, and occasionally urinating. Aggressive encounters between rival males were often associated with dung piles. When two groups met the two dominant males might approach each other, smell the dung pile together, both defecate on it (in turn or simultaneously), smell the pile, and then fight or separate. In the Pryor Mountain study area most dung piles were found on or near trails or in proximity to water holes, i.e. highly frequented sites. Feist and McCullough postulated that for this non-territorial population dung piles and urination sites had visual and olfactory importance in terms of dominance relationships, associations, and information on hormonal state which persisted over time. Miller and Denniston (1979) observed interband dominance hierarchy in a population of horses studied in the Red Desert of Wyoming. Dung piles might serve also as communication posts between such components of a population.

Flehmen occurred most frequently during the early part of the breeding season. Most observations involved a male response to female genitalia, urine, or faeces (53 per cent of $n=55$), but males also flehmened in response to their own urine.

In the same study area, Turner, Kirkpatrick, and Perkins (1981) also quantified eliminative behaviour in stallions. They defined eliminative behaviour as the behavioural response of stallions to urination and/or defecation by other horses which was simultaneously noticeable by the stallion. Elimination behaviour as such was not performed by females or immature animals. The response of stallions to adult female elimination varied seasonally. The average frequency of response ($n$=172) was 43.4 per cent. However, frequency of response peaked in the breeding season (May—93 per cent; June—89 per cent) and was very low in November—February (1 per cent). The average number of eliminations per hour did not differ between months. The seasonality of stallion eliminative behaviour correlated positively and strongly with the breeding season and seasonal increase in stallion plasma androgens and mare plasma steroids. Stallions, under laboratory conditions, have responded appropriately to oestrous behaviour by mares that have been ovariectomized. Thus sex steroids from adrenal sources may mediate oestrous behaviour in female horses (Asa, Goldfoot, Garcia, and Ginther 1980).

Rubenstein's (1978) study of territorial harem groups on Shakleford Banks indicates that eliminative behaviour in that locale is similar to that of non-territorial horses but that spatial placement of dung piles is different. Large dung piles occur both on territory boundaries and within territories. Encounters between rival males involve the same ritualized sequence of defecations, dung-sniffing, and aggression as described by Feist and McCullough (1975).

## Discussion

Perrisodactyls emerged in the Palaeocene and radiated in the Eocene and Oligocene. During the latter period they were abundant in America and Eurasia. Competitive pressure from artiodactyls which possess an efficient foregut digestive system may have led to the reduction or removal of perrisodactyls from many herbivore niches (Kingdon 1979). The remaining species appear to have either (i) specialized in high-fibre diets or (ii) evolved large body size as a strategy to avoid ruminant competition (Janis 1976). Janis postulates that perrisodactyls became cellulose eaters and caecal fermenters during the Palaeocene when their body size was relatively small. Perrisodactyls, unlike ruminants, are not limited by rumen size and passage rates, and hence can process more herbage and survive on a poorer quality (higher fibre) diet.

Primitive perrisodactyls were probably small forest-dwelling herbivores (Colbert 1969). Today the tapirs are considered to be the most primitive members of the order. In fact they are considered little changed from the Eocene fossils. They are typically solitary, live in overlapping home-ranges, and feed on browse. Eisenberg and Lockhart (1972) have suggested that the tendency of forest ungulates to be solitary may be partly due to the effect of poor visibility upon group communication and coordination. Perhaps a more critical factor is the spatial and temporal availability of food. In dense forest, food is more widely scattered and not quickly renewable, hence predisposing tapirs to be widely distributed (Jarman 1974; Laurie 1978). In any case, tapirs appear to be highly dependent on scent marks for locating conspecifics. Anatomically they have very large skeletal nasal openings (Willoughby 1974) and their long proboscis is flexible and sensitive. Although they do not appear to have scent glands they do scrape their dung and also wallow in mud in which they have defecated and urinated. These excretory odours probably leave scent-trails as individuals move through the dense vegetation.

The most 'primitive' rhino, the Sumatran rhino, also lives in dense tropical forests. Similar to tapirs they are primarily solitary, live in overlapping home-ranges, and are browsers. Squirt-urination and wallowing in mud-defecation sites are their major modes of scent communication. Javan and Indian rhino both have pedal scent glands which may be important for leaving scent trails. Indian rhino males are territorial and dung-piles are a common phenomenon. Javan and Indian rhinos are found in more open forests and alluvial floodplains and compared with Sumatran rhinos are more commonly found in aggregations of two or more. Black rhino are primarily browsers and are found in more open habitat. Depending on their denisty they may have a spacing system of over-lapping home-ranges or what appear to be territories inhabited by males in a dominance hierarchy. Black rhinos of both sexes squirt-urinate, utilize dung piles, and dung scrape. The latter behaviour appears to enable individuals to discriminate between scents of conspecifics and to track their dung trails appropriately. Among white rhinos some major departures from the general perrisodactyl condition occur. The white rhino has a high genetic variability (blood protein) and may be in an active stage of evolution (Osterhoff and Keep 1970, in Kingdon 1979), possibly because its adaptation to a diet of grass may be relatively recent. In any case they live in an open habitat, have a relatively evenly distributed and abundant food supply, eat grass, and males live in territories. Creating dung piles and scraping and squirting urine are the prerogatives of the dominant and breeding territorial male only. Females and subordinate males do not engage in this type of display. For rhinos, scent remains a major mode of communication and perhaps a prime regulator of their social existence (Kingdon 1979).

Equids are capable of great plasticity in their social behaviour and mating systems. Their social organization ranges from stable harem bands in mesic habitats where grazing is the main mode of feeding, to territorial males and small stable groups of mother and offspring in more arid habitat, where browse is the predominant food source. Dung piling and pawing is commonplace, and in territorial populations dung heaps may be localized along boundaries. Responses to dung and urination are made primarily by dominant males. Rival males may engage in defecation rituals associated with aggressive displays. Dominant males will typically cover female excreta and hence potentially announce their presence and mask the female's scent. Hormonal levels in the urine corresponding to ovulation may provide important cues to a female's reproductive status. However, adrenal steroids may play an equally important role. Ovariectomized and seasonally anoestrous horses display oestrous behaviour. 'Sexual activity outside the breeding season, made possible by the absence of complete control of steroids from the ovaries over oestrous behaviour, could contribute to the maintenance of social bonds between the harem stallion and his mares' (Asa *et al.* 1980).

Among perrisodactyls dung piles, scent trails (due to dung scraping or pedal scent glands), and urine marks (on the ground or sprayed on to vegetation) provide important olfactory information on location, social, and reproductive status for mobile members of, often low-density, populations.

## References

Antonius, O. (1951). *Die Tigerpferde.* Schops, Frankfurt.

Asa, C. S., Goldfoot, D. A., Garcia, M. C., and Ginther, O. J. (1980). Sexual behavior in ovariectomized and seasonally anovulatory pony mares (*Equus caballus*). *Horm. Behav.* 14, 46–54.

Bonney, S. and Crotty, M. J. (1978). Breeding the mountain tapir, *Tapirus pinchaque. Intern. Zoo. Yrbk*, Vol. 19 (ed. P. J. S. Olney) pp. 198–200.

Borner, M. (1977). A field study of the Sumatran rhinoceros, *Dicerorhinus sumatrensis* (Fisher 1814): ecology and behavior, conservation situation in Sumatra. Ph.D. thesis, Basel University.

Cave, A. J. E. (1966). The preputial glands of *Ceratotherium. Mammalia* 30, 152–9.

— and Aumonier, F. J. (1964). Lymph node structure in *Rhinoceros unicornis. J. R. microsc. Soc.* 83, 251–3.

Colbert, E. H. (1969). *The evolution of the vertebrates.* Wiley, New York.

Eisenberg, J. F. and Kleiman, D. G. (1972). Olfactory communication in mammals. *A. Rev. Ecol. Syst.* 3, 1–32.

— and Lockhart, M. (1972). An ecological reconnaissance of Wilpattu National Park, Ceylon. *Smithson. Contr. Zool.* No. 101, 1–118.

Estes, R. D. (1972). The role of the vomeronasal organ in mammalian reproduction. *Mammalia* 36, 316–41.

Feist, J. D. (1971). Behavior of feral horses in the Pryor Mountain Wild Horse Range. M.Sc. thesis, University of Michigan, Ann Arbor.

— and McCullough, D. (1975). Reproduction in feral horses. *J. Reprod. Fert.* Suppl. 23, 13–18.

— — (1976). Behavior patterns and communication in feral horses. *Z. Tierpsychol.* 41, 337–71.

Frame, G. W. (1980). Black rhinoceros (*Diceros bicornis* L.) sub-population on the Serengeti Plains, Tanzania. *Afr. J. Ecol.* 18, 155–66.

Fraser, A. F. (1968). *Reproductive behaviour in ungulates.* Academic Press, London.

Goddard, J. (1966). Mating and courtship of the black rhinoceros (*Diceros bicornis* L.) *E. Afr. wildl. J.* 4, 69–75.

— (1967). Home range, behavior, and recruitment rates of two black rhinoceros populations. *E. Afr. wildl. J.* 5, 133–50.

Groves, C. P. and Kurt, F. (1972). *Dicerorhinus sumatrensis. Mamm. Sp.* No. 21, pp. 1–6.

Hitchins, P. M. (1971). Preliminary findings in a radio-telemetric study on the black rhinoceros in Hluhluwe Game Reserve, Zululand. *Symp. in Biotelemetry*, Pretoria, 1971. CSIR, Pretoria.

Hunsaker, D. II, and Hahn, T. C. (1965). Vocalization of the South American tapir, *Tapirus terrestris. Anim. Behav.* **13**, 69–74.

Janis, C. (1976). The evolutionary strategy of the equidae and the origins of rumen and caecal digestion. *Evolution* **30**, 757–74.

Jarman, P. J. (1974). The social organisation of antelope in relation to their ecology. *Behaviour* **48**, 215–67.

Joubert, E. (1972). The social organization and associated behavior in the Hartmann zebra (*Equus zebra hartmannae*). *Madoqua* Ser. I, No. 6, pp. 17–56

— and Eloff, F. C. (1971). Note on the ecology and behaviour of the black rhinoceros *Diceros bicornis* Linn. 1758 in South West Africa. *Madoqua* **3**, 5–53.

Kingdon, J. (1979). *East African mammals*, Vol. III, Part B. Academic Press, London.

Klingel, H. (1969*a*). Zur soziologie des Grevyzebras. *Zool. Anz.* Suppl. **33**, 311–16.

— (1969*b*). Reproduction in the plains zebra, *Equus burchelli beohmi*. Behavior and ecological factors. *J. Reprod. Fert.* Suppl. **6**, 339–45.

— (1972). Social behaviour of African Equidae. *Zool. Africana* **7**, 175–86.

— (1975*a*). Die soziale organization der Equiden. *Verh. dt. zool. Ges.* 71–80.

— (1975*b*). Social organization and reproduction in equids. *J. Reprod. Fert.* Suppl. **23**, 7–11.

— (1977). Observations on social organizations and behavior of African and Asiatic wild asses (*Equus africanus* and *E. hemionus*). *Z. Tierpsychol.* **44**, 323–31.

Laurie, W. A. (1978). The ecology and behavior of the greater one-horned rhinoceros. Ph.D. thesis, University of Cambridge.

McCort, W. D. (1980). The feral asses (*Equus asinus*) of Ossabaw Island, Georgia. Ph.D. thesis, Pennsylvania State University.

Miller, R. and Denniston II, R. H. (1979). Interband dominance in feral horses. *Z. Tierpsychol.* **51**, 41–7.

Moehlman, P. D. (1974). Behavior and ecology of feral asses (*Equus asinus*) Ph.D. thesis, University of Wisconsin.

— (1979). Behavior and ecology of feral asses (*Equus asinus*). *Nat. Geog. Soc. Res. Rep.* 1970 projects, pp. 405–11.

Mukinya, J. G. (1973). Density, distribution, population structure and social organization of the black rhinoceros in Masa Mari Game Reserve. *E. Afr. wildl. J.* **11**, 385–400.

Osterhoff, D. R. and Keep. M. E. (1970). Natural variation in the blood protein of white and black rhinoceros. *Lammergeyer* **11**, 50–3.

Overall, K. L. (1980). Coatis, tapirs and ticks: a case of mammalian interspecific grooming. *Biotropica* **12**, 158.

Owen-Smith, N. (1973). The behavioral ecology of the white rhinoceros. Ph.D. thesis, University of Wisconsin.

— (1975). The social ethology of the white rhinoceros *Ceratotherium simum* (Burchell 1817). *Z. Tierpsychol.* **38**, 337–84.

Pollok, F. T. and Thom, W. S. (1900). *Wild sports of Burma and Assam*. Hurst and Blackeit, London.

Richter, W. von (1966). Untersuchungen über angeborene Verhaltensweisen des Schabrackentapirs (*Tapirus indicus*) und des Flachlandtapirs (*Tapirus terrestris*). *Zool. Beitr.* 12, 67-159.

Rubenstein, D. I. (1978). Islands and their effects on the social organization of feral horses. In *ABS Symp. in Social Behavior on Islands* (ed. R. Wallace).

Schaffer, J. (1940). *Die Hautdrüseorgane der Säugetiere.* Urban und Schwarzenberg. Berlin.

Schauenberg, P. (1969). Contribution a l'etude du Tapir Pinchaque, *Tapirus pinchaque* Roulin 1829. *Revue suisse Zool.* 76, 211-56.

Schenkel, R. and Schenkel-Hulliger, L. (1969a). The Javan rhinoceros (*Rhinoceros sondaicus* Desmarest) in Udjong Kulon Nature Reserve, its ecology and behavior. *Acta trop.* 26, 98-133.

—— —— (1969b). *Ecology and behavior of the black rhinoceros (Diceros bicornis L.).* Verlag Paul Parey, Hamburg.

Schneider, K. M. (1936). Zur Fortpflanzung, Aufzucht und Jugendenwicklung des Schabrakentapirs. *Zool. Gart.* 8, 83-96.

Sontag, W. A. von Jr. (1974). Beobachtungen an gemeinsam gehaltenen Flachlandtapiren (*Tapirus terrestris*) und Capybaras (*Hydrochoerus hydrochaeris*) im Zurcher Zoo. *Zool. Gart. Lpz.* 44, 317-23.

Strien, N. J. van (1974). *Dicerorhinos sumatrensis* (Fisher). The Sumatran or two-horned Asiatic rhinoceros. A study of literature. *Meded. Landb Hoogesch. Wageningen* 74, 1-82.

Terwilliger, V. J. (1978). Natural history of Baird's tapir on Barro Colorado Island, Panama Canal Zone. *Biotropica* 10, 211-20.

Trumler, E. (1958). Beobachtungen an den Bohmzebras des 'George-v.-Opel Freigeheges fur Tierforschung e.V' Kronberg im Taunus, I. Das Paarungsverhalten. *Säugetierk. Mitt.* 6, 1-48.

Turner, J. W., Kirkpatrick, J. F. and Perkins, A. (1981). Elimination behavior in feral horses. *Symp. on Wild Equids*, University of Wyoming.

Welsh, D. (1975). Population, behavioral and grazing ecology of the horses of Sable Island, Nova Scota. Ph.D. thesis, Dalhousie University.

Williams, K. D. and Petrides, G. A. (1980). Browse use, feeding behavior, and management of the Malayan Tapir. *J. wildl. Mgmt.* 44, 489-94.

Willoughby, D. P. (1974). *The empire of equus.* Barnes, London.

Woodward, S. L. (1979). The social system of feral asses (*Equus asinus*). *Z. Tierpsychol.* 49, 304-16.

Wroughton, R. C. and Davidson, W. M. (1918). Bombay Natural History Society's Mammal survey of India, Burma, and Ceylon. Rpt. No. 29 Pegu. *J. Bombay nat. Hist. Soc.* 25, 481-572.

# 14 The even-toed ungulates: order Artiodactyla

Sources, behavioural context, and function of chemical signals

L. M. GOSLING

## Introduction

The transmission of information by odour and taste is profoundly important in the social behaviour of artiodactyls. In this chapter I attempt to review the sources of these signals, and some of the more common behaviours used to transmit or receive them, and, most important, to explore the contribution that data from the artiodactyls can make in understanding their adaptive significance.

The main sources of odours are faeces, urine, saliva, vaginal secretions, and the specialized skin glands which are developed in exceptional profusion in this group. These are briefly described on pp. 550–66. Because man has a relatively poor sense of smell, most of our knowledge about olfactory signals is based on descriptions of the way that individuals produce scents and the way they respond to them. These observations are clearly biased towards behaviours that are easily seen and to responses that occur immediately after a smell is detected. Fortunately, many artiodactyls use distinctive behaviours to place scent marks in the environment, on their own bodies, and on conspecifics. These behaviours, and the marks, provide important clues as to function and they form a major part of the descriptions of signalling behaviour by individuals that comprise pp. 566–81.

The idea that mammals respond to external analogues of hormones, as envisaged in the concept of 'pheromones' (Karlson and Lüscher 1959) has been criticized by Beauchamp, Doty, Moulton, and Mugford (1976). However, it continues to influence research on mammalian olfactory signalling. All investigations of the composition of social odours in the artiodactyls have attempted to isolate one or a few 'active' or 'major' constituents. At variance with this approach are observations that odours and glandular secretions are highly complex and that many animals combine two or more marking substances to produce even greater complexity. An attempt to resolve this problem is one objective of the functional analysis on pp. 581–607. Another is to examine the consequences of the fact that, uniquely among social signals, scent-marks provide a spatial and historical record of individual behaviour.

## The sources of signal odours

### The glands

*Vestibular nasal glands*    Vestibular nasal glands have been found in marsh deer, *Blastocerus dichotomus* (Jacob and von Lehmann 1976) and pampas deer, *Ozotoceros bezoarticus* (Langguth and Jackson 1980). They consist of a pair of flattened oval sacs which open through a narrow canal onto the outer rim of each nostril. In one female pampas deer the glands measured 2.4 × 1.5 cm and 0.5 cm deep (Langguth and Jackson 1980). The tissue consists mainly of enlarged sebaceous glands with a few apocrine glands near the neck of the sac (Jacob and von Lehmann 1976).

*Interramal glands*    Interramal glands (Fig. 14.1) are present in all mouse deer and chevrotains (the genera *Tragulus* and *Hyemoschus*). They consist of thickened pads of skin overlying the mandibles, which contain enlarged sebaceous and apocrine sudoriferous glands. A number of long, presumably tactile, hairs protrude from the surface. The glands are largest in *T. napu* and *T. javanicus*. In males of the latter species the oval pad of glandular tissue measures 4.8 × 2.2 cm and is 1.4 cm deep. The glands are similar in size and microscopic structure in both sexes except in *T. javanicus* where that of the female is smaller (Pocock 1919; Schaffer 1940; Dubost 1975; Ralls, Barasch, and Minkowski 1975).

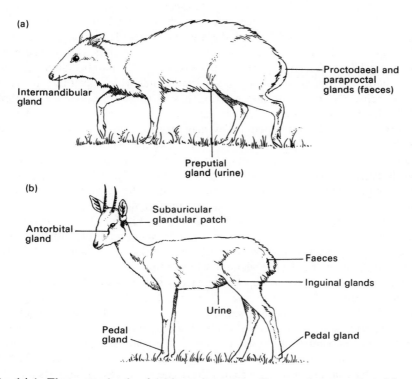

Fig. 14.1. The scent glands of African chevrotain, *Hyemoschus aquaticus* (a) and oribi, *Ourebia ourebia* (b). Chevrotains use small quantities of faeces and urine to mark their ranges and male oribi mark their territories with concentrations of faeces known as dung piles.

An interramal gland has also been found in steenbok, *Raphicerus campestris*. It consists of a thickened glandular layer in the upper two-thirds of the dermis and contains a dense concentration of enlarged sebaceous and apocrine glands. As in the chevrotains it is present and active in both sexes but larger in the male (Cohen and Gerneke 1976). The secretion consists of a whitish flaky substance that attaches itself to objects rubbed over the gland.

Pigs, *Sus scrofa*, are sometimes said to have an interramal gland but the tissue of the structure is similar to that in the surrounding skin and Schaffer (1940) believes it to have a purely tactile function. Haltenorth (1963) includes *Sus* and a number of cervids that possess similarly debatable structures in a list of animals possessing interramal glands but these are excluded from Table 14.1.

*Frontal glands*    Frontal glands which occur on the forehead are known in only three genera but this may be because they often lack an easily recognizable external structure. Only muntjac, *Muntiacus muntjak*, have clearly defined glands. These consist of two deep grooves which extend in a 'V' shape from just above the eye to the base of the antler pedicels. The grooves have thick, muscular lips which are under muscular control (Schaffer 1940). There are numerous large sebaceous glands at the surface and a number discharge into single hairs. Further below there are apocrine glands which open into the roots of the hairs. The odour of the secretion is said to be '*tres prenante*' (Dubost 1971). Impala, *Aepyceros melampus*, have no such structure although Jarman (1979) suggests that territorial males may develop two glandular skin folds on the forehead. Otherwise the skin of this area has a normal appearance but produces a copious oily secretion (Jarman 1979). Histological examination reveals large multilobular sebaceous glands. The secretion has a conspicuous, strong odour and is produced over the entire body, although in smaller quantities than on the forehead. High-ranking, non-territorial males (in male groups) also produce the secretion. A number of artiodactyls, for example hartebeest, *Alcelaphus buselaphus* and wildebeest, *Connochaetes taurinus*, rub their foreheads onto objects as if they were scent marking (Fig. 14.8) and a histological survey might show that frontal glands are common.

Male roe deer, *Capreolus capreolus*, have a similar glandular area but there are no reports of forehead grooves. Schumacher (1936) describes the greatly enlarged sebaceous and apocrine sudoriferous glands and their seasonal activity. Roe deer have a strongly seasonal cycle of social behaviour and their forehead glands reach maximum secretory activity in the summer when they compete with other males for access to females and mate. This sort of cycle, which is paralleled by changes in testes weight (Short and Mann 1966), suggests that the glands are under androgenic control (Adams 1980). The forehead skin of black-tailed deer, *Odocoileus hemionus*, has well developed apocrine glands but only minute sebaceous glands (Quay and Müller-Schwarze 1970).

*Antorbital glands*   Antorbital glands lie in a lachrymal depression just in front of the eye (Figs. 14.1 and 14.2). They are extremely common (Table 14.1) and it is tempting to think that this position is ideal for visual control during object marking (pp. 566–72). As described by Pocock (1910) the form of the gland is highly variable, ranging from simple tissue masses to deeply invaginated structures that are under muscular control. The glands of Uganda kob, *Kobus kob*, are of the first type, simple flattened structures, oval in dorsal view, which lack any central duct or invagination. In male Coke's hartebeest the gland is a solid tissue about 3.5 × 3.0 cm in area and 1.5 cm deep. The gland of adult females is considerably smaller although apparently active. There is a clear central depression, c. 2 mm in diameter, which branches into fine ducts in the

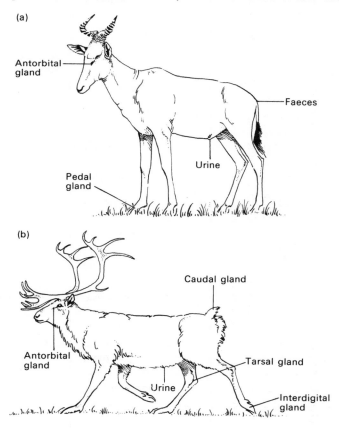

Fig. 14.2. The scent glands of Coke's hartebeest, *Alcelaphus buselaphus cokei* (a) and reindeer, *Rangifer tarandus* (b). Male hartebeest also mark their territories with faeces and urine and reindeer urinate on to the tarsal glands in the behaviour known as 'rub-urination'.

Table 14.1. Artiodactyl skin glands. The table indicates the presence (+) of each gland in the genera listed. Question marks (?) are shown in cases where marking has been observed but no glandular structure described. The genera accepted are a compromise between those given by Haltenorth (1963), Ansell (1971), and Walker (1975). The primary sources of information about the glands are the detailed and accurate observations by Pocock (1910, 1918a, b, c, d, e, 1919, 1923); these accounts are amplified from the reviews by Schaffer (1940). Haltenorth (1963), and from a large number of publications that deal with one or a few species; these are all cited in the bibliography. The table does not include every published reference to the occurrence of particular glands since some are clearly inaccurate; Pocock mentions many instances where early observers incorrectly stated the presence or absence of particular glands. In spite of an enormous literature the table is incomplete, although to differing extents: for example, the occurrence of antorbital, interdigital, and inguinal glands is probably quite well known while more diffuse, or otherwise less obvious, glandular areas, such as those of the forehead and prepuce will undoubtedly prove more common than indicated.

| | Vestibular nasal | Interramal | Frontal | Antorbital | Suborbital | Subauricular | Postcornual | Occipital | Dorsal | Ischiadic | Caudal (infra/circum) | Circumanal | Proctodaeal | Prevulval | Preputial | Inguinal | Carpal | Tarsal | Metatarsal | Interdigital | Ungulicular |
|---|---|---|---|---|---|---|---|---|---|---|---|---|---|---|---|---|---|---|---|---|---|
| *Tragelaphus* | | | | | | | | | | | | | | | | + | | | | | + |
| *Boocercus* | | | | | | | | | | | | | | | | | | | | | |
| *Taurotragus* | | | | | | | | | | | | | | | | | | | | | + |
| *Boselaphus* | | | | + | | | | | | | | | | | | | | | | | + |
| *Tetracerus* | | | | + | | | | | | | | | | | | | | | | | + |
| *Cephalophus* | | | | + | | | | | | | | | | | | | | | | + | |
| *Sylvicapra* | | | | | | | | | | | | | | | | + | | | | + | |
| *Redunca* | | | | | | + | | | | | | | | | | + | | | | | |
| *Kobus* | | | | + | | | | | | | | | | | | + | | | | | |

*Addax*
*Hippotragus*
*Oryx*
*Connochaetes*
*Alcelaphus*
*Damaliscus*
*Aepyceros*
*Antilope*
*Antidorcas*
*Procapra*
*Litocranius*
*Gazella*
*Ammodorcas*
*Oreotragus*
*Madoqua*
*Doreatragus*
*Ourebia*
*Raphicerus*
*Neotragus*
*Pelea*
*Pantholops*
*Saiga*
*Naemorhedus*
*Capricornis*
*Oreamnos*
*Rupicapra*
*Capra*
*Ammotragus*
*Hemitragus*
*Pseudois*
*Ovis*
*Budorcas*
*Ovibos*

Table 14.1 (*cont.*)

The glands

| | Vestibular nasal | Interramal | Frontal | Antorbital | Suborbital | Subauricular | Postcornual | Occipital | Dorsal | Ischiadic | Caudal (infra/circum) | Circumanal | Proctodaeal | Prevulval | Preputial | Inguinal | Carpal | Tarsal | Metatarsal | Interdigital | Unguicular |
|---|---|---|---|---|---|---|---|---|---|---|---|---|---|---|---|---|---|---|---|---|---|
| **Suidae** | | | | | | | | | | | | | | | | | | | | | |
| *Potamochoerus* | | | | + | | | | | | | | | | | | | + | | | + | |
| *Sus* | | | | + | | | | | | | | | + | | + | | + | | | | |
| *Phacochoerus* | | | | + | | | | | | | | | | | | | | | | | |
| *Hylochoerus* | | | | | + | | | | | | | | | | | | | | | | |
| *Babyrousa* | | | | | | | | | + | | | | | | + | | | | | | |
| *Tayassu* | | | | + | | | | | | | | | | | | | | | | | |
| **Hippopotamidae** | | | | | | | | | | | | | | | | | | | | | |
| *Hippopotamus* | | | | | | | | | | | | | | | | | | | | | |
| *Choeropsis* | | | | | | | | | | | | | | | | | | | | | |
| **Camelidae** | | | | | | | | | | | | | | | | | | | | | |
| *Camelus* | | | | | | | | + | | | | | | | | | | | | | |
| *Lama* | | | | | | | | | | | | | | | | | | | | + | |
| *Vicugna* | | | | | | | | | | | | | | | | | | | | + | |
| **Tragulidae** | | | | | | | | | | | | | | | | | | | | | |
| *Hyemoschus* | | + | | | | | | | | | | + | + | | + | | | | | | |
| *Tragulus* | | + | | | | | | | | | | | | | | | | | | | |
| **Cervidae** | | | | | | | | | | | | | | | | | | | | | |
| *Moschus* | | | | + | | | | | | | + | + | | + | + | | | | | + | |
| *Muntiacus* | | | + | + | | | | | | | | | | | | | | | | + | |
| *Elaphodus* | | | | + | | | | | | | | | | | | | | | | + | |
| *Dama* | | | | | | | | | | | | | | | + | | | | + | + | |

*Axis*
*Cervus*
*Elaphurus*
*Odocoileus*
*Blastocerus*
*Ozotoceros*
*Hippocamelus*
*Mazama*
*Pudu*
*Alces*
*Rangifer*
*Hydropotes*
*Capreolus*
**Giraffidae**
*Giraffa*
*Okapia*
**Antilocapridae**
*Antilocapra*
**Bovidae**
*Bos*
*Bubalus*
*Anoa*
*Syncerus*
*Bison*

gland and brings musty smelling secretion to the surface. Columns of smooth muscle run vertically through the glandular tissue, which is divided into a dorsal layer of enlarged sebaceous glands and a ventral layer of apocrine sudoriferous tubules (Gosling 1975).

The simplest invaginated glands consist of hemispherical depressions with little glandular activity. In the caribou, *Rangifer tarandus*, this depression is about 2.5 cm wide. Sebaceous and apocrine glands are scattered throughout the dermis but both are only slightly larger than in the surrounding skin; the sebaceous glands open into saccular hair follicles (Quay 1955). The glands of male reindeer, *Rangifer tarandus*, appear to be more active. They are larger than those of females and show more seasonal activity than other glands in this species. This variation is confined to the apocrine part of the secretory epithelium; glandular activity reaches a peak during the rut and is reduced in castrates, which suggests androgenic control (Mossing and Källquist 1981). The gland of male red deer, *Cervus elaphus*, is a simple sacculated structure whose inner lining is devoid of hair follicles or glandular tissue. The margins have well-developed sebaceous glands. The pouch of the gland is filled with a waxy plug for most of the year but in the mating season the plug is lost and a copious sticky fluid is produced which sometimes oozes onto the animal's cheek (Lincoln 1971). The glands of both male and female black-tailed deer contain a similar plug which according to Quay and Müller-Schwarze (1970) consists of accumulated orbital drainage, sloughed epidermal cornified cells and various adventitious materials. The lip of the structure contains slightly enlarged sebaceous and apocrine glands.

The most complex antorbital glands are deeply invaginated structures with muscular lips that are under voluntary muscular control. The lips are normally kept pressed together and are opened during marking or other forms of odour production (pp. 566-72). Sometimes, as in the fallow deer, *Dama dama*, the gland can be completely everted so that the inner surface is exposed as a hemispherical protrusion. The secretions of these glands, particularly those used in object marking, are often black and viscous. In dikdik, *Madoqua* spp, and steenbok this colour is caused by melanocytes which secrete into the enlarged sebaceous glands that form a large part of the gland (Richter 1971; Cohen and Gerneke 1976). The dik-dik gland is divided into a black dorsal layer of sebaceous glands plus melanocytes and a pale ventral layer of apocrine sudoriferous tubules. The products of both glands reach the surface through ducts in the centre of the gland (Richter 1971).

The glands of duikers, *Cephalophus* spp. and *Sylvicapra* spp. are unique, consisting of an elongated strip of bare skin studded with a series of pores which easily yield droplets of secretion when palpated (Pocock 1910). In transverse section the elongated antorbital gland of the red duiker, *Cephalophus natalensis*, is solid, flask-shaped, and enclosed in a connective-tissue capsule. The periphery of the gland consists of a layer of apocrine tubules and the centre of highly branched sebaceous glands. These lead to secretory ducts that open to one of the external pores arranged along the gland's surface (Mainoya 1978).

*Subauricular glandular patches* Subauricular patches (Fig. 14.1) occur in three distantly related genera (Table 14.1) but, despite this, are similar in appearance. In pronghorn, *Antilocapra americana*, oribi, *Ourebia ourebia*, and reedbuck, *Redunca arundinum*, there is a patch of black, almost hairless skin near the base of each ear. It is present in male pronghorn only and consists of a concentration of sebaceous and apocrine glands which are seasonally active; in the summer small globules of secretion can be seen on the fine hairs of the gland (Moy 1970). The glands are present in both male and female oribi. They also occur in both sexes in reedbuck but their presence is variable according to locality; in some areas they occur on all adults while in others their occurrence is irregular (Claussen and Jungius 1973). Reedbuck subauricular patches contain enlarged sebaceous glands. Similar patches have been reported for waterbuck, *Kobus defassa* (Dollmann 1931) but they are probably attributable to the occasional presence of a patch of cornified skin caused by scratching with the hind feet.

*Postcornual and occipital glands* Postcornual and occipital glands both occur in the skin overlying the base of the skull. Postcornual glands are restricted to the Caprini and Rupicaprini except for the Asian gazelles whose taxonomic affinities are uncertain. Best known and described by Pocock (1910) and Schaffer (1940) are the large postcornual glands of male chamois, *Rupicapra rupicapra*. These are present in both sexes but are larger in the male and, in this sex, become distinctly swollen during the mating season. The surface of the enlarged gland is purple with short, sparse hairs and deep intersecting grooves. The occipital glands of male camels, *Camelus dromedarius*, show similar seasonal variation and Yagil and Etzion (1980) found a clear correlation between serum androgen levels and those of the glandular secretion. During the rut the gland produced a copious watery secretion which had a strong foetid odour.

*Dorsal and ischiadic glands* Dorsal glands are discrete glandular structures in the centre of the back. They are found in collared and white-lipped peccaries, *Tayassu tajacu* and *T. albirostris* and in springbok, *Antidorcas marsupialis*, and pronghorn antelope. The gland of the collared peccary is located c. 15 cm anterior to the base of the tail and consists of a raised area of skin c. 5 X 8 cm in area (Sowles 1974). It is composed of both sebaceous and sudoriferous glands that empty into a common storage sac (Epling 1956). Secretion is expelled through a nipple-like protruberance at the centre of the gland.

The dorsal glands of both springbok and pronghorn are surrounded by long erectile hairs. In the springbok the gland is at the anterior end of this hair patch, known as a 'dorsal fan'; it produces a secretion that has a strong sweet smell (Bigalke 1972). The gland of male pronghorn is situated about 28 cm anterior to the tail base. It is elliptical in shape (60 × 30 mm) and the glandular base reaches a maximum depth, at the centre of the structure, of just over 4 mm. The sebaceous layer comprises less than a millimetre of this depth; individual glands are generally small and are grouped around hair follicles. The bulk of the gland consists of tightly coiled apocrine tubules with large dilated lumina. Airborne odour could be produced as the hairs of the gland are erected (Moy 1971). The gland appears to be active at an early age: when two young pronghorn were raised together in captivity, the eight-day-old female intensely licked the gland of the five-day-old male (Müller-Schwarze and Müller-Schwarze 1972).

The ischiadic, or rump, glands occur in the centre of each white rump patch of both male and female pronghorn. Glandular activity is only visible externally from a waxy secretion at the base of the hairs which has a pungent odour. A layer of small lobulated sebaceous glands overlies a deeper zone of sudoriferous tubules. A detailed study by Moy (1970) shows similar glandular activity in males, females, and in castrates, with little variation through the year.

*Caudal glands*    The caudal gland (Fig. 14.2) of caribou and reindeer consists of a deep layer of glandular skin which covers the ventral and lateral surfaces of the short hair-covered tail. The gland occurs in males, females, and calves and produces a pungent yellow exudate. Sebaceous glands in the tail are small and associated with hairs. Underneath is a deep layer of apocrine sudoriferous glands which empty into a funnel-like orifice at the base of a hair above the opening of a sebaceous gland. The layer of apocrine tubules reaches its maximum depth (c. 2 mm) on the underside of the tail near its tip. Odour may be discharged when the hairs are erected and the tail held vertical (Lewin and Stelfox 1967; Müller-Schwarze, Quay, and Brundin 1977). The caudal gland of red deer comprises the bulk of the tail and reaches about 140 g in both males and females (Schaffer 1940; Lincoln 1971). It is composed of modified sebaceous glands and shows no sign of seasonal changes in activity other than those attributable to changes in condition; castrated males have particularly large glands (Lincoln 1971).

The tail of male musk deer, *Moschus moschiferus*, is short, largely hairless, except for a large tuft at the end, and is normally concealed in the rump hair. It is highly glandular and two longitudinal grooves that extend from base to tip are filled with yellow, crumbly secretion (Pocock 1910).

*Anal and proctodaeal glands*    There is a complex system of glands in the terminal section of the rectum and both within, and immediately outside, the sphincter musculature. These have been reviewed by Schaffer (1940) and Ortmann (1960) but, according to the latter, the account is still far from complete. There are some problems of definition, particularly with the glandular pouches found between the base of the tail and the anus. These are usually called subcaudal pouches although the histological account by Ortmann (1960) suggests that they might be better regarded as a variant of the glands in the anal region.

The proctodaeal and paraproctal glands of the African chevrotain, *Hyemoschus aquaticus*, are particularly well known through the work of Dubost (1975). The proctodaeal glands form a continuous sleeve which ranges between 4 and 5 mm deep, being deeper in males. Most of this consists of greatly enlarged sebaceous glands. In males the apocrine sudoriferous layer is thin and sometimes absent while in females the sudoriferous layer is up to half the thickness of the sebaceous layer. Two spherical paraproctal glands, 7-10 mm in diameter, form discrete structures within the proctodaeal sleeve. Each contains a central reservoir which opens into the proctodaeal space through a single duct. The structures consist entirely of sebaceous glands, and sebum is forced into the central reservoir by contractions of surrounding muscles. Presumably the faeces are coated with secretion as they pass through and distend the proctodaeum.

In other species specialized glandular fields become more dense outside the anus. In roe deer the cutaneous zone and that next to the anus are characterized by densely packed sebaceous glands, while outside these are gradually replaced by apocrine glands; towards the base of the tail there is a solid field of apocrine glands (Ortmann 1960). There is a similar area of apocrine glands in red deer (Schumacher 1943) and a massive concentration of these glands in the anal and circumanal area of the giraffe which seems interesting in view of the absence of other cutaneous glands in this species (Ortmann 1960). Ortmann also describes the subcaudal pouches in *Capra ibex*, *C. hircus*, *Hemitragus*, *Pseudois*, *Ammotragus*, and *Ovis*. In the first of these the pouch is about the size of the ball of a thumb. It has a concentration of sebaceous glands at each rim and, at the rim nearest the tail, this layer is underlaid by a vast field of apocrine sudoriferous glands. The bottom of the pouch has a continuous layer of very small sebaceous glands and is heavily vascularized.

*Preputial glands*　Preputial glands have been adequately described in only a few species but are probably more common than suggested in Table 14.1. The prepuce of rutting fallow deer becomes swollen with enlarged sebaceous glands and the epidermis covered with keratinous villi-like processes. At this time the voided urine and the hair tuft at the end of the penis sheath have a pungent odour which is absent in bladder urine (Kennaugh, Chapman, and Chapman 1977). The preputial gland of male chevrotain also consists mainly of sebaceous glands which emerge through numerous tiny apertures arranged in a circle around the opening of the penis sheath (Dubost 1975). As in the fallow deer, secretion is added to the urine during urination. Sharpe's grysbok, *Raphicerus sharpei*, have a discrete sac-shaped gland whose secretion empties into a duct that opens anterior to the urethra (Ansell 1964). The well-known gland of the musk deer is anatomically similar but reaches 'the size of an orange' (Lederer 1950).

*Inguinal glands*    Inguinal glands (Fig. 14.3) consist of either shallow, oval-shaped depressions or deep pockets which lie in the sparsely haired area near the mammae. They always number one or two pairs and they sometimes contain a tuft of long hair, as in oribi (Fig. 14.3). Inguinal glands often produce copious and strongly smelling secretion (Pocock 1910). That of Uganda kob is a yellow waxy substance with a pungent smell (Buechner and Schloeth 1965). Pocock (1910) described the odour of secretion from the glands of the Persian gazelle, *Gazella subgutturosa*, as similar to the urine of mice; other glands smell of sour milk or faeces.

Fig. 14.3.  Inguinal glands of Uganda kob, *Kobus kob* (*left*), and oribi, *Ourebia ourebia* (*right*). (Redrawn from Pocock (1910); semi-diagrammatic.)

A number of inguinal glands have been examined histologically. Those of male Thomson's gazelle, *Gazella thomsoni*, have a superficial layer of sebaceous glands (1.9 mm deep) lining the glandular sac and a layer of apocrine glands (1 mm deep) (Mainoya 1980). Both discharge into the lumen of the gland and produce a yellow waxy secretion. The sebaceous glands of females are non-ramified and scattered in the dermal zone and the apocrine-gland layer is thinner (0.3 mm thick) and apparently less active than that of males. The inguinal glands of sheep have similar layers of sebaceous and apocrine glands (Schaffer 1940) but the apocrine glands appear to be absent in reedbuck (Claussen and Jungius 1973).

*Glands on the legs*    Most leg glands occur in the skin overlying the tarsal joint and the metatarsus. The rare exceptions include the glands associated with the false hooves in the four-horned antelope, *Tetraceros quadricornis* (Fig. 14.4; Pocock 1910) and the small, flask-shaped glands that form lines on the posterior surface of the carpal section of the leg in pigs (Schaffer 1940).

Tarsal glands are usually visible as a tuft of long hair on the medial side of the tarsal joint (Fig. 14.2). In the species examined they contain enlarged sebaceous and apocrine sudoriferous glands and enlarged erector pili muscles. In the centre of the glands of black-tailed deer the muscles reach the considerable length of 1.3–3.2 mm (Quay and Müller-Schwarze 1970). The dermal layer, containing hair follicles, muscles, and sebaceous glands, varies between 1.1 and 2.6 mm; the layer of sudoriferous tubules beneath is usually less than a millimetre. The structure of the gland is similar in white-tailed deer, *O.virginianus* (Quay 1959) but in the caribou the sudoriferous layer is much thicker in females (1.45 mm) and thinner in males (0.35 mm) (Quay 1955). This sex difference is not apparent in reindeer (Mossing and Källquist 1981). Reindeer tarsal glands appear to maintain a similar level of glandular activity throughout the year. The visible secretion of tarsal glands often takes the form of a powdery or flaky substance that collects amongst the hair or adheres to its surface (Pocock 1910; and others). In black-tailed deer there are specialized cuticular scales along the hairs of the tarsal gland which may enhance the retention of secretion and release of odour; such specialized hairs are called osmotrichia (Müller-Schwarze, Volkman, and Zemanek 1977).

The metatarsal gland of black-tailed deer is externally visible as an elongated patch of long hair, with a central keratinized ridge which lies about half way down the outer surface of the metatarsal section of the leg. In impala (Fig. 14.4) the gland lies above the false hooves and the conspicuous dark hair lies

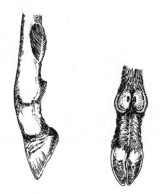

Fig. 14.4. Metatarsal gland of impala, *Aepyceros melampus* (*left*), and unguicular glands (in the false hooves) of four-horned antelope, *Tetraceros quadricornis* (*right*). (Redrawn from Pocock (1910).)

in two distinct zones on either side of the central ridge (Pocock 1910). The glandular tissue of black-tailed deer is characterized by greatly enlarged apocrine glands which form a layer 1.3 to 1.8 mm deep. A smaller quantity of enlarged sebaceous glands are found at the surface near to the central ridge (Quay and Müller-Schwarze 1970). The metatarsal glands of red deer contain well-developed sebaceous and apocrine glands and are visible as a tuft of hair about 2.5 cm in diameter (Lincoln 1971). Metatarsal glands are generally of similar appearance in males and females and seem to show little seasonal variation in activity (e.g. black-tailed deer: Quay and Müller-Schwarze 1970; red deer: Lincoln 1971).

*Pedal (or interdigital) glands*  Pedal glands (Fig. 14.5) consist of a variety of depressions or invaginations which lie between the two digits and open either anteriorly, usually just above the hooves, or ventrally between the base of the hooves. Ventral openings are rare but occur in dik-dik (Fig. 14.5) and in a number of duikers (Pocock 1910). Anterior openings are very common if highly variable. Apart from the deeply invaginated examples in Fig. 14.5 there

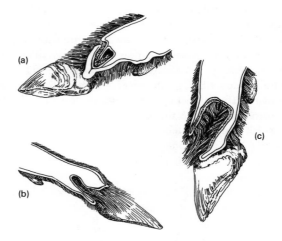

Fig. 14.5. Sagittal sections of the feet of goral, *Naemorhedus goral* (fore-foot (a); dik-dik *Madoqua phillipsi* (hind-foot) (b); and black wildebeest, *Connochaetes gnou* (fore-foot) (c) to show the pedal glands. (Redrawn from Pocock (1910).)

are shallow depressions such as those of chamois, deep, but open-fronted, depressions such as those of the oribi and tiny pits in reedbuck (Pocock 1910). The glandular skin is sometimes almost bare, or with a covering of sparse, short hair and is sometimes covered with long, thick hairs that project out of the gland (e.g. wildebeest, Fig. 14.5). Most glands have long hairs which arise around the orifice or rim of the gland and extend down onto or between the hooves. These hairs sometimes become thickly coated with secretion. Pedal glands often produce large quantities of secretion of various colour and consistency: the interdigital glands on the fore-feet of male hartebeest produce a copious, viscous black secretion, while, according to Pocock (1910) the gland of a rhebok, *Pelea capreolus*, 'was filled with semifluid, pale-coloured, evil-smelling secretion, recalling exactly the odour of dog's dung'.

Pedal glands are sometimes restricted to the fore-feet (e.g. hartebeest), sometimes to the hind-feet (e.g. reindeer), and are sometimes found on all four (e.g. dik-dik: Pocock 1910; black-tailed deer: Quay and Müller-Schwarze 1970). Like other mammalian scent glands the pedal glands are composed of apocrine, sudoriferous, and sebaceous glands but the way in which these components are arranged varies between species. The large, sock-shaped glands of pronghorn antelope consist mainly of large, lobulated and actively secreting sebaceous glands; a few apocrine glands surround these and empty into the sebaceous ducts (Moy 1971). In contrast, the interdigital gland of the red duiker consists mainly of large nests of coiled apocrine gland tubules with small sebaceous glands in association with hair follicles in the upper part of the dermis (Mainoya 1978). Layers of richly lobulated sebaceous glands and of apocrine gland tubules are both well developed in caribou hind-foot pedal glands although the apocrine layer is deeper (Quay 1955). In the fore-feet the sebaceous glands and their erector pili muscles are no larger than in unmodified skin but the apocrine layer is similar to that of the hind-feet. In black-tailed deer there are well-defined layers of complex, multilobulated sebaceous glands and sudoriferous glands as in the glands of both fore- and hind-feet. In fully grown animals the layers are respectively about 1.3 and 1.4 mm deep. The glands contain slightly enlarged erector pili muscles which probably have the function of expelling glandular secretion (Quay and Müller-Schwarze 1970).

Pedal glands are usually similar in males and females (e.g. caribou: Quay 1955) and they do not become more active in breeding seasons. Mossing and Källquist (1981), in a study of reindeer, found the greatest depth of sudoriferous epithelium in castrated males. The amount of sebaceous tissue does not vary seasonally in reindeer and that of sudoriferous tissue is greater in the late winter than in the mating season. Both types of gland are well represented in the pedal glands of male roe deer and they remain large through the year (Adams and Johnson 1980).

*Other sources of social odours*

The occurrence of social odours from saliva has been inferred in a number of mammals. Amongst ungulates the best-known case concerns the submaxillary and parotid salivary glands of male domestic pigs. These produce the steroids androstenol (Patterson 1968) and various androstenes (Booth 1980) which have a signal role in reproductive behaviour. Saliva is used in scent-marking in European wild boar, *Sus scrofa* (Beuerle 1975) and could have a signal odour in species such as caribou and reindeer: males perform a distinctive licking display next to females during precopulatory behaviour (Bergerud 1973) and a conspicuous breath odour develops in response to increasing androgen levels during the rut (Mossing and Damber 1981).

The relatively unspecialized glands of the body surface might sometimes provide a source of signal odour. The enlarged sebaceous glands on the back and sides of male pigs contain androstenol and androstenone and produce a characteristic odour; they are less active in castrated males and females (Stinson and Patterson 1972). Seasonal changes in the activity of sebaceous glands over wide areas of body skin are common and in some cases they appear to be under androgenic control. The glands between the horns, and on the neck and shoulder of male goats, become greatly enlarged during the rut. This increase does not occur in other parts of the body, or in females or castrated males (McEwan Jenkinson, Blackburn, and Proudfoot 1967).

Vaginal secretions are undoubtedly a source of odours that are detected by males during reproductive behaviour although most evidence for artiodactyls is circumstantial. However, mucous from the vulva of an oestrous pig, when smeared on a dummy, will sometimes induce mounting by a recalcitrant male (Booth 1980).

Faeces and urine often elicit behavioural responses and are often used in scent-marking (see pp. 572-4) but it is not known to what extent their signal value depends on the addition of glandular secretion, such as that of preputial glands to urine and proctodaeal glands to faeces. Urine sometimes has a strong odour during the rut in species which are not known to have preputial glands (e.g. reindeer: Mossing and Damber 1981).

## Marking and odour production

Olfactory and gustatory signals appear to be transmitted in two main ways. The substances containing the signal are either placed on some object so that detection by other animals (or the same animal) can occur later, or they are produced in such a way that the recipient can detect them directly. Examples of the first case are when animals mark objects in their territory, or themselves, with glandular secretion, and of the second when animals release airborne odour that releases flight by group members or when females produce substances in their urine that allows the detection of oestrus when tasted by males. Information about the first sort of signalling is more comprehensive because marking behaviour is often easily observed while odour production can usually be detected only when, and if, there is an immediate response.

The aim of the section that follows is to describe marking behaviour and odour production, using what observational data are available, and thus to provide the information that will be placed in a functional context in the next section.

### Marking objects

*Manufacture of marking sites*    The extent to which artiodactyls manufacture marking sites is unknown but it occurs unequivocally in at least one antelope: territorial male oribi bite off tall grass culms to an apparently optimum height, before marking the tips with the copious black secretion of the antorbital gland

(Fig. 14.6; Gosling 1972). Two other cases that may be comparable are those of territorial male pronghorn which nibble at sites on shrubs before marking with the subauricular gland (Kitchen 1974; Kitchen and Bromley 1974), and of male chevrotains, *Hyemoschus aquaticus*, which smell plants at about their shoulder height, eat the foliage, then mark the leafless stem tips with the inter-mandibular gland (Dubost 1975) (Fig. 14.7). Finally, territorial male Coke's hartebeest, *Alcelaphus buselaphus cokei*, reduce dwarf *Acacia drepanolobium* trees to bare stumps by vigorous thrashing with the horns, and the tip of the stump is marked with the antorbital gland and rubbed against the skin of the forehead and between the horns (Gosling 1972, 1975). However, while a marking site is clearly produced in this case there are other possible interpretations (see p. 574).

Fig. 14.6. Manufacture of a scent-marking site: a territorial male oribi bites off a tall grass culm (*upper*) and marks the tip with the viscous black secretion of its antorbital gland. (Redrawn from Gosling (1972).)

Fig. 14.7. Chevrotain, *Hyemoschus aquaticus*, marking the tip of a plant stem with the secretion of the intermandibular gland. (Drawn from a photograph in Dubost (1975).)

*Marking with glandular secretion*   Object marking consists of rubbing the secretion of a specialized gland, or some other substance with equivalent signal value, onto objects such as plants or rocks. Of the glands listed in Table 14.1, at least seven (interramal, frontal, antorbital, subauricular, postcornual, occipital, and dorsal) are regularly used to mark objects. Most of the rest are never used in this way. Other substances used in object marking include saliva and, with some problems of definition, urine and faeces. In the artiodactyls excreta are not placed on prominent objects, as in many carnivores, but instead constitute the object itself (see pp. 572–5).

Chevrotains and steenbok, althogh distantly related, use the interramal gland in a similar way. After sniffing a site such as a vertical plant stem they extend the muzzle and rub the gland on to its apex (Fig. 14.7) (Dubost 1975; Cohen and Gerneke 1976). In captivity, mouse deer, *Tragulus napu*, use the gland to mark food and water containers and straw on the ground (Ralls *et al.* 1975).

The unique frontal gland of the muntjac is generally used to mark grass tussocks, stones, and sometimes the flat ground. Individuals typically sniff a site, retract the muzzle towards the throat, and lean forward to press the forehead against the ground. The frontal (and antorbital) gland is rubbed backward and forward, and both remain dilated for a short while after the head is raised (Dubost 1971). Marking with the frontal gland by territorial male impala is more typical of those species with diffuse glandular tissue over the whole forehead (see pp. 552). Impala tilt the head forward and rub the forehead against a plant with gentle vertical movements. The ears are directed backwards and the eyes are half closed (Jarman 1979). This sort of marking, which also occurs in roe deer (Kurt 1968), is very common in a wide range of bovids and cervids (Fig. 14.8) but in most it is not known whether or not the forehead skin contains specialized glandular tissue.

Fig. 14.8. A territorial male Coke's hartebeest, *Alcelaphus buselaphus cokei*, rubs the skin of the forehead and pedicel on to the tip of a stump. Such stumps, of the dwarf tree *Acacia drepanolobium*, are also used for antorbital gland marking and are created by thrashing actions of the horns. (Drawn from a photograph.)

Marking with the secretion of the antorbital gland is possibly the best known form of scent marking; the illustration of male black-buck, *Antilope cervicapra*, marking the end of a twig in Hediger's early paper on territory demarcation (Hediger 1949) is typical. The head movement involved is almost invariably a deft inclination towards the site, bringing the gland into contact, followed by small nodding movements as the secretion is applied. The contact between the gland and the site is variable, depending largely on the great variety in the structure of antorbital glands in different species. Coke's hartebeest usually rub the entire dorsal surface of the gland against the 1–3-cm wide apex of the broken stumps of small trees and leave a musty odour but no visible secretion (Gosling 1975). The related topi, *Damaliscus korrigum*, has a deeper invagination in the centre of the gland and lowers its head so that the tips of low herbs and forbs enter the orifice where they are anointed with a gelatinous secretion. Thomson's gazelle (Fig. 14.9), dik-dik, and blackbuck are among the species whose pocket-shaped glands open as they approach the tip of a stem or twig and close over it, leaving a visible smear of secretion. Antorbital gland marking has only recently been recognized in some species even though the associated behaviour was well known. Sowls (1974) believed that the collared peccary illustrated in Fig. 14.10 was rubbing its cheek in dorsal gland secretion which had previously been smeared on the site. However, peccaries have a large and active antorbital gland and the animal illustrated was probably using it to mark (Byers 1978). Artiodactyls often mark more than once during a bout and sometimes alternate the left and right gland (Walther 1979).

Many bovids and cervids thrust their horns and antlers into herbaceous or

Fig. 14.9. A territorial male Thomson's gazelle, *Gazella thomsoni*, marks the tip of a grass culm with the black secretion of the antorbital gland. (Drawn from a photograph.)

Fig. 14.10. A collared peccary, *Tyassu tajacu*, marking a stump with the secretion of the dorsal gland (*left*) and with the antorbital gland secretion. (Redrawn from Sowls (1974).)

woody vegetation then gently 'weave' the head from side-to-side and/or 'thrash' the vegetation with vigorous sweeping or nodding movements. Male Grant's gazelle, *Gazella granti*, retract the muzzle, lower the horns close to the ground, and then rhythmically weave the head back and forth through low vegetation (Walther 1965). This species has functional antorbital glands but does not use them to mark small objects as in the case of Thomson's gazelle. 'Weaving' may be a functional equivalent. Adult male blackbuck open their antorbital glands during 'weaving' and sometimes a male stops to mark a plant stem in the specific manner described above (Dubost and Feer 1981). In contrast, territorial male impala thrash bushes with great vigour. The horns are lowered to a horizontal position then whipped up through the bush to the vertical (Jarman 1979). The evidence for an olfactory role is that the glandular forehead makes contact with the vegetation and that the male often sniffs the site after sweeping the horns upward. The sequence of thrash down/sweep up/sniff may be repeated several times. Another case of such generalized application of secretion is the striking behaviour known as 'preaching': male sambar, *Cervus unicolor*, and chital, *Axis axis*, stand upright on their hind legs and wipe their face, and prominent antorbital glands, on the leaves of overhanging branches (Schaller 1967).

But, why do these animals involve the horns and antlers rather than just wiping the face on to the vegetation? The main theoretical possibilities are (i) that the disturbed vegetation is a visual advertisement for the scent mark (see pp. 574-5); (ii) that the behaviour creates a site that is more suitable for marking (see pp. 566-7); (iii) that the horns, and other parts of the head become marked with secretion; and (iv) that the animal marks itself not only with its own secretion but also with the odour of the crushed vegetation; these last possibilities are considered on pp. 575-8).

Subauricular, postcornual, and occipital glands are all used in object marking. Male pronghorn sniff and sometimes nibble the tip of a plant stem (at about gland height) then extend the head and rub the subauricular patch (Kitchen 1974; Kitchen and Bromley 1974). No marking use is known for the subauricular patches of reedbuck or oribi. Mountain goats, *Oreamnos americanus*, and chamois both mark by drawing vegetation, such as a thin branch or herb stem, across the postcornula gland (Geist 1963; Hediger 1949). The occipital or poll gland of the camel, *Camelus bactrianus*, is rubbed against fences and piles of branches in captivity and marked objects become darkly stained (Wemmer and Murtaugh 1980).

The dorsal glands of springbok and pronghorn antelope are not used for object marking but collared peccaries mark objects such as tree stumps (Fig. 14.10; Sowls 1974). The marked surface acquires a conspicuous brown coloration and 'short streams' of secretion are often visible.

The leg glands are not employed in any specific marking activity but some occur in positions which press against the ground during lying and could thus scent-mark bedding sites. This interpretation can be applied to metatarsal glands, such as those of impala, the unguicular glands of the four-horned antelope (Fig. 14.4), and the carpal glands of pigs, which are located posteriorly on the fore-legs (Schaffer 1940) and pressed against the ground during lying with the legs outstretched. Inguinal glands could also be pressed against the ground while lying. This interpretation is not applicable to the tarsal glands, which occur on the medial (inner) surface of the tarsal joint (Pocock 1910).

Pawing, that is drawing the forefoot backwards in contact with the ground, is sometimes performed with great vigour and it has been suggested that secretion from the pedal gland is transferred to the substrate (e.g. Walther 1979). This interpretation is only possible for glands in the fore-feet; artiodactyls never paw with the hind-feet even though pedal glands in hind-feet are common

(Pocock 1910; Schaffer 1940). A problem in implicating pawing as a marking behaviour is that the pedal glands do not seem structurally adapted to depositing secretion as the foot is moved backward. The opening of the gland is usually anterior (Fig. 14.5) which suggests that marks would be left on objects as the foot was drawn *forwards*. The obvious candidate for such marking behaviour is the movement of the foot during walking, with secretion being left on herbage that passes between the digits and is pulled directly over the surface of the gland. A few pedal glands have ventral openings, an example being the dik-dik (Fig. 14.5), which is one of the many antelopes which paws before defecating on well-defined dung piles. A link with pawing thus seems possible but it is difficult to imagine a less suitable substrate for a mark than the dry dusty soil that is typical of such sites.

Unequivocal observations of the use of saliva in object marking are rare. A well-documented case is that of rutting male wild boar, which anoint vertical sticks and tree trunks with viscous saliva (Beuerle 1975).

*Marking with faeces and urine*    Faeces and urine are rarely used to mark objects in the artiodactyla. Rather they, and particularly faeces, are used to create conspicuous objects that have olfactory significance. These are usually dung piles, concentrations of faeces that accumulate during prolonged use, which often measure several metres in diameter. The concentrated faeces may bear the odour of the proctodaeal and anal glands (Ortmann 1960); odours that are a product of the food ingested and digestive processes (including breakdown by the protozoa and bacteria that constitute the 'gut flora' of ruminants) could also have signal value. Urine might have similar intrinsic signal value in addition to the specialized additions of preputial glands.

Dung piles are constructed by distantly related species and most consistently by territory owners. Territorial male hartebeest defecate almost entirely on dung piles at their territory boundaries. These piles measure up to 5 m in diameter and often contain thousands of faecal pellets. Typically, hartebeest nose the faeces, paw (sometimes vigorously with alternate feet), kneel, rub the forehead, face glands, and horns in the dung pile, stand up, step forward, and defecate on the spot that was nosed and pawed. Sometimes kneeling and head rubbing are omitted and sometimes horning the ground and urination inserted (Gosling 1975). Similar sequences are performed by many antelopes. Wildebeest sometimes lie down and roll during the sequence (Estes 1969). Dik-dik (Fig. 14.11), Thomson's gazelle, pronghorn antelope and blackbuck are among the many that do not kneel or lie: their sequence is usually nose/paw/urinate/defecate (Walther 1964, 1968; Kitchen 1974; Hendrichs 1975; Dubost and Feer 1981). Male pigmy hippopotamus, *Choeropsis liberiensis*, spread their faeces by rapid tail movements, and simultaneously spray a jet of urine backwards; the muscular tail appears to be adapted for this purpose, since it is less developed in the female, which does not defecate in this way (Hediger 1949).

Fig. 14.11. A territorial male dik-dik, *Madoqua kirki*, sniffs at a dung pile (*upper left*), paws, urinates, and defecates (*lower right*). This sequence of actions occurs with a high degree of consistency. (Drawn from photographs.)

Faeces and urine are frequently augmented by the secretion of specialized glands. The case of muntjac demonstrates the potential complexity of the odour signals that could result. Pausing in their movements, individuals sniff a point on the ground, rub their frontal and antorbital gland on it, then step forward to defecate or urinate on the same spot (Dubost 1971). Dik-dik also mark with the antorbital gland near to dung piles (Tinley 1969). One large mark generally occurs close to the dung pile and the link between the two forms of marking is confirmed by experimentally moving suitable twigs nearer to a dung pile than the existing marking site; in every case the subjects prefer to mark the site that is nearest to the defecation point (Gosling, unpublished).

In a few cases pairs or groups of animals contribute to dung piles. Dik-dik territories are occupied by an adult male and female which have a long-term pair-bond. These add to dung piles together and in a consistent fashion. The female approaches the dung pile first and, without sniffing or pawing, defecates then urinates. The male follows and sniffs, paws, urinates, and defecates (Fig. 14.11) (Hendrichs 1975). Larger groups appear to use the same dung piles when their membership remains similar for long periods. Female vicuña,

*Vicugna vicugna*, and their offspring live in such groups within the territory of a single male and both males and females defecate on well-defined dung piles: both smell the dung, make 'kneading' movements with the fore-feet then defecate and/or urinate (Koford 1957; Franklin 1974).

The striking postures adopted by animals as they urinate and defecate at dung piles also have clear significance as visual displays. For example, the urination and defecation postures of gazelles are conspicuously exaggerated (Walther 1968) (Fig. 14.12). These postures sometimes occur without the production of faeces or urine (muntjac: Dubost 1971; vicuña: Franklin 1974; and others).

Fig. 14.12. Linked urination and defecation by a territorial male Thomson's gazelle, *Gazella thomsoni*. Such postures have a visual display function and sometimes occur without the production of urine or faeces. (Drawn from photographs in Walther (1979).)

## Visual advertisement of scent marks

A number of behaviours that occur in association with scent-marking yield conspicuous visual features which might attract the attention of other animals and thus 'advertise' the mark. The most common are scraping bark, thrashing vegetation, and pawing the ground. There is no direct evidence for a visual advertisement role although field experiments would be relatively simple.

As mentioned on pp. 569–71 many artiodactyls thrash bushes or grass with their horns or antlers. These behaviours produce broken and torn foliage which could act as a visual signal that a scent mark is present. Thrashing commonly occurs in the presence of opponents and might reinforce a link between the visual and olfactory stimuli. However, in this context it is usually interpreted as a threat. Schaller (1967) gives examples: male chital pause while circling an opponent to thrash tufts of grass and shrubs with their antlers; on one occasion a male 'belabored a *Butea* shrub so violently . . . that the leaves flew high into the air'. Barasingha, *Cervus duvauceli*, also jab and thrash grass and bushes in similar situations. Thrashing may also be a form of 'redirected attack': one young male barasingha thrashed a bush after losing two sparring contests (Schaller 1967).

The pale appearance of wood that is exposed when bark is scraped away, often appears to be used to advertise scent marks. Roe deer thrash small trees with their antlers and, in the case of conifers, often leave a completely barked white stem which is marked with the forehead gland (Kurt 1968). Male muntjac remove patches of bark from trees or shrubs by scraping with the lower incisors. After 3–10 scrapes the animal rubs the hair-covered pedicel of the antlers on the site then sniffs and scrapes again. This sequence is repeated up to 30 times (Dubost 1971; Barrette 1977) and a conspicuous bare area of wood is produced. Saliva and, presumably, wood sap are rubbed on the skin of the pedicel while glandular secretion from the pedicels (if any exists) are rubbed onto the wood (Dubost 1971). Female elk, *Cervus canadensis*, similarly smell the trunk of a sapling or branch, before removing an area of bark by scraping with the incisors. Afterwards they rub the muzzle and chin on the exposed surface then rub these parts on their flank. Males perform a similar sequence except that the bark is removed by their antlers (Graf 1956). As with foliage thrashing, bark removal by horns or antlers might sometimes be a threat but this possibility can be excluded in the case of bark stripping with the incisors.

The role of pawing (and more rarely horning the ground) could be clarified by more detailed observation. Functional possibilities include depositing pedal-gland scent, removing previous marks, spreading the mark created by the subject and creating a visual advertisement. The first two seem improbable: marking with the pedal gland is unlikely (see pp. 564–5) and only two artiodactyls are known to paw after defecation (ibex: Walther 1979 and musk deer: Frädrich 1966). However, pawing before defecation is extremely common (Walther 1979) and all pawing creates a visual mark: often this is impressive with scratch marks from the hooves, an excavated depression, and scattered earth and faeces (e.g. wildebeest, Estes 1969; hartebeest, Gosling 1975); in dik-dik dung piles the depth of soil that has been loosened by pawing and mixed with faeces reaches 10 cm and these disturbed patches can be seen at considerable distances (L. M. Gosling and M. Petrie, unpublished).

*Self-marking*

Self-marking using glandular secretion, faeces, and urine is very common and a number of species also anoint their bodies with foreign substances, including mud and wood sap, in a way that seems comparable to 'scent rubbing' in carnivores (e.g. Rieger 1979).

Antorbital gland secretion is used for self-marking both directly, by rubbing the gland onto the body, and indirectly by rubbing the body on marked objects. Hartebeest rub their antorbital glands on to the sides of the body (Fig. 14.13; Gosling 1982) and musk ox mark their outstretched fore-leg (Pedersen 1958).

Fig. 14.13. Self-marking. A territorial male Coke's hartebeest, *Alcelaphus buselaphus cokei*, rubs the antorbital gland on to its side.

Most artiodactyls that mark objects with their antorbital glands also rub other parts of the head on the site: peccaries rub their cheeks while marking (Sowls 1974; Byers 1978) and hartebeest rub their forehead and sometimes other parts of the face on the end of marked stumps (Gosling 1975). Tsessebe, *Damaliscus lunatus*, mark the tip of short grass stalks with the sticky, gelatinous secretion of the antorbital gland, then rub the head and horns on the site. In the related blesbok, *D. dorcas*, this results in the accumulation of thick layers of secretion in the grooves on the horns (Joubert 1972). In other cases these behaviours occur before marking: Schaller (1967) saw a male barasingha rub its neck up and down a sapling, then gently wipe its antlers and afterwards both antorbital glands over the same area. If self-marking is involved then the animal may mark itself with the odour already present at the marking site; the theoretical consequences of this interpretation are considered in pp. 593–8. Animals that 'weave' their face and horns or antlers rhythmically back and forth through vegetation (see pp. 569–71) probably transfer secretion to the vegetation and comprehensively anoint their heads with its odour.

A self-marking behaviour provides the most easily interpreted use of an interdigital gland. Both male and female reindeer lift a hind-foot, insert an antler tip into the invaginated gland and then rub the tips on the inguinal area or tarsal joint (Espmark 1977). In those species with pedal glands in their hind-feet, secretion could be transferred to those parts of the body surface that can be reached during scratching. However, this seems unlikely because the foot moves in an anteroposterior direction in scratching while most interdigital glands have anterior openings (Fig. 14.5).

Many, perhaps most, male ungulates impregnate part of their pelage with urine (Hediger 1944). Goats and their relatives are particularly well known for this behaviour which seems to be largely responsible for their strong smell during mating periods. Ibex, *Capra ibex*, and markhor, *C. falconeri*, lower their hind-quarters and spray urine from an extended penis so that the fore-legs, chest, belly, and throat become saturated (Hediger 1950; Geist 1971; Schaller and Mirza 1971). Male chamois wet large areas of their belly and flanks by shaking their body while urinating, and both dibatag, *Ammodorcas clarki*, and camels, *Camelus dromedarius*, splash urine onto their bodies using their tails (Coblentz 1976; Walther 1979). In seasonally breeding species self-marking is often restricted to the mating period: male red deer spray the underside of the body with jerky movements of the penis and, at the time of peak reproductive activity, the urethral opening, brisket, and inside of the fore-legs become darkly stained with strongly smelling urine (Lincoln 1971).

In the behaviour known as 'rub-urination' males spray urine on to their tarsal glands while the opposing glands on the hind-legs are rubbed against each other. Whether this behaviour should be regarded as adding urine to a glandular secretion on the animal's body or adding secretion to a urine mark on the ground, or both, is a matter of conjecture. The area reached by the urine varies between species: male reindeer spray urine in a loosely defined jet which wets the underparts in general as well as the tarsal glands (Espmark 1964, and personal communication). The urine jet appears to be more accurately directed towards the glands in black-tailed deer (Müller-Schwarze 1971). A behaviour which may be analagous occurs in male eland, *Taurotragus oryx*. These large antelopes urinate on the ground, rub their foreheads in the urine and then rub the muddy mixture on to trees (Walther 1966; Hillman 1976); it is not known if the forehead is glandular in this case. Urine is also combined with the secretion of the preputial gland in fallow deer and the belly and legs of rutting males become stained through repeated spray urination (Kennaugh *et al.* 1977).

A number of species lie or roll in faeces and urine. Coke's hartebeest, tesessebe, blesbok, and bontebok, *Damaliscus dorcas*, territorial males lie in dung piles (Joubert 1972; Du Plessis 1972; David 1973; Gosling 1975). Wildebeest lie down then roll during the sequence of behaviours associated with defecation at dung piles (Estes 1969). Hog deer, *Axis porcinus*, squirt urine onto the ground before lying down on it (Schaller 1967). These behaviours reach their most extreme development in wallowing. Wallows are muddy pools that are sometimes excavated by pawing and digging with horns and antlers. Male barasingha dig in swampy areas by thrusting the antler tips into the ground, then suddenly jerking

their heads up so that earth is thrown into the air; digging sometimes continues
after the males lie down in the excavated wallow (Schaller 1967). In some cases
urine, and possibly ejaculate, is added to the mud and the animals then lie,
sometimes partly submerged, in the strongly smelling mixture. During the rut,
male moose, *Alces alces*, dig wallows by alternate pawing with the fore-feet. In
between bouts of pawing they squat and urinate, sometimes up to ten times
during the excavation (Geist 1963). Male red deer dig with the fore-feet and
antlers and, while they do so, release spurts of a clear fluid (probably urine)
from their unsheathed penis. Like barasingha, red deer continue to scrape with
the antlers while lying in the wallow (Müller-Using and Schloeth 1967). Elk,
which lie in wallows for hours, also rub their antorbital glands on to the rim of
the pit (Struhsaker 1967). Rutting male bison, *Bison bison*, which 'wallow'
in dry, dusty places where the surface is broken by horning, as well as in wet
mud holes, urinate as they paw at the site and also during and after lying down
and rolling (McHugh 1958). Most of these cases appear to be elaborate forms
of self-marking for individual males but wallowing and dust bathing also have
body-care functions; female bison also wallow but they do not usually urinate
while doing so (McHugh 1958). Communal wallowing also occurs, for example
by mixed- and single-sex groups of red deer, outside the rut (Müller-Using and
Schloeth 1967; Gossow and Schürholz 1974).

### Marking conspecifics

A number of males mark females during precopulatory behaviour. Antorbital
gland secretion is used to mark the back, rump, or flank in gerenuk, *Litocranius
walleri* (Fig. 14.14; Backhaus 1958; Walther 1958; Leuthold 1971), Lichtenstein's hartebeest, *Alcelaphus lichtensteini* (Wilson 1966), and fallow deer
(Chapman and Chapman 1975). Male mouse deer, *Tragulus napu* and *T. javanicus*,
and male steenbok mark females with intermandibular gland secretion in the

Fig. 14.14. A male gerenuk, *Litocranius walleri*, marks a female with antorbital gland secretion during precopulatory behaviour. (Drawn from a photograph in Walther (1966).)

neck, back, and rump (Ralls *et al.* 1975; Cadigan 1972; Cohen and Gerneke 1976) and Gray's waterbuck, *Kobus megaceros*, saturate their own underparts by 'spray-urination' then rub their dripping chin and throat on the head or rump of a female (Walther 1966). The males of the other species, including most goats, which similarly spray their underparts (reviewed by Coblentz 1976) could transfer urine to the back and haunches of females during mounting and copulations. Male wild boar mark the back and flanks of females with saliva during courtship (Beuerle 1975). Sometimes marking is reciprocal: male and female Maxwell's duiker, *Cephalophus maxwelli*, press their large antorbital glands together (Ralls 1971) and various members of peccary groups, including both males and females, rub their cheeks on to each other's rump and dorsal gland (Sowls 1974); this behaviour may also involve marking with the antorbital gland.

Adult males sometimes mark each other during agonistic encounters. Male wildebeest rub the antorbital gland onto the rump of their opponent during boundary encounters (Estes 1969) and captive male Maxwell's duiker press their antorbital glands together before and during bouts of fighting (Ralls 1974). Male mouse deer, *Tragulus napu*, mark other adult males and also male offspring (Ralls *et al.* 1975).

Females mark their offspring in a number of species including Zebra duiker, *Cephalophus zebra* (Frädrich 1964), and muntjac (Dubost 1971). Female Maxwell's duiker mark each other by rubbing their antorbital glands together (Aeschlimann 1963; Ralls 1974) and female wildebeest rub their antorbital gland on to another's rump (Estes 1969).

*The production of substances that are directly smelled or tasted by conspecifics*

The vestibular nasal glands of marsh deer produce mainly low volatility lipids which casts doubt on the possibility of their discharging airborne odour. Jacob and von Lehmann (1976) suggest that the secretion might serve as the substrate for the product of another sudoriferous apocrine gland but it is not clear how this would occur. Langguth and Jackson (1980) watched deer touching noses and suggested that they were smelling the nasal gland. However, this behaviour is almost universal in the artiodactyls (Schloeth 1956, and others) most of which do not possess these glands. It may prove difficult to discriminate between the effects of the odour of nasal glands and of the other odours that are available near the nose and mouth. The saliva of male pigs contains the steroids androstenol and androstenone, which are produced by the submaxillary gland and which have a functional role in mating (Patterson 1968; Melrose, Reed, and Patterson 1971). Saliva is produced in large quantities during agonistic encounters between European wild boar (Beuerle 1975) and might be detected during 'yawning', which is common in conflict situations, and during nose-to-nose contact.

In those species with antorbital glands that are pocket-shaped and under muscular control it is common for the gland to be dilated in reproductive and agonistic contexts. For example, male muntjac dilate their prominent antorbital gland while closely pursuing females during precopulatory behaviour (Dubost 1971), male blackbuck and black-tailed deer dilate their glands in agonistic encounters (Schmied 1973; Brownlee, Silverstein, Müller-Schwarze, and Singer 1969) and red deer males open their glands during threat roaring (Lincoln 1971). Subordinate male impala sniff the forehead glandular area when it is presented by dominant males (Jarman 1979) although in this case the production of airborne odour is probably not under muscular control. Territorial male pronghorn similarly present their subauricular glandular patches to females during reproductive behaviour (illustrated by Gilbert 1973). In all of these cases involving glands on the head, the odours produced are also used for object marking.

The dorsal and ischiadic glands are used in a similar way by a few distantly related species, generally when the animals involved are alarmed. The odour of the dorsal gland can be clearly distinguished after a band of peccaries has fled and one individual has visibly ejected secretion to a height of 'about one foot' (Neal 1959). These observations provide a marked contrast to the use of the gland in object marking and when the gland of a conspecific is directly sniffed (Sowls 1974; see pp. 569–71); in these contexts the secretion clearly lacks any alarming effect. Springbok and pronghorn antelope both possess long white erectile hairs on the back and rump which become coated with the secretion of the dorsal gland and, in the pronghorn, the ischiadic glands. When pronghorn are startled, the rump hair is raised 'with a jerk that made the patch flash in the sun' (Seton 1927). At the same time Seton detected the musky odour of glandular secretion '20 or 30 yards down the wind'. Caribou and reindeer use their glandular tails in an analagous way. When alarmed, the long hairs around the gland are erected and the tail held stiffly upright as they flee (Lewin and Stelfox 1967; Müller-Schwarze, Quay, and Brundin 1977). The visual display of springbok and pronghorn is sometimes enhanced by spectacular stiff-legged, bounding gaits known as 'stotting' or 'pronking' (Bigalke 1972; Walther 1968, 1981). Stotting also occurs in a variety of antelopes that lack associated odour production, such as gazelles and hartebeest (Walther 1968; Gosling 1975). It is common in play but, in spite of reservations expressed by Walther (1981), it is most easily explained as a form of antipredator behaviour (see pp. 583–4).

There are a number of situations in which the secretion of inguinal glands appear to be smelled or tasted. Inguinal nuzzling of females by males is most conspicuous. Male Uganda kob thrust the muzzle deeply between the hind legs of a female and lick both inguinal glands and mammae; more rarely they approach the glands from in front of one hind leg, as in sucking (Buechner and Schloeth 1965). Inguinal nuzzling occurs in other species including lesser kudu, *Tragelaphus imberbis* (Walther 1958), waterbuck (Spinage 1969) and nyala, *Tragelaphus ongasi* (Anderson 1980). Of these, inguinal glands are present in kob and lesser kudu and absent in waterbuck and nyala.

Juveniles probably smell the odour of inguinal gland secretion when suckling but these glands are also common in males. Many ungulates smell and nuzzle their own inguinal regions but this also occurs in species which lack inguinal glands (e.g. hartebeest, Gosling 1975).

Evidence for odour production by metatarsal and tarsal glands is confined to the cervids and consists of observations of glandular activity during social encounters, responses by conspecifics to the glands and gas–liquid chromatography of glandular secretion. Adult male black-tailed deer erect and rhythmically move the hair tuft overlying the tarsal gland during a threat posture that includes lowering the head, laying back the ears and opening the antorbital gland. This behaviour is often followed by rub-urinating (Brownlee *et al.* 1969). Black-tailed deer also sniff, and sometimes lick, the tarsal glands of other deer both in captive groups and in the wild; in a group of fawns such inspections occur up to 5.5 times per hour (Müller-Schwarze 1971). When the odour of a gland is altered by the application of strange secretion or of a distillate of the secretion dissolved in petroleum ether, other deer approach and show prolonged sniffing and licking; extracts from tarsal hair release more frequent responses than extracts of the gland (Müller-Schwarze 1971). In an experiment with young male roe deer, Broom and Johnson (1980) found a well-defined response to the odour of both male and female metatarsal gland secretion and virtually no response to the secretion of the forehead and pedal glands.

Many artiodactyls smell the genitalia and anal region of conspecifics both in species that use faeces and urine in object marking and those that do not, and in species with specialized glands in these areas (preputial glands, anal glands, etc.) and those where they are absent. An example is that of territorial male hartebeest which use faeces to mark their territories and also smell the anal region of opponents during agonistic encounters (Gosling 1974, 1975). Most, probably all, artiodactyls smell the urine or vulva of females during reproductive interactions and some males produce odours (which remain unidentified) that have long-term 'priming' effects on female reproduction. These will be discussed on pp. 601–4 and the odours used by females to recognize their young on pp. 604–6.

## Context and function of olfactory signals

### Scents and marks as social signals

The olfactory signals of artiodactyls must be expected to have the general properties of all signals in that they cause changes in the behaviour of the responding animals that are to the advantage of the signaller. On average there should also be an advantage to the reactor but this is clearly not the case for particular signals under all circumstances. There are many examples of signals that appear to subvert a response to the signaller's benefit, irrespective of the consequences for the reactor (Dawkins and Krebs 1978). Cases from the artiodactyls which conform with this general framework are developed in the sections that follow. With one exception, there are no grounds for suspecting that olfactory signals differ in any fundamental way from signals that are detected by the other senses. The exception, which is frequently mentioned (for example by Mykytocwycz 1972) is that communication can take place through the medium of scent marks in the absence of the signalling animal.

It is often assumed that mammalian scent communication occurs through responses to one or a few signal compounds which are analagous to hormones. This belief, which originates in studies of insect 'pheromones', has been criticized by Beauchamp *et al.* (1976) on a variety of grounds which include the fact that most mammalian olfactory signals are extremely complex. Enough is known of the chemical composition of artiodactyl secretions (and of other sources of odour) to extend this generalization to the group but little is known of the functional significance of this complexity.

The techniques of analysis that are chosen determine, to a large extent, what is found in a given substance and too little is known about the operation of chemical signals to provide firm guidelines for what is appropriate in a particular case. For example, it is not known whether or not only volatile substances should be considered and whether lipids, which require different analytical techniques, have signal value of their own or are merely 'vehicles' for other substances. In spite of these problems some secretions have been partially analysed. These include the secretions of the preputial gland of musk deer (Lederer 1950), the tarsal gland of mule deer, *Odocoileus hemionus* (Brownlee *et al.* 1969), the subauricular gland of pronghorn antelope (Müller-Schwarze, Müller-Schwarze, Singer, and Silverstein 1974), the interdigital and caudal gland of reindeer (Brundin, Andersson, Andersson, Mossing, and Källquist 1978), and the pedal gland of blesbok (Burger, le Roux, Garbers, Spiers, Bigalke, Pachler, Wessells, Christ, and Maurer 1976; Burger *et al.* 1977). In a few cases, isolated and identified components have been shown to elicit particular behavioural responses but, partly because of the considerable technical difficulties, it has not been demonstrated whether these components are the only, or even the main, active constituents. Müller-Schwarze and his co-workers (Brownlee *et al.* 1969; Müller-Schwarze 1969, 1971) found that one lactone from the secretion of the tarsal gland of black-tailed deer elicited more licking by conspecifics than other isolated compounds but that the addition of further compounds caused a greater response. In addition, the distillate that remained after extraction of these compounds also elicited some licking. Recent theoretical advances in the evolution of behaviour suggest that there will be selection for both signals and responses to vary according to social context. However, the possibility that the signal value of an odour might be changed by varying the relative concentration of a number of compounds is excluded by attempts to identify single active components. To date, the investigations of odour signals and responses have been demonstrations of physiological capabilities rather than of function.

The importance of the complexity of odours is underlined by the common occurrence of behaviours which combine the products of odour-producing organs (some examples are given on pp. 573-4). One *general* explanation of these behaviours is that complex odours are less easily replicated and may thus be less ambiguous. Complex odours also provide an opportunity to vary a signal by altering the proportions of the constituents in different contexts. Regardless of which hypothesis is considered it seems generally clear that, in the artiodactyls at least, selection has favoured odour complexity and heterogeneity.

*Self-orientation*

The suggestion that scent marks are used in self-orientation is a recurring theme in the literature. Even Hediger (1949), in a rarely cited section of his classic paper on territory demarcation, suggests that marks might help animals to orientate themselves. More recently Walther (1978*a*) concludes that marking by territorial male Thomson's gazelle seems mainly 'to be significant for the owner himself and for his orientation'. Non-territorial animals, particularly those in very large ranges, might also benefit from such self guidance. Lyall-Watson (1964) believes that scent marking 'serves to maintain the animal's familiarity with its environment . . . odour is added to specific visual landmarks both to familiarize the animal with new territory and to refamiliarize it with old terrain'. A specific problem for this hypothesis is that since animals visit marking sites regularly it is difficult to say whether they do so in order to mark (or inspect the site for the odour of conspecifics) or in order to orientate or familiarize themselves. More generally, 'familiarity' is essentially a subjective concept and thus difficult to test. For these reasons it might be more profitable to abandon the hypothesis until it is reformulated with specific predictions about the effect of scent-marking on the efficiency of exploiting particular resources (such as a dispersed food supply or escape routes from predators).

*Antipredator behaviour*

The most conspicuous use of odour in antipredator contexts is the discharge from glands on the back, rump, or tail in springbok, pronghorn antelope, caribou, reindeer, and peccaries. In all cases the odour is apparently discharged as the potential prey detect, and suddenly start to run away from, a predator. In the two bovids this discharge is accompanied by a visual display including pilo-erection around the gland and spectacular leaping gaits. Caribou and reindeer similarly erect their short, highly glandular tails when alarmed (Fig. 14.2(b) and see p. 560). The adaptive significance of odour production in this context may be related to that of the linked visual signals. However, these are a matter of controversy; for example, it is not known whether the signal is directed towards conspecifics (although they are known to react in a number of cases) or to the predator. Signallers might increase their fitness if the signal alerted kin which then had an increased chance of escape. However, animals that flee first are not always those that escape—the opposite may be the case: animals that flee early may elicit a hunting response. Thus the signaller might improve its own chance of escape by triggering an escape response in nearby conspecifics either because the predator is attracted to another animal or because it becomes confused by many fleeing prey. On the other hand the prey might signal to the predator that it is difficult to catch relative to other animals.

Of these possibilities some seem relatively implausible when the signal concerned is olfactory rather than visual and when the frequent lack of nearby kin, at least for males in species with polygynous mating systems, and known variation in predator hunting techniques are taken into account. The idea which best survives these considerations is that the function of the odour is to elicit flight by conspecifics so that the chance of the signaller being caught is diluted. The same hypothesis is advanced for the case of alarm calling by birds in flocks (Charnov and Krebs 1975) and is known elsewhere as the 'never break ranks' theory (Dawkins 1976). But, it is unlikely that this dilution effect is simple, because predators, for example, lions, *Panthera leo* (Schaller 1972) and hyaena, *Crocuta crocuta* (Kruuk 1972) often select animals that are in poor health or in poor condition. When an animal elecits flight in conspecifics it increases the chance that an individual in poorer condition than itself will flee and thus reduce the chance that it will itself be selected for serious pursuit. It seems likely that a predator would use visual cues to select prey animals because individual odours must often be confused during hunting attempts. It has been suggested elsewhere that the bounding leaps shown by some ungulates in antipredator contexts signal the good physical condition of the individual relative to others (Zahavi, cited in Dawkins 1976; Gosling and Petrie 1981). The position of the erectile hair around scent glands suggests that pilo-erection might provide visual reinforcement for the odour.

The hypothesis suggested requires that the animals receiving the signal detect an intrinsic property of the odour rather than a contextual property such as that advocated for competitor assessment in the following sections. Theoretically this odour could be learnt by association with predation attempts or it could be recognized innately. Which of these is the case is unknown but there is experimental evidence for an intrinsic effect in the response of black-tailed deer to the garlic-like odour of the metatarsal gland which is also produced in 'fear-inducing' situations: subjects avoid feeding when the scent is placed underneath a perforated food bowl (Müller-Schwarze 1971).

## Territoriality

Many animals gain an advantage by denying conspecifics access to a limited resource. The resource may be of direct value, for example food, or of indirect value, for example a food supply that attracts potential mates. Resource defence becomes most explicit in territorial behaviour, which is most usefully defined as dominance with a spatial reference and some degree of exclusivity. While within the territority, the owner is dominant over potential competitors for the defended resource. In the artiodactyls, territoriality is most commonly expressed by individual males although there are some cases of pairs of males and females occupying territories and of species where all breeding adults occupy individual territories (Leuthold 1977). A feature of territories that becomes important when considering scent marking is that boundaries are defined to varying degrees. Sometimes they are very specific, particularly when under continual threat from intruders. Hartebeest territorial males sometimes adopt a dry stream bed or a road as their boundary and, under some circumstances, respond immediately when, at a distance, they see intruders cross (Gosling 1974). In contrast, male red deer defend a loosely defined area around a group of females during their breeding season (Fraser Darling 1937).

Almost all territories of artiodactyls are scent marked. The present section will deal with two aspects of this behaviour. The first explores territory marking as an activity that, regardless of function, is subject to various economic constraints such as a finite supply of marking substance and of time available for marking. This approach is useful in considering questions such as whether or not marks are optimally distributed to intercept animals moving in and out of a territory. The results obtained from this approach do not constitute evidence for a particular function of marking, but they are most simply interpreted in the light of the hypothesis that marks are distributed to maximize the chance of detection by intruders. Other possibilities will be mentioned but most discussion of functional issues is reserved for pp. 593-8 where existing hypotheses are reviewed and where existing data are reinterpreted. However, the only assumptions required for the economic argument are that marking does have a function and that this involves detection of marks by animals that enter the territory.

*The economics of scent-marking in territories*    If a territory owner scent-marked in order to maximize the chance that another animal, for example a mate or a competing intruder, would detect the marks, then it should mark the territory at every point that these animals might visit. The absence of such comprehensive marking is almost certainly a consequence of constraints that operate on the marking animal. These constraints, such as a limited supply of glandular secretion and limited time available for marking, will generally become more critical as the animal attempts to defend a larger area. This way of considering marking behaviour can be regarded as an economic approach because it emphasizes what an animal can *afford* to do under particular environmental circumstances (Gosling and Petrie 1981; Gosling 1981). It is illustrated in Fig. 14.15 by a graphical model that considers the trade-off between the effectiveness of marking an area of increasing size and its food value.

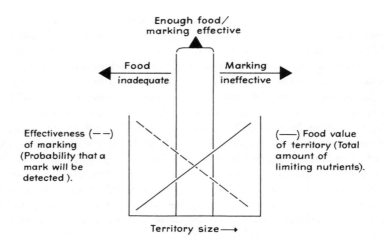

Fig. 14.15. A graphical model of the variation in marking effectiveness as territory size increases. The optimum size indicated could be modelled quantitatively if both variables were expressed in the same 'currency', preferably the fitness consequences for the marking animal. However, in this case it would be necessary to include other costs of territory occupancy, such as that of active defence. (From Gosling (1981).)

When marks are limited, territory owners can maximize the chance that they will be detected by placing them only where they are most likely to be found. They should avoid placing them where detection, although possible, is less likely. This consideration provides a framework for considering observed patterns of marking. It can be divided into behaviour that influences the conspicuousness of marks and patterns of marks that influence the chance of intercepting other animals.

There is good evidence that scent marks are placed at sites where they are likely to be detected by an animal that passes nearby. Most sites seem to be selected for their intrinsic conspicuousness, which is enhanced by a variety of visual advertisement. Examples of the first effect are the selection by male gerenuk of stiff twigs that protrude outside the surface of a bush for antorbital gland marking (Gosling 1981) and the use of old termite mounds for the construction of dung piles by territorial male blesbok (Coe and Carr 1978). I recently suggested (Gosling 1980) that the frequent observation of defecation by a number of species on common dung piles might be explained by interspecific scent communication. Another interpretation is that each animal makes the site more conspicuous with each addition and that subsequent animals simply take advantage of this when they place their own faeces on the pile. The wide range of behaviour that results in enhanced visual conspicuousness was reviewed above (pp. 574–5). Here it can be added that the advertisement of scent marks in this fashion can only be effective if the animal to whom the signal is addressed responds by approaching. While these animals could be manipulated into a response of this kind, even when it is to their disadvantage, such approaches could also occur because the scent mark contains information that is useful (see pp. 593-607).

Observed patterns of marks generally support the idea that territory owners place marks to maximize the chance of detection by another animal. For example, antorbital gland marking sites are sometimes at about head height and while this could be interpreted as simply reflecting the normal height of the gland, in fact, this is not the case: while active, animals move their heads over a wide height range. Browsing gerenuk feed at ground level and up to 1.75 m (when standing in the bipedal position) but 88 per cent of their marks are restricted to the narrow vertical range of 105 to 129 cm (Gosling 1981). Oribi mark at a similar height in relation to their head height and again the distribution is highly leptokurtic; in this case the distribution is determined by the marking animal which bites off stems at a suitable height before marking (Gosling 1972; Fig. 14.16). This height is quite different from the normal head height of active oribi, which spend most time grazing from short grass swards. Perhaps these marks are placed at a height where they have the best chance of detection by a walking animal.

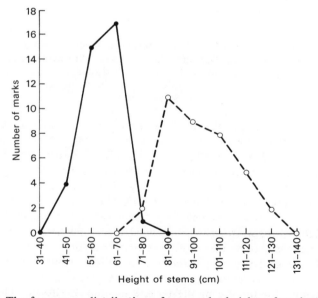

Fig. 14.16. The frequency distribution of grass culm heights after they have been bitten off and marked by territorial male oribi, *Ourebia ourebia* (●——●) compared with that of their original heights (o – – –o). 50-70 cm may be an optimum height for detection by walking intruders. (From Gosling (1972).)

This last interpretation receives some support from data which show that marks are often placed near to game trails. This was noted by Gilbert (1973) for the scent marks in pronghorn antelope territories and, in the case of gerenuk, 40 per cent of a sample of marks in one territory (*n* = 60) were at the edge of, or overhanging, a game trail (Gosling 1981). This tendency becomes even clearer in antelopes that live in closed habitats, such as the dik-dik; in an area of dense thicket in Tsavo National Park where I mapped dik-dik antorbital gland sites, nearly every mark was next to a trail (Fig. 14.17).

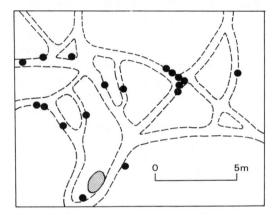

Fig. 14.17. A map of part of a dik-dik, *Madoqua kirki*, territory in *Commifera/ Sanseveria* scrub in Tsavo National Park, Kenya. Antorbital gland secretion marks (●), mainly on the tips of twigs, are restricted to trails and might thus intercept intruding animals. The stippled oval is a dung pile. (Gosling, unpublished observations.)

Territory owners might also be expected to mark on, or near, objects that would attract other animals. One obvious case is that of food: a mark placed on a preferred food plant would be more likely to be detected than one on a plant that was rarely eaten. This prediction has been tested for gerenuk and a relationship found between the frequency of marking and feeding on those plants where both occurred (Gosling 1981). However, there are exceptions: dead bushes or trees are marked very frequently and particular shrubs, for example *Calyptrotheca taitensis* are avoided even though they are a preferred food. The reasons may be that dead twigs provide an optimum site for marking, being stiff and generally free of concealing foliage, while the succulent and pliable twigs of *C. taitensis* are too unstable. Coe and Carr (1978) suggested that one reason for the use of the termite mounds as sites for blesbok dung piles is that the mounds are used for antipredator scanning; this interpretation can be extended to suggest that the scent mark is placed at a site which, for the same reason, is likely to be visited by a conspecific.

In an earlier section (pp. 571-2) it was suggested that the scent from pedal glands might be deposited passively as herbage was drawn between the digits while walking. If this is true then a territory owner would mark the herb layer at an intensity that precisely reflected its range use. Thus, seasonally preferred food, for example, would be marked most intensively. This could be important where grassland territories contain a number of discrete herb communities which are utilized at different seasons; one case is that of Coke's hartebeest, which has large and very active pedal glands and whose territories (that average 0.31 km$^2$) contain up to five grass communities which are utilized in a seasonal pattern (Gosling 1974). Marks left in the herb layer would presumably be detected during grazing. Müller-Schwarze, Källquist, Mossing, Brundin, and Andersson (1978) have shown that captive reindeer respond to pedal gland secretion by sniffing, licking, and searching the ground.

Because this form of marking is passive it differs from that predicted by the above formulation of an economic hypothesis: marking reflects the usage of a resource rather than being restricted to those parts that are most used. However, the pattern is explicable in economic terms if, unlike the substances used in active marking, the product of the pedal glands is super-abundant in relation to the amount of 'marking' (i.e. the number of steps) that is needed at particular levels of activity. Perhaps the waxy secretion, often found in the lumen of pedal glands (Pocock 1910 and others) has been selected as an odour 'vehicle' whose persistent qualities are adapted to the continual removal of small quantities as herbage brushes past. The characteristically large lumens of pedal glands (Fig. 14.5) could then be explained as a reservoir for this material.

In contrast to this hypothetical distribution of pedal gland scent, marks that are actively deposited by the head glands are placed on a minority of the available sites. Assuming that this is because marks are in limited supply then it is valid to compare observed spatial patterns of marking with the patterns that would be predicted from the economic argument. For example, if the pattern of marks is adapted to intercept animals entering a territory then the owner should mark the boundary of the territory. If it is an advantage for the animal entering to detect more than one mark then the owner should mark in two or more concentric 'circles' up to the limit of the marks available and the duration of their signal value. Since marks are limited, the animal should avoid marking except in this optimum pattern. Clearly such predictions do not apply to many field situations where details of topography and marking function would be needed to make the hypothesis realistic. However, they provide a general indication of the patterns that would be expected.

Until recently tests of these predictions have suffered from the technical problem that it is difficult objectively to describe observed patterns of marking. A number of observers report concentrations of marks around territory boundaries and some (e.g. Gilbert 1973; Walther 1978a; Jarman 1979) have seen behaviour that appears to be directed to this objective. However, marks also occur within territories and further information is needed to explore whether or not these are part of an integrated system of marking. The techniques available are statistical comparisons with distributions that would be expected by chance and a nearest-neighbour mapping technique that has recently been used to describe the pattern of marks in a gerenuk territory (Gosling 1981). The first procedure, used for the same data on gerenuk, yields the result that antorbital gland marks occur close together more frequently than if they were randomly distributed but it does not show the pattern that is responsible. This is revealed by taking a map of the marks, joining each mark to its nearest neighbour, then each group of marks, thus created, to *its* nearest neighbour, and so on until all marks are linked. The technique assumes functional significance in the proximity of marks but this is also implicit in the subjective judgement that marks are clustered around boundaries, or other objects, and the mapping procedure merely makes it explicit.

The technique has been used to investigate the pattern of antorbital gland marks in a gerenuk territory (Gosling 1981) and of the same type of marks in the territory of a Thomson's gazelle which was meticulously mapped by Walther (1978*a*). In both cases (Figs. 14.18 and 14.19) an important feature of the pattern revealed is a large ring of marks. In the gazelle this was known, from observation, to be the boundary of the males' territory. Although the circle

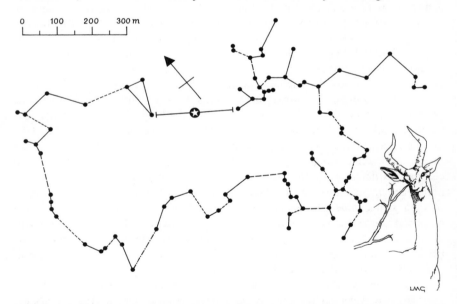

Fig. 14.18. The pattern of antorbital gland secretion marks in the territory of a gerenuk, *Litocranius walleri*, as revealed by the nearest-neighbour mapping technique described in the text. The broken line indicates the gap left when the mapping process is complete. (Redrawn from Gosling (1981).)

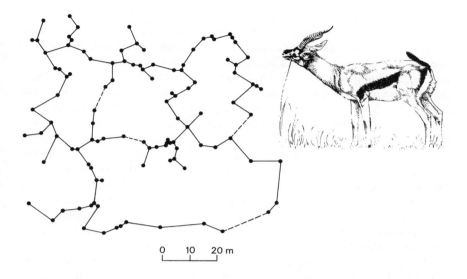

0    10    20 m

Fig. 14.19. The pattern of antorbital gland secretion marks in the territory of a Thomson's gazelle, *Gazella thomsoni*, revealed by applying the nearest-neighbour mapping technique, described in the text, to the map of marks obtained by Walther (1978*a*). The broken lines indicate gaps left when the mapping process is complete.

of marks made by the gerenuk was known to be the product of one territorial male it was smaller than territories mapped by Leuthold (1978*a*) and might have been a concentric circle within a ring of boundary marks (Gosling 1981). In either case the patterns of both species are broadly consistent with the predictions of the economic hypothesis. The existence of boundary marking receives further support from observations of territory owners that, in an apparently purposeful way, devote periods of time to marking repeatedly a section of their boundary. This behaviour has been seen in a number of species including impala and Thompson's gazelle (Jarman 1979; Walther 1978*a*). A case that demonstrates the time and effort that is sometimes involved is described by Gilbert (1973) for a territorial male pronghorn antelope. Prior to the marking episode the male was intermittently courting an oestrous female and periodically looking towards one part of the boundary of his territory, some 2 km away. Suddenly the male ran towards the boundary and arrived 13 minutes later. Over the next 15 minutes the male marked with the subauricular gland 20 times, horn thrashed four bushes, and showed the linked sniffing/pawing/urination/defecation sequence at four sites. It ran between each bout of marking activity and finally returned to the female group 32 minutes after departing.

The maps of gerenuk and Thomson's gazelle antorbital gland marking sites (Figs. 14.18 and 14.19) also reveal patterns of marks in addition to the major 'circles'. In the gazelle additional lines of marks divide the territory into sectors. These are small in the north, where the frequency of boundary encounters with neighbouring males was high (Walther 1978*a*) and large in the south. Perhaps the function of this arrangement is to maximize the number of times that animals entering the territory will encounter marks. Individuals that enter from the north would pass into a small 'enclosure' of marks which they must cross whatever their route. The same principle may apply to the south, but here the male marks at a lower intensity and the sectors are consequently larger. In the case of gerenuk there are a number of 'arms' of marks that radiate out from the main circle. If these link up with a line of boundary marks then the territory might be broken into sectors as in the gazelle. If a boundary line proves to be absent then the arms might represent a less 'costly' alternative since they would tend to intercept animals that walked obliquely through the periphery of the territory (Gosling 1981).

The constraints on marking can be inferred from the different patterns of marking with different substances. Faeces, and thus the dung piles used in territory marking, are clearly in a limited supply and the economic constraints on their use should be more severe than on antorbital gland secretion. This expectation is fulfilled by using the nearest neighbour technique with the dung piles (18 in number) shown in Walther's (1978*a*) map. This shows that dung piles are confined to the territory boundary and that they are most common where intrusion is most likely. Hendrichs (1975) found that dik-dik pairs only construct dung piles along those sections of their territory boundaries that adjoin neighbouring territories. The number of piles varies between 1.7 and 11.6 per 100 m of marked boundary and highest densities occur when a pair attempts to expand their territory. The complete absence of piles on boundaries that did not adjoin other territories may be linked to the mating system: dik-dik live in pairs which can persist for many years. There appear to be relatively few non-territorial animals and the continuous threat of intrusion by non-territorial competitors that characterizes the polygynous species is much reduced.

These qualitative judgements could be tested using a model that included the marking capacity of territorial males in relation to the duration of the signal value of each mark. Some of these data are already to hand. Leuthold (1978*b*) recorded an overall marking frequency of 2.24 per h so that in the 12 daylight hours a male could mark 27 times. There are 121 marks on the map shown (Fig. 14.18) and if these were all the marks in the territory the male could mark each site every 4.5 days. This value would be reduced if marking occurred at night. Walther (1978*a*) found 110 antorbital gland marking sites in the territory he examined and, in another study (Walther 1978*b*), found that territorial males marked an average 39.1 sites between sunrise (06.30) and nightfall (19.00). Males could thus mark each site every 2.8 days. Presumably there is an upper limit beyond which a mark becomes ineffective and it might be possible to determine this experimentally. If so, an elementary model of the economics of territory marking could be constructed. More complete models should consider factors such as the change in marking rate at different stages of territory occupancy (cf. Walther 1972) and, finally, the functional significance of the information that is conveyed by scent marks in territories.

*The function of scent-marking in territories*    Almost all territorial artiodactyls scent-mark and a functional link with area defence has been suspected for some time (early references are given by Johnson 1973). However, non-territorial animals also scent-mark and Ralls (1971), amongst others, has argued that territory marking is one manifestation of a high marking frequency by dominant animals and is a kind of threat. This line of thought has the advantage of considering what is clearly a continuum of marking behaviour; it has the disadvantage of de-emphasizing the unique spatial property of scent marks. Here I will argue the alternative that an understanding of the functional basis of scent-marking in territories provides the key to understanding object marking in a wider context. The present section is thus devoted to a brief review of existing ideas and the development of an interpretation (Gosling 1982) that places this form of marking in the theoretical framework of competitor assessment. In the sections that follow, this interpretation is extended to the general context of competition for mates, and to some aspects of sexual selection.

The explanations of territory demarcation by scent marks are legion. The earliest hypothesis, that marks deter intruders from entering an area (Uexküll and Kriszat 1934; Hediger 1949, 1950), has been refuted by numerous observations of intrusion into scent-marked territories (Grau 1976; Franklin 1980; and others). The idea that scent marks were involved in area defence was retained in the hypothesis that marking might intimidate intruders (e.g. Geist 1965; Mykytowycz 1965) and enhance the confidence of residents. Although originally proposed from data on rabbit territoriality (Mykytowycz 1973; Mykytowycz, Hesterman, Gambale, and Dudzinski 1976) this idea has been independently used to explain marking in ungulates: von Richter (1972) suggests that black wildebeest, *Connochaetes gnou*, gain 'self-assurance' from the presence of their own scent marks and Joubert (1972) states that tsessebe gain 'a sense of ownership or possession'. The main disadvantage of these ideas are that the psychological states proposed are subjective and cannot be tested. In addition, it is difficult to discriminate between the intimidating effect of scent marks on an intruder and the effect of entering an unfamiliar area (Johnson 1973).

Mykytowycz (1970) has also suggested that marks could indicate the territory owner's 'level of influence and readiness to defend the marked area'. This suggestion implies that marks possess intrinsic properties which convey information about the social status of individual animals. This may be true but most of the observations that attempt to test this idea (e.g. Mykytowycz *et al.* 1976) are more economically explained by the 'scent-matching' mechanism that is discussed below.

The views that marks in territories help orientate the movements of the owner, that they attract or stimulate mates and that they assist in pair bond formation are discussed on pp. 598–600 and 600–4. The hypotheses that territory marks assist in population regulation (Wynne-Edwards 1962; Aleksiuk 1968; and others) assumes the operation of group selection and the suggestion that marks assist owners to forage optimally (Henry 1979) is not applicable to the artiodactyls; neither of these will be discussed further here.

The 'scent-matching' theory depends on the theoretical propositions that residents have more to gain from retaining a territory than has an intruder in occupying it and that the costs of agonistic interactions can be reduced by prior assessment of opponents. The first proposition is based on such (unverified) factors as the greater benefits to an animal that is familiar with an area than one that does not know it (Dawkins and Krebs 1978). Since a resident has more to gain it will escalate encounters that potentially decide territory ownership more readily than an intruder. The costs of territory defence may be purely energetic but owners also risk injury or death (Gosling and Petrie 1981). These costs can be reduced by prior assessment of the likely outcome of any interaction (Maynard Smith and Price 1973; Parker 1974; Maynard Smith and Parker 1976; Zahavi 1975, 1977). Given these assumptions there will be an advantage for territory owners in allowing themselves to be identified, and scent-marking may be a part of this process. Intruders might recognize some intrinsic quality of the odour used or, more simply, they could compare the odour of an animal they meet, or the odour of a mark they see it make, with the memorized odour of the scent marks in the vicinity. These marks can only have been made by an animal that has occupied, and successfully defended, the area for a reasonable length of time so that if the scents *match* the animal concerned is probably the territory owner. Having identified the owner, and thus assessed it as a competitor that might escalate any ensuing agonistic encounter to a dangerous level, most intruders should withdraw. The minority that persisted in an encounter would be attempting to displace an identified owner. Thus, by marking, and allowing themselves to be assessed, territorial males reduce the time and energy spent in territory defence. While signals and responses need not be of mutual advantage (Dawkins and Krebs 1978) in this case the intruders also benefit by reducing the chance of costly encounters, except when they compete for ownership. Marking by owners may have evolved by subverting this advantage to intruders in detecting marks and using the information they contain.

If the assessment hypothesis were true, territory owners would be expected to (i) mark their territory in a way that maximizes the chance that marks will be detected by intruders; (ii) mark themselves with the substances used to mark the territory; (iii) remove or replace marks that do not match their own odour; and (iv) make themselves available for scent-matching.

The first of these predictions was explained in the previous section (pp. 585–593) where it was shown that, in general, marks are placed in positions that maximize their chance of detection. This conclusion can, of course, be used to support alternative hypothetical functions, for example, that this distribution allows residents to find their own marks and thus orientate themselves. Apart from any deficiencies in such interpretations (see p. 583) none explain the full range of behaviours predicted by the assessment hypothesis.

Many of the self-marking behaviours reviewed earlier (pp. 575–8) are consistent with the second prediction. Some examples were of non-territorial animals but, as will be discussed below (pp. 598–607) the assessment hypothesis may also apply to other contexts. Other cases, such as that of hartebeest which rub antorbital gland secretion onto their sides, were of territory owners. Self-anointing with foreign substances, best known as 'scent-rubbing' in carnivores (Rieger 1979), is also explicable by this hypothesis. If *matching* is the critical process in identification of the owner, rather than recognition of any intrinsic property of a secretion, then owners could anoint themselves with any odour present in the territory, particularly if it was area specific and likely to be detected by intruders. Thus wallowing by territorial male deer might combine the smell of urine with the soil odour of a particular site. When muntjac rub their antler pedicels on to areas of bare wood that they have scraped free of bark (Dubost 1971) they might anoint themselves with the combined odour of wood sap, saliva, and any glandular secretion from the pedicels. This interpretation might explain the observation that many animals rub parts of their body on to regularly used marking sites before marking themselves. When hartebeest rub their heads in dung piles before contributing their own faeces and rub the skin of their faces, on antorbital marking sites before marking they presumably anoint themselves with either the odour of any other animals that have used the site or the deteriorated odour of their own marking substances. If the matching principle is correct then the male should attempt to acquire these odours, particularly that of its own deteriorated marks since these might sometimes be in the majority.

This process could, of course, eventually be counter-productive, hence the attempts of some owners to remove or conceal previous marks. Perhaps self-anointing with a strange odour, and removing it, are opposing tendencies which can be optimized to produce the highest degree of scent-matching possible, consistent with economic constraints such as changes in the rate of decay of marking substances and in territory size at different seasons. Examples of the removal of previous marks are those of male oribi which bite off, and presumably eat, the tip of previously marked grass stems before remarking with antorbital gland secretion (Gosling 1972) and of male hartebeest that paw and scatter the most recent addition to a dung pile before adding fresh faeces (Gosling 1974, 1975).

Clearly the owner cannot be identified if it does not make itself available for 'scent-matching' during agonistic encounters. Such behaviour might consist only of approaching to a distance (or allowing approach to a distance) at which an odour or taste is detectable, or it might consist of one of the behaviours that seem specifically adapted to this function. One case is that of territorial male hartebeest which stand with the head high and deflected away as an intruder approaches. Sometimes the intruder withdraws with a lowered head but often it approaches and, in the encounter that follows slides its nose, with nibbling movements of the lips, down the immobile neck of the territorial male (Gosling 1974). It seems certain that the intruder smells and, possibly, tastes, pelage which is impregnated with faeces and antorbital gland secretion in addition to any local sebaceous secretion. Dominant male impala similarly tilt their fore-heads towards subordinates, which sniff the glandular area (Jarman 1979) and territorial male pronghorn antelope deflect their head to display the subauricular patch towards intruders (Kitchen 1974); in both cases the same glands are used for object-marking in territories.

Owners that mark during an agonistic encounter may do so to provide a particularly unambiguous link between themselves and the scent marks that the intruder has detected elsewhere in the territory. This behaviour might also be a device to provide the intruder with the owner's odour without approach-ing to a distance that could elicit potentially dangerous fighting. Marking is very common during agonistic encounters (e.g. Ralls 1971; Johnson 1973). Examples are forehead rubbing, antorbital gland marking, and defecation during encounters between hartebeest territory owners and intruders (Gosling 1974, 1975) and rub-urination by black-tailed deer and reindeer during agonistic encounters in the rut (Müller-Schwarze 1971; Mossing and Damber 1981).

The behaviour of intruders into territories is less well known than that of territory owners but some observations suggest that they behave in a way that would allow assessment using scent-matching. Some clearly seek out scent marks in the environment and smell or taste owners during agonistic encounters. Black-tailed deer entering a new area sniff many twigs that may carry scent from forehead rubbing, before engaging in any social encounter (Müller-Schwarze 1974). Subordinate male blackbuck sniff the antorbital glands of dominants during agonistic encounters (Schmied 1973) and a large number of species that use faeces in territory marking sniff the rump or anal region during agonistic behaviour (Schloeth 1956). No data exist on artiodactyls directly to test whether or not intruders respond appropriately after smelling owners: theoretically, low-status intruders should withdraw from an animal whose scent matches that of scent marks in the vicinity; high-status males should withdraw except when they are competing for territory ownership. The fact that most intruding hartebeest do withdraw from owners after behaviour that includes smelling and tasting is generally consistent with the hypothesis. A more specific case gives an indication of the importance of olfactory signals during attempts to take over territories. Animals that are likely to displace a territory owner are usually difficult to identify but, in the case of hartebeest, high-status males occupy and defend small temporary territories which are used as a base for intrusions into a neighbouring permanent territory. The intrusions consist largely of marking behaviour: the intruders walk from one dung pile to another and paw and defecate at each. The owner approaches, vigorously paws the faeces left by the intruder and defecates itself, before resuming the pursuit. Meanwhile the intruder has moved to another dung pile. Sometimes the owner overtakes the intruder and agonistic encounters occur. The intruder almost invariably withdraws and eventually returns to its temporary territory (Gosling 1974, 1975). Such behaviour appears to be an attempt to scent mark the permanent territory as part of a protracted attempt to assume ownership. These tactics have the advantage of not only displacing a male with little risk of injury but also of ensuring a comprehensive marked territory in advance of the critical early stages of ownership.

Neighbouring territorial males can be regarded as a particular class of high status, potential intruders and some parallels with the use of olfactory signals in interactions with other intruders might be expected. Neighbours differ from other intruders in presenting a continual threat through territory expansion, as envisaged in Huxley's (1934) 'rubber disk' theory of territory size. This idea has empirical support from the observation that exceptionally vigorous male hartebeest can sometimes expand their territories at the expense of neighbours (Gosling 1975). Given these premises the assessment hypothesis of territory marking would predict that owners should mark their territories where this threat is most frequent, mark themselves, and make themselves available for scent-matching to the neighbouring males. The form of boundary interaction in a wide range of ungulates strongly supports these predictions. Boundary interactions between neighbouring territorial male hartebeest occur at repeatedly used dung piles: in turn, the males sniff the dung pile, paw vigorously, kneel and rub their heads on the ground then stand and defecate. The males rub their faces onto their own shoulders and sides, smell and nibble the neck of their opponent, and sniff its anal region (Gosling 1974, 1975). Wildebeest have similar boundary encounters and, in addition to the above behaviour, lie down and roll in the dung pile (Estes 1969).

The need for an owner to match the odour of marks in its territory in order to maintain an equilibrium with its neighbours might explain the unusual behaviour of neighbours during competition for vacant hartebeest territories. Vacancies usually occur when owners are killed by predators. There is immediate competition for the territory between a number of high-status, non-territorial males before one of them, often a male that has previously occupied the territory, eventually establishes itself as the new owner. This process involves a very high frequency of marking with faeces and scent glands; some of this occurs while neighbouring territorial males, atypically, intrude deep into the disputed territory and initiate prolonged encounters with the new resident

(Gosling 1975). These intrusions may be a consequence of the discrepancy between the smell of the animal in occupancy and the smell of the marks in the territory. After a day or two, when the territory is comprehensively marked, the intrusions cease.

In a number of cases marking behaviour appears to have visual signal value in addition to its olfactory significance. Particularly clear examples are the exaggerated postures that accompany urination and defecation in many bovids (Walther 1979). In other instances marking seems too distant for any olfactory information to be transmitted: hartebeest territorial males sometimes paw, forehead rub, and defecate in response to seeing similar behaviour by a neighbour at distances of 30–100 metres (Gosling 1974); territorial males also behave in this way during interactions with intruders and after they have been expelled. Comparable behaviour occurs in the vicuña: Koford (1957) describes how a territorial male engaged in an agonistic encounter with a neighbour at their common boundary 'ran off to a dung pile as much as 50 yards distant, defecated, and then returned to posture even closer to his opponent'. Such behaviour may have been selected because it provides a clear link between the marking animal and the scent marks in its territory. In this sense there is an analogy with the scent-matching hypothesis although, like all visual signals, and in contrast to scent marks, marking displays are significant only in the immediate context.

*Competition for mates*

Trivers (1972) has argued that members of the sex that invests the least in each offspring will compete to mate with members of the higher investing sex. Amongst the artiodactyls, female investment is usually relatively high and polygyny the most common mating system. Male reproductive success is usually limited by the number of females that can be fertilized and female success by the number of offspring they can raise. As a consequence there is often intense competition between males, and selection for large body size and corresponding fighting ability. Males generally increase their chance of mating either by acquiring a territory which females visit (see pp. 584–5) or by competing for access to females in a group of males. Direct competition over receptive females occurs in some vertebrate groups but systems of this kind are unknown in the artiodactyls. In this group most males, such as those of oryx, *Oryx besia* and *O. gazella* (Walther 1978b, 1980) or bison McHugh 1958; Lott 1979, 1981), establish dominance relationships which probably determine prior access to receptive females. These interactions frequently involve a high frequency of scent marking and other forms of olfactory signalling.

Ralls (1971) reviewed the occurrence of object marking in animals that are not territorial and found a correlation between marking frequency and dominance ranking. She concluded that marking is a form of threat and that it is effective whether or not animals are territorial. This may be partially true because some marking behaviour involves behaviours such as 'horn thrashing' (pp. 569-71) which are particularly vigorous during agonistic encounters. However, a great deal of marking is done in isolation, and this, together with marking in agonistic encounters, is more easily explained from the principle of competitor assessment by scent matching as outlined in the previous section (pp. 593-8). If dominant animals mark as outlined in the previous section detected by subordinates, then the dominants could be identified by a comparison of their odour with that of the marks that form the pattern. The simplest 'characteristic pattern' would be that dominant males mark at a higher frequency than other males and, as documented by Ralls (1971), Shank (1972), and others; this occurs in most cases that have been investigated. Subordinates could thus compare the odour of the male with the memorized odour of the majority of marks in the vicinity in order to identify the dominant animal. It would also be predicted that dominant males should prevent subordinates from marking and remove or cover their marks. Schaller (1967) observed behaviour by barasingha which is consistent with this idea: the males of this species establish dominance relationships within mixed-sex groups during the breeding season and, in one case, a high-ranking male that had just wallowed drove away a younger male that briefly lay down in the same spot. As in the case of territoriality, dominant males should make their odour available for matching and this may explain the high frequency of marking during agonistic encounters. For example, there is repeated marking with the postcornual gland and pawing in 'rutting pits', during encounters between male mountain goats in the mating season (Geist 1963). Rutting male bison urinate and wallow during encounters with competitors (McHugh 1958).

In groups of competing males, individuals may have prior knowledge of, or learn, each other's identity in encounters that determine dominance relationships. Individuals are recognized by their smell in other contexts (pp. 593-8) and it seems quite possible that odour is one sort of information used in this process. The extreme complexity of the substances produced by various glands and the mixtures of substances used in self-marking (pp. 575-8) must facilitate production of an unique odour. In contexts where immediate recognition of individuals (of known relative dominance) is possible, males could use this information to assess competitors and decide whether or not to escalate an encounter that decides access to a receptive female. Individual knowledge of the fighting ability of competitors, using complex individual odours to reduce ambiguity in recognition, is thus a theoretical alternative to the scent-matching hypothesis. It should occur where males meet frequently, for example in groups with consistent membership. Where males meet less frequently, selection should favour assessment by scent-matching.

Given the operation of scent-matching, there is little fundamental difference between the use of scent-marking for competitor assessment in territorial or hierarchial mating systems. Ralls (1971) also makes this suggestion although she reaches a different functional conclusion. Territoriality is an expression of dominance in space and may simply occur when it is economically possible to defend a particular resource, such as a food supply that attracts females (Davies 1978; Gosling and Petrie 1981). Intruders are the equivalent of subordinates in multi-male groups and both are potentially capable of disrupting mating behaviour. The chance of disruption may be reduced in both cases by providing a means of accurate competitor assessment.

On this assumption it becomes particularly critical for the males to mark the immediate area during sexual behaviour. Thus, a male blackbuck that was alone in its territory bounded towards two females that it saw approaching, circled them, then stopped and marked with antorbital gland secretion (Schaller 1967). Male caribou frequently rub-urinate while 'tending' oestrous females (Lent 1965). After smelling the vulva of a female, territorial male hartebeest some-times, paw and defecate, or paw, lie down, and rub the face glands on the ground (Gosling 1975). Male warthog, *Phacochoerus aethiopicus*, sniff the urine of females in oestrus then urinate on the spot (Frädrich 1974) and European wild boar urinate in the same context and also mark branches with saliva (Beuerle 1975). Such behaviour may ensure that there are scent marks in the vicinity should a competitor approach.

This behaviour becomes most explicit when females are directly marked using glandular secretion or urine (see pp. 578–9). It would be interesting to determine to what extent females limit this behaviour, since in doing so they could exert choice: when marked by a particular male there would theoretically be an increased chance of mating with it. One case demonstrates an active female role: female moose seek out wallows which contain the urine of rutting males and compete with other females for exclusive wallowing rights (Altmann 1959). Reciprocal marking, as occurs between male and female Maxwell's duiker (p. 579) may be linked to the monogamous mating system. Mutual marking also occurs in collared peccaries which live in large permanent bands. In this species females are usually dominant over males; they initiate most courtship and mate with a number of males when in oestrus (Sowls 1974). An understanding of intrasexual competition for mates in these two species is needed before it would be possible to make specific predictions about the role of scent-marking in competitor assessment.

Further work is also required to interpret the marking of male opponents by wildebeest and duikers (p. 579). However, selection might favour this behaviour because such marks are uniquely available to the competitor. Marking opponents might thus be equivalent to object-marking during agonistic encounters which, as argued above, may provide marks which competitors can compare with those in the vicinity that are typical (in frequency or pattern) of high-status males.

*Mating*

As introduced on p. 598 polygynous mating systems predominate in the artio-dactyla. Under these conditions males compete to maximize the number of females that they can successfully fertilize during their lifetime. The production of offspring by females is less variable and most females have the opportunity for choice between males. For example, female hartebeest ranges include 20-30 male territories (Gosling 1974). In such systems there should be selection for females to exercise considerable choice and to prefer males that are successful in competing with others; by mating with such animals the female would ensure that its offspring would inherit any genetic component of competitive ability (Fisher 1930).

In spite of this theoretical position it is often assumed that females have no role in mate choice. An example is the assumption of a primary male role in the detection of oestrus. There are no data to test the prediction that females should signal their receptivity only to high-quality males. Females do increase the frequency of urination during oestrus (for example in domestic pigs, Quinto 1957) but the link with male quality has not been investigated.

There is extensive documentation of the ability of male artiodactyls to detect oestrus by smelling vaginal secretions and by smelling and/or tasting their urine (e.g. Fraser 1968; Estes 1972; Adams 1980). Many males routinely check the odour of females. For example, territorial male hartebeest wait near the boundaries of their territories as females enter, then wheel around and nose the vulva of each as they pass (Gosling 1975). They also smell females as they stand up at the end of the resting period of the day. The females of many species that are approached in this way stop and urinate (Walther 1964, 1968). The subsequent behaviour of the male varies within and between species: some smell the urine as it falls to the ground, some immerse their nose in the stream and some sniff the saturated ground after urination is complete. The next phase is almost universal amongst the artiodactyls: the male lifts its head, raises its nose, and retracts the upper lip in the behaviour known as 'flehmen' (Fig. 14.20). 'Flehmen' may also follow after a male has nosed or licked the female's vulva. In spite of conflicting interpretation (e.g. Dagg and Taub 1970) the most convincing explanation of this behaviour is that the nares are closed by the muscular contraction associated with the retraction of the upper lip so that inhaled air passed through the incisive ducts to the vomeronasal organ (Estes 1972). This organ contains a sensory epithelium that may be involved in detecting the endocrine breakdown products that characterize oestrus in female urine. Direct evidence for this process is lacking in ungulates but fluorescence has recently been found in the vomeronasal organs of guinea-pigs after the subjects were given free access to urine containing non-volatile fluorescent dye (Wysocki, Wellington, and Beauchamp 1980). Flehmen is absent, or barely perceptible, in a few artiodactyls (e.g. chevrotains: Dubost 1975, and hartebeest:

Fig. 14.20. A territorial male Uganda kob, *Kobus kob*, shows the 'flehmen' response after smelling the urine of a female. (Redrawn from Walther (1966).) Flehmen probably facilitates passage of volatiles to the vomeronasal organ and allows the male to identify females that are in an oestrous condition.

Backhaus 1959; Gosling 1975) and Estes (1972) suggests that these species can close the nares without using the labial musculature.

It is also generally assumed that females are manipulated by male-priming pheromones, the substances that trigger neuroendocrine and endocrine activity. These are best known for their medium-term effects on the reproduction of female rodents (reviewed by Bruce 1966; Whitten 1966; Bronson 1971; see Chapter 8). Less critical data are available for the artiodactyls and nearly all reports are confined to the domesticated forms. The evidence suggests that male odour hastens and synchronizes the oestrous cycles of sheep, goats, and pigs (Sinclair 1950; Schinckel 1954; Watson and Radford 1960; Brooks and Cole 1970). However, these effects are very variable. Brooks and Cole showed that puberty in young female pigs may be hastened by introducing a male but, if the male is introduced too early, puberty may be delayed. A group of female goats exposed to a vasectomized male before the start of the breeding season bred *later* on average than a control group (Shelton 1960). Some of this variation might be more easily understood on the assumption that females could respond flexibly to males of varying quality.

Priming effects have not been demonstrated in any wild artiodactyls but they are used to explain the production of strong male odours during breeding seasons. Coblentz (1976) suggests that the odour caused by spray urination in goats is to synchronize oestrus and Kennaugh *et al.* (1977) interpret the production of strongly smelling preputial gland secretion by rutting fallow deer in the same way. Perhaps the males of these species signal their status in this fashion and take advantage of female preference for high-quality males. Females could discriminate on the basis of an intrinsic property of the odour (as suggested by Coblentz 1976), or they could match the odour of the male with that of scent marks in the vicinity. Dominant males mark most frequently (Ralls 1971) and so nearby marks are likely to have been made by an animal of proven competitive ability. As in the case of territory defence (pp. 593–8), the advantages to the marking animal are a consequence of allowing itself to be identified. In a sense the male subverts the advantage to intruders and receptive females in, respectively, avoiding and mating with a high-quality male.

The hypothesis of mate choice by scent-matching yields predictions that are similar to those advanced for competitor assessment: (i) males should mark their territory or that part of a female's range where they are dominant, or potentially dominant, in competition for oestrous females; (ii) males should mark themselves with the same substances, except where these are otherwise detectable; (iii) they should make themselves available to the female for 'scent-matching'; and (iv) they should remove, or cover, the marks of other males, or prevent them from marking. Females should compare the marks in the environment and of a courting male and select males where these scents match. These predictions cannot be adequately tested with existing data but a number of observations lend them some credibility. Perhaps most important is that dominant and/or territorial males often display glands that are used in marking in pre-copulatory behaviour. For example, male pronghorn antelope present the sub-auricular patch and male muntjac dilate the antorbital gland (Bromley and Kitchen 1974; Dubost 1971) while courting. Female pronghorn sometimes approach the displaying male and sniff the subauricular patch and this is sometimes followed by copulation (64 seconds later in one case) (Bromley and Kitchen 1974). A number of female artiodactyls sniff male genitalia or their anal region during courtship (e.g. hartebeest: Gosling 1975; blesbok: Novellie 1979; impala: Jarman 1979; bison: Lott 1981) and might thus detect the odour of faeces, urine and other glandular secretions that are mixed with urine during scent-marking. Female ungulates often lick males during precopulatory behaviour (e.g. barasingha: Schaller 1967; roe deer: Kurt 1968; Eld's deer, *Cervus eldi*: Blakeslee, Rice, and Ralls 1979) and could thus taste or smell any locally produced odour or any caused by self-marking.

Perhaps the best-known response by female ungulates to male odour is that of female pigs, when in oestrus, to the odour of male saliva. The source of the odour is the submandibular salivary gland (Patterson 1968) and its active components androstenol and androstenone. These sometimes elicit lordosis in the female, which allows the male to mount and copulate (Melrose *et al.* 1971). The functional significance of variation in the occurrence of this response has not been investigated. The lordosis response can also be elicited by mixtures of boar urine and preputial gland fluid (Melrose *et al.* 1971) or by preputial fluid alone (Signoret 1970), none of which contain the steroids found in the salivary glands (Patterson, quoted in Beauchamp *et al.* 1976).

Apart from this case the role of scent signals during mating behaviour is largely speculative. There are several instances of responses to particular sources of odour in addition to those mentioned above. The apparent response of male bovids to the inguinal glands of females is particularly interesting because of its occurrence in the extended postcoital behaviour of Uganda kob. Territorial males approach females while they stand immobile after copulation and lick and nuzzle both the inguinal pouches and the mammae (Buechner and Schloeth 1965). Buechner and Schloeth suggest that this behaviour may cause uterine contractions that promote the transport of spermatozoa. Some doubt is cast on this idea by observations of inguinal nuzzling in the *pre*copulatory behaviour of lesser kudu (Walther 1958), waterbuck (Spinage 1969), and nyala (Anderson 1980). Another possibility is that males employ the immobilization effect of touching the inguinal region which may have its main functional significance in the contexts of suckling and possibly also of consuming the faeces and urine of very young calves. I used this response to control the movements of a tame hartebeest calf and later saw a female hartebeest use the same device to stop its calf from walking away. The immobilization response has obvious significance in precopulatory behaviour and may occur in kob postcoital behaviour because of the extremely small territories in this species (Buechner 1960; Leuthold 1966) and the consequent possibility that an oestrous female could be mated by a neighbouring male if allowed to depart immediately. In this interpretation the inguinal glands play only a fortuitous role. It may be that their functional significance is in their airborne odour which can be detected during any reasonably close encounter; as in other cases the importance of a particular secretion may be missed if the observer admits only responses that occur after the gland in question has been directly smelled or licked.

*Parent–offspring behaviour*

Most newborn artiodactyls are mobile soon after birth and, since female reproductive success is usually determined by preferential investment in their own offspring, it is important for the mother to learn the identity of its offspring as quickly as possible. This may be particularly important in highly gregarious species, such as wildebeest, where the accidental separation of mother and offspring can be a major cause of juvenile mortality (Estes and Estes 1979). There is considerable evidence that female ungulates do rapidly learn the identity of their offspring and that odour is the primary, sometimes the only, factor involved (Grau 1976; Stoddart 1980). This learning process, often called 'imprinting' probably occurs as the female licks the neonate free of embryonic membranes and fluid and then continues to groom it up to the time of first suckling; these behaviours are reviewed by Lent (1974). Female musk ox show the 'flehmen' response to licking their newborn calves (Lent 1974) which suggests the involvement of the vomeronasal organ. Most mothers pay particular attention to the anus and genitalia and a number, for example female hartebeest (Gosling 1959) eat the faeces and urine that are produced in response to this stimulation. The concentrations of skin glands near the anus (Ortmann 1960) might produce odours that are learnt by the mother.

The critical period for learning varies between species. In domestic goats it occurs within five minutes and young that are substituted immediately after birth are accepted. If removed and cleaned elsewhere the mother rejects her offspring when it is returned three hours later (Hersher, Moore, and Richmond 1958; Klopfer, Adams, and Klopfer 1964; Klopfer and Klopfer 1968). Six hours may be required for this process in the pronghorn (Prenzlow, in Lent 1974). Most ungulates smell their offspring before allowing them to suck and reject offspring that are not their own. Female hartebeest sniff calves that are about the size of their own and drive away strangers: usually they threaten with dipped horns but sometimes they butt or hook and could certainly cause serious injury (Gosling 1975). Similar behaviour has been seen in wildebeest (Estes and Estes 1979), feral goats (Rudge 1970), rocky mountain goats, *Oreamnos americanus* (Lentfer 1955), and fallow deer (Gilbert 1968). Female reindeer sometimes allow strange calves to suck but they eventually identify them, apparently from their odour, and drive them away (Espmark 1971).

Artiodactyls can be broadly divided into those where the young closely follow their mother for the early weeks of life and those that lie concealed in long vegetation (reviewed by Lent 1974): wildebeest calves are classic 'followers' and press closely against their mother's side as soon as they become mobile (Estes and Estes 1979); hartebeest calves lie hidden in long grass for much of the first week or two of life and emerge only for brief periods when they are suckled and when the female eats their faeces and urine (Gosling 1969). Presumably it is particularly critical for the females to learn their identity before calves 'lie out' for the first time. Little information is available on the time at which calves first depart but one hartebeest calf lay down 7 m from its mother, 80 minutes after birth, in a fashion that seemed to be a precusor of 'lying out' (Gosling 1969). A roe deer fawn, observed by Espmark (1969), twice showed an early tendency to 'lie-out', in scrub cover, only 3 and 5 m from its mother. These incidents occurred 5 hours 1 minute and 8 hours 8 minutes after birth. At 10 hours 52 minutes the fawn concealed itself in a similar way, 5 m from its mother, and when the female grazed away neither took any apparent notice of the other. The interval before the calf finally departs may be partly determined by a trade-off between the advantages of rapid concealment from predators and the need for the mother to learn the offspring's odour.

A related conflict may occur between the advantage to the calf in producing an odour that allows it to be recognized by its mother and the chance of detection by predators, many of which have a keen sense of smell. This may be resolved by the production of odours which can be detected only at close range: most ungulate females return to the general area where their calf is hidden and call, whereupon the calf leaves its hiding place and approaches (for example, in a number of gazelles, Walther 1968). If olfactory recognition occurs at the same range in this context as in the cases cited earlier, then the mother would only confirm the identity of its calf when it is very close (perhaps within a metre). This limitation becomes obvious when hartebeest calves are slightly older and join others of about the same age which rest together during the day. Jackals sometimes kill young hartebeest and occasionally they rush into a juvenile group of this kind and attempt to grab a calf. In the mêlée that follows calves run about searching for their mothers and the females smell each calf that approaches; most are rejected, sometimes violently, and new predation opportunities are thus created (Gosling 1975).

Young artiodactyls learn the identity of their mothers more slowly and sometimes attempt to suck not only other females but also males. Kurt (1968) believes that the critical period for this imprinting process extends over two or three weeks. Walther (1969) reports that an abandoned blesbok calf appeared to imprint on him during a bout of sniffing his bare feet 36 hours after birth; during this process the calf showed distinct lip-smacking movements. Before imprinting it was bottle fed by a number of people but afterwards it would accept a bottle from strange people only when it was wrapped in a piece of Walther's clothing. The odour of the mothers' inguinal glands must be very obvious to calves while sucking but it is not known if they provide a means of identification.

In a few cases there are indications that vocalizations are used by mothers and their offspring to recognize each other (Espmark 1971) but, in general, odour is more important. Reindeer females also appear to use the odour produced by the caudal gland to identify their calves (Müller-Schwarze, Quay, and Brundin 1977). There are no data to suggest that any artiodactyls can innately recognize the odour of their kin; the available evidence demonstrates that individual odours are learnt.

### Interactions within social groups

Comprehensive marking of conspecifics within social groups is rare in the artiodactyls compared with animals such as the social mongooses. However, some cases do occur, particularly in the pigs. Peccaries, for example, mark fellow members of their group with antorbital gland secretion and simultaneously anoint themselves with the odour of the other animal's dorsal gland. This behaviour may be reciprocated or unreciprocated and it involves all ages and sexes within the groups. Communal wallowing which occurs in pigs, such as wart-hog (Frädrich 1967) and deer, for example red deer (Gossow and Schürholz 1974) might sometimes have the same effect, regardless of whether the odour acquired is from urine or glandular secretion, or from the odour of the soil itself. In either case the smell would be similar for all members of the group. Scents that were characteristic of a group could also be obtained by sequential object marking, by lying on a substrate impregnated with secretion, faeces, or urine, or simply by resting in body contact.

The function of acquiring and dispensing particular odours within groups may be analagous to those suggested for marking mates and offspring. Depending on details of social behaviour, social groups, such as that of the peccary, probably contain potential mates and animals that are more closely related than animals of other groups. An individual's reproductive success thus depends to a large extent on the well-being of all members of its group and their competitive success in relation to other groups. These advantages obviously depend on the ability to recognize group members, and selection should thus favour the evolution of signals that unambiguously transmit this information. These arguments can be extended to groups whose members are not closely related if it is accepted that once an animal has joined a group (for any reason) it is advantageous for it to remain there rather than to move to another. Under these circumstances, where animals could sometimes gain antipredator or food-finding advantages, it might also be advantageous to recognize whether or not a nearby animal was a member of the group, and also to advertise group membership. 'Scent-matching' with the subject's own odour or with the odour marks in the home-range of the group, provides the most economical mechanism for such recognition.

## Conclusions

It is often inferred that the responses of mammals to odours are a consequence of innate recognition of one or a few 'active' compounds. Such beliefs are a result of uncritical application of the 'pheromone' concept, in a form that is appropriate for most invertebrates, to a group where more complex and flexible signalling must be expected. The artiodactyls demonstrate the inadequacy of existing theory. While there are one or two extensively cited examples of pheromonal effects, notably the lordosis response of female pigs to the steroids androstenol and androstenone (Melrose *et al.* 1971), these are very rare and, in clear contradiction, there is widespread evidence of selection for extreme complexity in chemical signals, not only within particular glandular secretions, but also in complex marking behaviour where a number of subtances are combined. Two major alternatives have been discussed here. The first involves contextual learning of particular odours. For example, young springbok might learn to associate the odour of dorsal-gland secretion with the appearance of a predator (pp. 583-4) and to respond by flight. Simpler cases are of a female learning the odour of its calf (pp. 604-6) or of members of a kin-related group learning the odour of other members (pp. 606-7). The second alternative is based on the fact that it is often unnecessary to invoke either innate or learnt recognition of particular odours. The fact that scent marks provide a historical and spatial record of an individual's behaviour give particular properties to odour signals that are not possible with those involving sound, vision, or touch. Where the record is characteristic of a particular social class any other animal can identify the marking animal as a member of the class by comparing its scent with that of the scent-marks. This concept is most simply illustrated using the case of scent-marking in territories (pp. 584–5). Only an animal that is able to occupy and defend a territory for a reasonable period can mark it comprehensively and so intruders can distinguish a territory owner from all other animals by comparing its odour with that of marks in the vicinity. The advantage to the resident is that it reduces the cost of territory defence: most intruders withdraw because owners are generally prepared to escalate agonistic encounters in defence of their territories (Gosling 1982).

Similar processes of recognition by scent-matching may be involved in such processes as kin recognition and mate choice (pp. 601–6) and it is important to note that the same odours may be used in these different functional contexts. For example, male pronghorn antelope deflect the head and simultaneously discharge scent from the subauricular gland; in response, subordinate males withdraw but females approach and smell the gland as part of pre-copulatory behaviour (Müller-Schwarze 1974; Bromley and Kitchen 1974).

Scent-matching, as an hypothesis, has a number of advantages over those involving the innate recognition of single compounds. It seems more credible in view of the observed tendency to manufacture extremely complex odours and explains the common absence of any immediate response to scent marks. However, a more interesting link between signal complexity and assessment theory is raised by the possibility of the evolution of deceitful signals. If high status confers a fitness advantage, why do low-status animals not imitate the signals of high status to their own advantage? There are a number of possibilities: there may be short-term advantages in being subordinate (Rohwer and Ewald 1981) or it may be dangerous to deceive because animals of higher quality may sometimes test the signaller. Alternatively, selection may favour signals that are difficult to mimic as in the case of competitive roaring by male red deer which is potentially exhausting (Clutton-Brock and Albon 1979). Scent-marking, in the case of territory marking at least, may be similarly expensive: owners must invest heavily in time and energy in order to mark an area comprehensively (Gosling 1975, 1981; Walther 1978a). Such signals could also be linked to a characteristic that only high-status animals possess, for example, a particular odoriferous substance. However, most appropriate compounds will also be available to other high-status competitors. This problem might be avoided by the observed tendency to create extremely complex odours which are virtually impossible to replicate (even by the owner). Such odours may signal high status only when they match scent-marks in the environment that could only have been made by a high-status animal. In this hypothesis the context determines the meaning of the signal. Competitors identify high-status animals using the historical and spatial record of individual behaviour that is the unique property of olfactory signals.

# References

Adams, M. G. (1980). Odour-producing organs in mammals. *Symp. zool. Soc. Lond.* **45**, 57–86.

— and Johnson, E. (1980). Seasonal changes in the skin glands of roe deer (*Capreolus capreolus*). *J. Zool., Lond.* **191**, 509–20.

Aeschlimann, A. (1963). Observations sur *Philantomba maxwelli* (Hamilton-Smith), une antilope de la forêt éburnée. *Acta trop.* **20**, 341–68.

Aleksiuk, M. (1968). Scent-mound communication, territoriality, and population regulation in beaver (*Castor canadensis* Kuhl). *J. Mammal.* **49**, 759–62.

Altmann, M. (1959). Group dynamics of Wyoming moose during the rutting season. *J. Mammal.* **40**, 420–4.

Anderson, J. L. (1980). The social organisation and aspects of behaviour of the nyala *Tragelaphus angasi* Gray, 1849. *Z. Tierpsychol.* **45**, 90–123.

Ansell, W. F. H. (1964). The preorbital, pedal and preputial glands of *Raphicerus sharpei* Thomas, with a note on the mammae of *Ourebia ourebia* Zimmermann. *Arnoldia* **18**, 1–3.

— (1971). *The mammals of Africa. An identification manual*, Part 15: *Order Artiodactyla*. Smithsonian Institution Press, Washington.

Backhaus, D. (1958). Beitrag zur Ethologie der Paarung einiger Antilopen. *Zuch thyg., Berl.* **2**, 281–93.

— (1959). Beobachtungen über das Freileben von Lelwel-Kuhantilopen (*Alcelaphus buselaphus lelwel*, Heuglin 1877) und Gelegenheitsbesbachtungen an Sennar-Pferdeantilopen (*Hippotragus equinus bakeri*, Heuglin 1863). *Z. Säugetierk.* **24**, 1–34.

Barrette, C. (1977). Scent marking in captive muntjacs, *Muntiacus reevesi*. *Anim. Behav.* **25**, 536–41.

Beauchamp, G. K., Doty, R. L., Moulton, D. G., and Mugford, R. A. (1976). The pheromone concept in mammalian chemical communication: a critique. In *Mammalian olfaction, reproductive processes, and behavior* (ed. R. L. Doty) pp. 143–60. Academic Press, London.

Bergerud, A. T. (1973). Movement and rutting behavior of caribou (*Rangifer tarandus*) at Mount Albert, Quebec. *Can. Fld Nat.* **87**, 357–69.

Beuerle, W. (1975). Freilanduntersuchungen zum Kamff- und Sexualverhalten des europäischen Wildschweines (*Sus scrofa* L.). *Z. Tierpsychol.* **39**, 211–58.

Bigalke, R. C. (1972). Observations on the behaviour and feeding habits of the springbok, *Antidorcas marsupialis*. *Zool. Afr.* **7**, 333–59.

Blakeslee, C. K., Rice, C. G., and Ralls, K. (1979). Behavior and reproduction of captive brow-antlered deer, *Cervus eldi thamin* (Thomas, 1918). *Säugetierk. Mitt.* **27**, 114–27.

Booth, W. D. (1980). Endocrine and exocrine factors in the reproductive behaviour of the pig. *Symp. zool. Soc. Lond.* **45**, 289–311.

Bromley, P. T. and Kitchen, D. W. (1974). Courtship in the pronghorn (*Antilocapra americana*). In *The behavior of ungulates and its relation to management* (ed. V. Geist and F. R. Walther) pp. 356–64. IUCN, Morges.

Bronson, F. H. (1971). Rodent pheromones. *Biol. Reprod.* **4**, 344–57.

Brooks, P. H. and Cole, D. J. A. (1970). The effect of the presence of a boar on the attainment of puberty in gilts. *J. Reprod. Fert.* **23**, 435–40.

Broom, D. M. and Johnson, E. (1980). Responsiveness of hand-reared roe deer to odours from skin glands. *J. nat. Hist.* **14**, 41–7.

Brownlee, R. G., Silverstein, R. M., Müller-Schwarze, D., and Singer, A. G.

(1969). Isolation, identification, and function of the chief component of the male tarsal scent in black-tailed deer. *Nature, Lond.* **221**, 284–5.

Bruce, H. M. (1966). Smell as an exteroceptive factor. *J. anim. Sci.* **25**, 83–9.

Brundin, A., Andersson, G., Andersson, K., Mossing, T., and Källquist, L. (1978). Short-chain aliphatic acids in the interdigital gland secretion of reindeer (*Rangifer tarandus* L.), and their discrimination by reindeer. *J. chem. Ecol.* **4**, 613–22.

Buechner, H. K. (1960). Territorial behavior in Uganda kob. *Science, NY* **133**, 698–9.

— and Schloeth, R. (1965). Ceremonial mating behaviour in Uganda kob (*Adenota kob thomasi* Neumann). *Z. Tierpsychol.* **22**, 209–25.

Burger, B. V., le Roux, M., Garbers, C. F., Spiers, H. S. C., Bigalke, R. G., Pachler K. G. R., Wessells, P. L., Christ, V., and Maurer, K. H. (1976). Studies on mammalian pheromones. I. Ketones from the pedal gland of the bontebok (*Damaliscus dorcas dorcas*). *Z. Naturforsch.* **31**, 21–8.

— — — — — — — — (1977). Studies on mammalian pheromones II. Further compounds from the pedal gland of the bontebok (*Damaliscus dorcas dorcas*). *Z. Naturforsch.* **32**, 49–56.

Byers, J. A. (1978). Probable involvement of the preorbital glands in two social behavioural patterns of the collared peccary, *Dicotyles tajacu*. *J. Mammal.* **59**, 855–6.

Cadigan, F. C. (1972). A brief report on copulatory and perinatal behaviour of the lesser Malayan mouse deer (*Tragulus javanicus*). *Malay. Nat. J.* **25**, 112–16.

Chapman, D. and Chapman, N. (1975). *Fallow deer: their history, distribution and biology*. The Lavenham Press, Lavenham.

Charnov, E. L. and Krebs, J. R. (1975). The evolution of alarm calls: altruism or manipulation? *Am. Nat.* **109**, 107–12.

Claussen, C. P. and Jungius, H. (1973). On the topography and structure of the so-called glandular subauricular patch and the inguinal gland in the reedbuck (*Redunca arundinum*). *Z. Säugetierk.* **38**, 97–109.

Clutton-Brock, T. H. and Albon, S. A. (1979). The roaring of red deer and the evolution of honest advertisement. *Behaviour* **69**, 145–70.

Coblentz, B. E. (1976). Functions of scent-urination in ungulates with special reference to feral goats (*Capra hircus* L.). *Am. Nat.* **110**, 549–57.

Coe, M. J. and Carr, R. D. (1978). The association between dung middens of the blesbok (*Damaliscus dorcas phillipsi* Harper) and mounds of the harvester termite (*Trinervitermes trinervoides* Sjostedt). *S. Afr. J. wildl. Res.* **8**, 65–9.

Cohen, M. and Gerneke, W. H. (1976). Preliminary report on the intermandibular cutaneous glandular area an the infraorbital gland of the steenbok. *J. S. Afr. vet. med. Ass.* **47**, 35–7.

Dagg, A. I. and Taub, A. (1970). Flehmen. *Mammalia* **34**, 686–95.

David, J. H. M. (1973). The behaviour of the bontebok, *Damaliscus dorcas dorcas* (Pallas 1766), with special reference to territorial behaviour. *Z. Tierpsychol.* **33**, 38–107.

Davies, N. B. (1978). Ecological questions about territorial behaviour. In *Behavioural ecology: an evolutionary approach* (ed. J. R. Krebs and N. B. Davies) pp. 317–50. Blackwell, Oxford.

Dawkins, R. (1976). *The selfish gene*. Oxford University Press.

— and Krebs, J. R. (1978). Animal signals: information or manipulation? In *Behavioural ecology: an evolutionary approach* (ed. J. R. Krebs and N. B. Davies) pp. 282–309. Blackwell, Oxford.

Dollmann, J. G. (1931). Development of auricular 'glandular' patches in the water-buck. *Proc. Linn. Soc., Lond.* **144**, 86–7.

Dubost, G. (1971). Observations éthologiques sur le muntjak (*Muntiacus muntjak* Zimmermann 1780 et *M. reevesi* Ogilby 1839) en captivité et semi-liberté. *Z. Tierpsychol.* **28**, 387–427.

— (1975). Le comportement du Chevrotain africain, *Hyemoschus aquaticus* Ogilby (Artiodactyla, Ruminantia). La signification écologique et phylo-génétique. *Z. Tierpsychol.* **37**, 403–48.

— and Feer, F. (1981). The behavior of the male *Antilope cervicapra* L., its development according to age and social rank. *Behaviour* **76**, 62–127.

Du Plessis, S. S. (1972). Ecology of blesbok with special reference to productivity. *Wildl. Monogr.* **30**, 1–70.

Epling, G. P. (1956). Morphology of the scent gland of the javelina. *J. Mammal.* **37**, 246–8.

Espmark, Y. (1964). Rutting behaviour in reindeer (*Rangifer tarandus* L.). *Anim. Behav.* **12**, 159–63.

— (1969). Mother–young relations and development of behaviour in roe deer (*Capreolus capreolus* L.). *Viltrevy* **6**, 461–540.

— (1971). Individual recognition by voice in reindeer mother–young relation-ship. Field observations and playback experiments. *Behaviour* **40**, 295–301.

— (1977). Hindleg–head-contact behaviour in reindeer. *Appl. anim. Ethol.* **3**, 351–65.

Estes, R. D. (1969). Territorial behavior of the wildebeest (*Connochaetes taurinus* Burchell, 1823). *Z. Tierpsychol.* **26**, 284–370.

— (1972). The role of the vomeronasal organ in mammalian reproduction. *Mammalia* **36**, 315–41.

— and Estes, R. K. (1979). The birth and survival of wildebeest calves. *Z. Tierpsychol.* **50**, 45–95.

Fisher, R. A. (1930). *The genetical theory of natural selection.* Clarendon Press, Oxford.

Frädrich, H. (1964). Beobachtungen zur Kreuzung zwischen Schwarzrücken-ducker, *Cephalophus dorsalis* Gray, 1846, und Zebraducker, *Cephalophus zebra* Gray, 1838. *Z. Säugetierk.* **29**, 46–51.

— (1966). Einige Verhaltensbeobachtungen am Moschustier (*Moschus moschi-ferus* L.). *Zool. Gart., Lpz.* **33**, 65–78.

— (1967). Das Verhalten der Schweine (Suidae, Tayassuidae) und Flusspferde (Hippopotamidae). *Handb. Zool.* VIII, 10 (26), 1–44.

— (1974). A comparison of behaviour in the suidae. In *The behaviour of ungulates and its relation to management* (ed. V. Geist and F. R. Walther) pp. 133–43. IUCN, Morges.

Franklin, W. L. (1974). The social behavior of the vicuña. In *The behavior of ungulates and its relation to management* (ed. V. Geist and F. R. Walther) pp. 477–87. IUCN, Morges.

— (1980). Territorial marking behavior by the South American vicuna. In *Chemical signals: Vertebrates and aquatic invertebrates* (ed. D. Müller-Schwarz and R. M. Silverstein) pp. 53–66. Plenum Press, New York.

Fraser, A. F. (1968). *Reproductive behaviour in ungulates.* Academic Press, London.

Fraser Darling, F. (1937). *A herd of red deer.* Oxford University Press.

Geist, V. (1963). On the behaviour of the North American moose (*Alces alces andersoni* Peterson 1950) in British Columbia. *Behaviour* **20**, 377–416.

— (1965). On the rutting behavior of the mountain goat. *J. Mammal.* **45**, 551–68.

— (1971). *Mountain sheep: a study in behavior and evolution.* University of Chicago Press.

Gilbert, B. K. (1968). Development of social behaviour in the fallow deer (*Dama dama*). *Z. Tierpsychol.* **35**, 867–76.

— (1973). Scent marking and territoriality in pronghorn (*Antilocapra americana*) in Yellowstone National Park. *Mammalia* **37**, 25–33.

Gosling, L. M. (1969). Parturition and related behaviour in Coke's hartebeest, *Alcelaphus buselaphus cokei* Gunther. *J. Reprod. Fert.* Suppl. **6**, 265–86.

— (1972). The construction of antorbital gland marking sites by male oribi (*Ourebia ourebia*, Zimmermann, 1783). *Z. Tierpsychol.* **30**, 271–6.

— (1974). The social behaviour of Coke's hartebeest (*Alcelaphus buselaphus cokei*). In *The behaviour of ungulates and its relation to management* (ed. V. Geist and F. R. Walther) pp. 485–511. IUCN, Morges.

— (1975). The ecological significance of male behaviour in Coke's hartebeest, *Alcelaphus buselaphus cokei,* Günther. Unpublished Ph.D. thesis, University of Nairobi.

— (1980). Defence guilds of savannah ungulates as a context for scent communication. *Symp. zool. Soc. Lond.* **45**, 195–212.

— (1981). Demarkation in a gerenuk territory: an economic approach. *Z. Tierpsychol.* **56**, 305–22.

— (1982). A reassessment of the function of scent marking in territories. *Z. Tierpsychol.* **60**, 89–118.

— and Petrie, M. (1981). The economics of social organization. In *Physiological ecology: an evolutionary approach to resource use* (ed. C. R. Townsend and P. Calow) pp. 315–45. Blackwell, Oxford.

Gossow, H. and Schürholz, G. (1974). Social aspects of wallowing behaviour in red deer herds. *Z. Tierpsychol.* **34**, 329–36.

Graf, W. (1956). Territorialism in deer. *J. Mammal.* **37**, 165–70.

Grau, G. A. (1976). Olfaction and reproduction in ungulates. In *Mammalian olfaction, reproductive processes, and behavior* (ed. R. L. Doty) pp. 219–41. Academic Press, London.

Haltenorth, Th. (1963). Klassification der Säugetiere: Artiodactyla. *Handb. Zool.* VIII, **1** (18), 1–167.

Hediger, H. (1944). Die bedeutung von miktion und defäkation bei wildtieren. *Schweiz. Z. Psychol. Anwend.* **3**, 170–82.

— (1949). Säugetier-territorien und ihre Markierung. *Bijdr. Dierkd.* **28**, 172–84.

— (1950). *Wild animals in captivity.* Butterworth, London.

Hendrichs, H. (1975). Changes in a population of dik-dik, *Madoqua (Rhynchotragus) kirki* (Günther 1880). *Z. Tierpsychol.* **38**, 55–69.

Henry, J. D. (1979). The urine marking behaviour and movement pattern of red foxes (*Vulpes vulpes*) during a breeding and post-breeding period. In *Chemical signals: vertebrates and aquatic invertebrates* (ed. D. Müller-Schwarze and R. M. Silverstein) pp. 11–27. Plenum Press, New York.

Hersher, L. A., Moore, U., and Richmond, J. B. (1958). Effect of post-partum separation of mother and kid on maternal care in the domestic goat. *Science, NY* **128**, 1342–3.

Hershkovitz, P. (1958). The metatarsal glands in white-tailed deer and related forms of the neotropical region. *Mammalia* **22**, 537–46.

Hillman, J. C. (1976). The ecology and behaviour of free-ranging eland (*Tauro-*

*tragus oryx* Pallas) in Kenya. Unpublished Ph.D. thesis, University of Nairobi.

Huxley, J. S. (1934). A natural experiment on the territorial instinct. *Br. Birds* 27, 270-7.

Jacob, J. and von Lehmann, E. (1976). Bemerkungen zu einer Nasendrüse des Sumpfhirsches, *Odocoileus (Dorcelaphus) dichotomus* (Illiger, 1811). *Säugetierkd. Mitt.* 24, 151-6.

Jarman, M. V. (1979). Impala social behaviour: territory, hierarchy, mating, and the use of space. *Adv. Ethol.* 21, 1-93.

Johnson, R. P. (1973). Scent marking in mammals. *Anim. Behav.* 21, 521-35.

Joubert, S. C. J. (1972). Territorial behaviour of the tsessebe (*Damaliscus lunatus* Burchell) in the Kruger National Park. *Zool. Afr.* 7, 141-56.

Karlson, P. and Lüscher, M. (1959). 'Pheromones': a new term for a class of biologically active substances. *Nature, Lond.* 183, 55-6.

Kennaugh, J. H., Chapman, D. I., and Chapman, N. G. (1977). Seasonal changes in the prepuce of adult fallow deer (*Dama dama*) and its probable function as a scent organ. *J. Zool., Lond.* 183, 301-10.

Kitchen, D. W. (1974). Social behavior and ecology of the pronghorn. *Wildl. Monogr.* 38, 1-96.

— and Bromley, P. T. (1974). Agonistic behavior of territorial pronghorn bucks. In *The behaviour of ungulates and its relation to management* (ed. V. Geist and F. R. Walther) pp. 365-81, IUCN, Morges.

Klopfer, P. H., Adams, D. K., and Klopfer, M. W. (1964). Maternal imprinting in goats. *Proc. natn. Acad. Sci. USA* 52, 911-14.

— and Klopfer, M. W. (1968). Maternal 'imprinting' in goats: the fostering of alien young. *Z. Tierpsychol.* 25, 862-6.

Koford, C. B. (1957). The vicuña and the puna. *Ecol. Monogr.* 27, 153-219.

Kruuk, H. (1972). *The spotted hyaena. A study of predation and social behaviour.* University of Chicago Press.

Kurt, F. (1968). Das Sozialverhalten des Rehes (*Capreolus capreolus* L.). Mamm. Depicta, Paul Parey, Hamburg and Berlin.

Langguth, A. and Jackson, J. (1980). Cutaneous scent glands in pampas deer *Blastoceros bezoarticus* (L., 1758). *Z. Säugetierkd.* 45, 82-90.

Lederer, E. (1950). Odeurs et parfums des animaux. *Fortschr. Chem. org. Nat-Stoffe* 6, 87-153.

Lent, P. C. (1965). Rutting behaviour in a Barren-ground caribou population: *Anim. Behav.* 13, 259-65.

— (1974). Mother–infant relationships in ungulates. In *The behaviour of ungulates and its relation to management* (ed. V. Geist and F. R. Walther) pp. 14-55. IUCN, Morges.

Lentfer, J. W. (1955). A two year study of the rocky mountain goat in the Crazy Mountains, Montana. *J. Wildl. Mgmt.* 19, 417-29.

Leuthold, W. (1966). Variations in territorial behaviour of Uganda kob, *Adenota kob thomasi* (Neumann, 1896). *Behaviour* 27, 214-57.

— (1971). Freilandbeobachtungen an Giraffengazellen (*Litocranius walleri*) in Tsavo-Nationalpark, Kenya. *Z. Säugetierk.* 36, 19-37.

— (1977). *African ungulates. A comparative review of their ethology and behavioural ecology.* Springer, Berlin.

— (1978a). On the ecology of the gerenuk, *Litocranius walleri* (Brooke, 1878). *J. anim. Ecol.* 47, 471-90.

— (1978b). On social organization and behaviour of the gerenuk *Litocranius walleri* (Brooke 1878). *Z. Tierpsychol.* 47, 194-216.

Lewin, V. and Stelfox, J. G. (1967). Functional anatomy of the tail and associated behavior in woodland caribou. *Can. Fld Nat.* **81**, 63–6.

Lincoln, G. A. (1971). The seasonal reproductive changes in the red deer stag (*Cervus elaphus*). *J. Zool., Lond.* **163**, 105–23.

Lott, D. F. (1979). Dominance relations and breeding rate in mature male American bison. *Z. Tierpsychol.* **49**, 418–32.

— (1981). Sexual behavior and intersexual strategies in American bison. *Z. Tierpsychol.* **56**, 97–114.

Lyall-Watson, M. (1964). The ethology of food-hoarding in mammals—with special reference to the green acouchi *Myoprocta pratti* Pocock. Unpublished Ph.D. thesis, University of London. [Cited in Kleiman (1966).]

McEwan Jenkinson, D., Blackburn, P. S., and Proudfoot, R. (1967). Seasonal changes in the skin glands of the goat. *Br. vet. J.* **123**, 541–9.

McHugh, T. (1958). Social behavior of the American buffalo (*Bison bison bison*). *Zoologica, NY* **43**, 1–40.

Mainoya, J. R. (1978). Histological aspects of the preorbital and interdigital glands of the red duiker (*Cephalophus natalensis*). *E. Afr. wildl. J.* **16**, 265–72.

— (1980). Observations on the histology of the inguinal glands of the Thomson's gazelle, *Gazella thomsoni. Afr. J. Ecol.* **18**, 277–80.

Maynard Smith, J. and Parker, G. A. (1976). The logic of asymmetric contests. *Anim. Behav.* **24**, 159–75.

— and Price, G. R. (1973). The logic of animal conflict. *Nature, Lond.* **246**, 15–18.

Melrose, D. R., Reed, H. G. B., and Patterson, R. C. S. (1971). Androgen steroids associated with boar odour as an aid to the detection of oestrus in pig artificial insemination. *Br. vet. J.* **127**, 497–502.

Mossing, T. and Damber, J-E. (1981). Rutting behaviour and androgen variation in reindeer (*Rangifer tarandus* L.). *J. chem. Ecol.* **7**, 377–89.

— and Källquist, L. (1981). Variation in cutaneous glandular structures in reindeer (*Rangifer tarandus*). *J. Mammal.* **62**, 606–12.

Moy, R. F. (1970). Histology of the subauricular and rump glands of the pronghorn (*Antilocapra americana* Ord). *Am. J. Anat.* **129**, 65–88.

— (1971). Histology of the forefoot and hindfoot interdigital and median glands of the pronghorn. *J. Mammal.* **52**, 441–6.

Müller-Schwarze, D. (1969). Complexity and relative specificity in a mammalian pheromone. *Nature, Lond.* **223**, 525–6.

— (1971). Pheromones in black-tailed deer (*Odocoileus hemionus columbianus*). *Anim. Behav.* **19**, 141–52.

— (1974). Social functions of various scent glands in certain ungulates and the problems encountered in experimental studies of scent communication. In *The behaviour of ungulates and its relation to management* (ed. V. Geist and F. R. Walther) pp. 107–13. IUCN, Morges.

— Källquist, L., Mossing, T., Brundin, A., and Andersson, G. (1978). Responses of reindeer to interdigital secretions of conspecifics. *J. chem. Ecol.* **4**, 325–35.

— and Müller-Schwarze, C. (1972). Social scents in hand reared pronghorn (*Antilocapra americana*). *Zool. Afr.* **7**, 257–71.

— — Singer, A. G., and Silverstein, R. M. (1974). Mammalian pheromone: identification of active component in the subauricular scent of the male pronghorn. *Science, NY* **183**, 860–2.

— Quay, W. B., and Brundin, A. (1977). The caudal gland in reindeer (*Rangifer tarandus* L.): its behavioural role, histology, and chemistry. *J. chem. Ecol.* **3**, 591–601.

— Volkman, N. J. and Zemanek, K. F. (1977). Osmetrichia: specialized scent hair in black-tailed deer. *J. ultrastruct. Res.* **59**, 223–30.

Müller-Using, D. and Schloeth, R. (1967). Das Verhalten der Hirsche (Cervidae). *Handb. Zool.* VIII, **10** (18), 1–60.

Mykytowycz, R. (1965). Further observations on the territorial function and histology of the submandibular cutaneous (chin) glands in the rabbit, *Oryctolagus cuniculus* (L.). *Anim. Behav.* **13**, 400–12.

— (1970). The role of skin glands in mammalian communication. In *Advances in chemoreception*, I. *Communication by chemical senses* (ed. J. W. Johnston, D. G. Moulton, and A. Turk) pp. 327–60. Appleton-Century-Crofts, New York.

— (1972). The behavioural role of the mammalian skin glands. *Naturwissenschaften* **59**, 133–9.

— (1973). Reproduction in mammals in relation to environmental odours. *J. Reprod. Fert.* Suppl. **19**, 433–46.

— Hesterman, E. R., Gambale, S., and Dudzinski, M. L. (1976). A comparison of the effectiveness of the odors of rabbits, *Oryctolagus cuniculus*, in enhancing territorial confidence. *J. chem. Ecol.* **2**, 13–24.

Neal, B. J. (1959). A contribution on the life history of the collared peccary in Arizona. *Am. Midl. Nat.* **61**, 177–90.

Novellie, P. A. (1979). Courtship behaviour of the blesbok (*Damaliscus dorcas phillipsi*). *Mammalia* **43**, 263–74.

Ortmann, R. (1960). Die Analregion der Säugetiere. *Handb. Zool.* VIII 3(7), 1–68.

Parker, G. A. (1974). Assessment strategy and the evolution of fighting behaviour. *J. theor. Biol.* **47**, 223–43.

Patterson, R. L. S. (1968). Identification of 3α-hydroxy-5α-androst-16-ene as the musk odour component of boar submaxillary salivary gland and its relationship to the sex odour taint in pork meat. *J. Scient. Fdn. Agric.* **19**, 434–8.

Pedersen, A. (1958). *Der Moschusochs*. Zeimsen, Wittenberg-Lutherstadt.

Pocock, R. I. (1910). On the specialised cutaneous glands of ruminants. *Proc. zool. Soc. Lond.*, 840–986.

— (1918a). On some external characters of ruminant Artiodactyla. Part I. The Cephalophinae, Neotraginae, Oreotraginae, and Madoquinae. *Ann. Mag. nat. Hist.* **1**, 426–35.

— (1918b). On some external characters of ruminant Artiodactyla. Part II. The Antilopinae, Rupicaprinae, and Caprinae, with a note on the penis of the Cephalophinae and Neotraginae. *Ann. Mag. nat. Hist.* **2**, 125–44.

— (1918c). On some external characters of ruminant Artiodactyla. Part III. The Bubalinae and Oryginae. *Ann. Mag. nat. Hist.* **2**, 214–25.

— (1918d). On some external characters of ruminant Artiodactyla. Part IV. The Reduncinae (Cervicaprinae) and Aepycerinae. *Ann. Mag. nat. Hist.* **2**, 367–74.

— (1918e). On some external characters of ruminant Artiodactyla. Part V. The Tragelaphinae. *Ann. Mag. nat. Hist.* **2**, 440–8.

— (1919). On the external characters of existing chevrotains. *Proc. zool. Soc. Lond.* **1919**, 1–11.

— (1923). On the external characters of *Elaphurus*, *Hydropotes*, *Pudu*, and other Cervidae. *Proc. zool. Soc. Lond.* **1923**, 181–207.

Quay, W. B. (1955). Histology and cystochemistry of skin gland areas in the caribou, *Rangifer*. *J. Mammal.* **36**, 187–201.

— (1959). Microscopic structure and variation in the cutaneous glands of the deer, *Odocoileus virginianus. J. Mammal.* **40**, 114–28.

— and Müller-Schwarze, D. (1970). Functional histology of intergumentary glandular regions in black-tailed deer (*Odocoileus hemionus columbianus*). *J. Mammal.* **51**, 675–94.

Quinto, M. G. (1957). The breeding habits of Berkshire swine. *Phillipp. Agric.* **41**, 319–26.

Ralls, K. (1971). Mammalian scent marking. *Science, NY* **171**, 443–9.

— (1974). Scent marking in captive Maxwell's duikers. In *The behaviour of ungulates and its relation to management* (ed. V. Geist and F. R. Walther) pp. 114–32. IUCN, Morges.

— Barasch, C., and Minkowski, K. (1975). Behavior of captive mouse deer, *Tragulus napu. Z. Tierpsychol.* **37**, 356–78.

Richter, J. (1971). Untersuchungen an Antorbitaldrüsen von Madoqua (Bovidae, Mammalia). *Z. Säugetierk.* **36**, 334–42.

Rieger, I. (1979). Scent rubbing in carnivores. *Carnivore, Seattle* **2**, 17–25.

Rohwer, S. and Ewald, P. W. (1981). The cost of dominance and advantage of subordination in a badge signaling system. *Evolution* **35**, 441–54.

Rudge, M. R. (1970). Mother and kid behaviour in feral goats (*Capra hircus* L.). *Z. Tierpsychol.* **27**, 687–92.

Schaffer, J. (1940). *Die Hautdrüsenorgane der Säugetiere.* Urban and Schwarzenberg, Berlin.

Schaller, G. B. (1967). *The deer and the tiger.* University of Chicago Press.

— (1972). *The Serengeti lion.* University of Chicago Press.

— and Mirza, Z. B. (1971). On the behaviour of the Kashmir markhor (*Capra falconeri cashmirensis*). *Mammalia* **35**, 548–66.

Schinckel, P. G. (1954). The effect of the presence of the ram on the ovarian activity of the ewe. *Aust. J. agric. Res.* **5**, 465–9.

Schloeth, R. (1956). Zur Psychologie der Begegnung zwischen Tieren. *Behaviour* **10**, 1–80.

Schmied, A. (1973). Beiträge zu einem Aktionssystem der Hirschziegenantilope (*Antilope cervicapra* Linne 1758). *Z. Tierpsychol.* **32**, 153–98.

Schumacher, S. (1936). Das stirnorgan des rehbockes (*Capreolus capreolus* L.), ein bisher unbekanntes Duftorgan. *Z. Mikrosk.-anat. Forsch.* **39**, 215–30.

—— (1943). Die Duftorgane (Hautdrüsenorgane) unseres Haarwidles. *Z. Jagdkunde* **5**, 41–64.

Seton, E. T. (1927). *Lives of game animals,* Vol. III. Doubleday, New York.

Shank, C. C. (1972). Some aspects of social behaviour in a population of feral goats (*Capra hircus* L.). *Z. Tierpsychol.* **30**, 488–528.

Shelton, M. (1960). Influence of the presence of a male goat on the initiation of estrous cycling and ovulation of Angora does. *J. anim. Sci.* **19**, 368–75.

Short, R. V. and Mann, T. (1965). Androgenic activity in a seasonally breeding animal, the roe buck (*Capreolus capreolus*). *J. Endocr.* **31**, 19–20.

Signoret, J. P. (1970). Swine behavior in reproduction. Symposium Proceedings 70-0. Effect of Disease and Stress on Reproductive Efficiency in Swine, pp. 28–45. Extension Service, University of Nebraska College of Agriculture.

Sinclair, A. N. (1950). A note on the effect of the presence of rams on the incidence of oestrus in maiden merino ewes during spring mating. *Aust. vet. J.* **26**, 37–9.

Sows, L. K. (1974). Social behaviour of the collared peccary *Dicotyles tajacu*

(L.). In *The behaviour of ungulates and its relation to management* (ed. V. Geist and F. R. Walther) pp. 144–65. IUCN, Morges.

Spinage, C. A. (1969). Naturalistic observations on the reproductive and maternal behaviour of the Uganda defassa waterbuck *Kobus defassa ungandae* Neumann. *Z. Tierpsychol.* **26**, 39–47.

Stinson, C. G. and Patterson, R. L. S. (1972). $C_{19}-\Delta^{16}$ steroids in boar sweat glands. *Br. vet. J.* **128**, 61–2.

Stoddart, D. M. (1980). *The ecology of vertebrate olfaction.* Chapman and Hall, London.

Struhsaker, T. T. (1967). Behavior of elk (*Cervus canadensis*) during the rut. *Z. Tierpsychol.* **24**, 8–114.

Tinley, K. L. (1969). Dikdik, *Madoqua kirki*, in South West Africa: notes on distribution, ecology and behaviour. *Madoqua* **1**, 7–33.

Trivers, R. L. (1972). Parental investment and sexual selection. In *Sexual selection and the descent of man* (ed. B. Campbell) pp. 136–79. Aldine–Atherton, Chicago.

Uexküll, J. V. and Kriszat, G. (1934). *Streifzüge durch die Umwelten von Tieren und Menchen.* Berlin. [Quoted in Hediger (1944).]

von Richter, W. (1972). Territorial behaviour of the black wildebeest. *Zool. Afr.* **7**, 207–31.

Walker, E. P. (1975). *Mammals of the world*, 3rd ed. Johns Hopkins University Press, Baltimore.

Walther, F. R. (1958). Zum Kamp- und Paarungsverhalten einiger Antilopen. *Z. Tierpsychol.* **15**, 340–80.

— (1964). Einige Verhaltensbeobachtungen an *Gazella thomsoni* im Ngorongoro-Krater. *Z. Tierpsychol.* **21**, 871–90.

— (1965). Verhaltensstudien an der Grant gazelle (*Gazella granti* Brooke, 1872) im Ngorongoro-Krater. *Z. Tierpsychol.* **22**, 167–208.

— (1966). *Mit Horn und Huf.* Paul Parey, Berlin.

— (1968). *Verhalten der Gazellen.* Ziemsen, Wittenberg Lutherstadt.

— (1969). Ethologische Beobachtungen bei der küntslichen Aufzucht eines Blesbokkalbes (*Damaliscus dorcas philippsi* Harper 1939). *B. Zool. Garten* **36**, 191–215.

— (1972). Territorial behaviour in certain horned ungulates, with special reference to the examples of Thomson's and Grant's gazelles. *Zool. Afr.* **7**, 303–7.

— (1978*a*). Mapping the structure and the marking system of a territory of the Thomson's gazelle. *E. Afr. wildl. J.* **16**, 167–76.

— (1978*b*). Quantitative and functional variations of certain behaviour patterns in male Thomson's gazelle of different social status. *Behaviour* **65**, 212–40.

— (1978*c*). Behavioural observations on oryx antelope (*Oryx beisa*) invading Serengeti National Park, Tanzania. *J. Mammal.* **59**, 243–60.

— (1979). Das Verhalten der Horträger (Bovidae). *Handb. Zool.* VIII, **10**(30), 1–184.

— (1980). Aggressive behaviour of oryx antelope at water holes in the Etosha National Park. *Madoqua* **11**, 271–302.

— (1981). Remarks on behaviour of springbok, *Antidorcas marsupialis* Zimmermann, 1780. *Zool. Gart., NF* **51**, 81–103.

Watson, R. H. and Radford, H. M. (1960). Influence of rams on the onset of oestrus in Merino ewes in the spring. *Aust. J. agric. Res.* **2**, 65–71.

Wemmer, C. and Murtaugh, J. (1980). Olfactory aspects of rutting behavior in the Bactrian camel (*Camelus bactrianus ferus*). In *Chemical signals: Vertebrates and aquatic invertebrates* (ed. D. Müller-Schwarze and R. M. Silverstein) pp. 107–24. Plenum Press, New York.

Whitten, W. K. (1966). Pheromones and mammalian reproduction. *Adv. reprod. Physiol.* **1**, 155–77.

Wilson, V. J. (1966). Observations on Lichtenstein's hartebeest, *Alcelaphus lichtensteini*, over a three-year period, and their response to various tsetse control measures in eastern Zambia. *Arnolia* No. 15, Vol. 2, 1–14.

Wynne-Edwards, V. C. (1962). *Animal dispersion in relation to social behaviour.* Oliver and Boyd, Edinburgh.

Wysocki, C. J., Wellington, J. L., and Beauchamp, G. K. (1980). Access of urinary non volatiles to the mammalian vomeronasal organ. *Science, NY* **207**, 781–3.

Yagil, R. and Etzion, Z. (1980). Hormonal and behavioural patterns in the male camel (*Camelus dromedarius*). *J. Reprod. Fert.* **58**, 61–5.

Zahavi, A. (1975). Mate selection—a selection for a handicap. *J. theor. Biol.* **53**, 205–14.

— (1977). Reliability in communication systems and the evolution of altruism. In *Evolutionary ecology* (ed. B. Stonehouse and C. Perrins) pp. 253–9. Macmillan, London.

# 15 The carnivores: order Carnivora

DAVID W. MACDONALD

## Introduction

The order Carnivora contributes both insight and frustration to the study of mammalian social odours: insight because carnivores are not only bountifully supplied with odorous glands, and are conspicuous in their deployment of urine and faeces, but also because, historically, interest in these sources of scent has led to a relatively substantial literature; frustration, because despite the passage of time and effort, sadly few carnivoran social odours can be succinctly translated in terms of their functional significance. Nevertheless, with mounting interest in behavioural ecology one may be optimistic that field studies and naturalistic experiments will increasingly pull together fragments of knowledge. Perhaps the greatest hope for success lies in a better understanding of carnivore societies, and hence of the ways in which social odours may function within these societies. At the moment, undeveloped ideas about mammalian societies make the task of understanding olfactory communication akin to that of attempting a translation from a foreign language based on the order and abundance of words, but with little inkling of their (chemical) spelling and even less of their meaning.

Aspects of scent-marking by carnivores have been reviewed by Schaffer (1940), Kleiman (1966), Ewer (1973), and mentioned in the wider reviews of Ralls (1971), Eisenberg and Kleiman (1972), and Johnson (1973). More recently, Gorman (1980) has discussed the functions of some of the glandular secretions produced by carnivores, while I considered their patterns of marking with urine and faeces (Macdonald 1980). In particular, the magnitude of intraspecific variation in patterns of scent-marking is striking. Variation in the nature of social behaviour between species and populations of carnivores can be seen as adaptations to their ecological circumstances (e.g. Kruuk 1976; Macdonald 1983), and olfactory communication is one component of social organization that is the subject of that adaptation. Consequently, in this chapter I will try to interpret the functions of carnivores' social odours in terms of social pressures within ecological frameworks. Recognition of a scent mark, as distinct from elimination, is especially difficult regarding urine and faeces; Kleiman (1966) suggested that the following criteria could be used: (i) orientation to specific objects, (ii) elicitation by familiar landmarks and novel stimuli, and (iii) frequent repetition on the same object. To this useful list I have found it helpful to add, especially in the case of urine, (iv) often involves small volumes, irrespective of posture. In addition to urine and faeces various skin glands produce social odours. The ones dealt with most completely here are summarized in Tables 15.1 and 15.2.

Table 15.1. Sources of social odours in the Carnivora

| Family | Anal pouch and auxiliary anal glands | Anal sacs | Subcaudal glands | Perineal glands | Supracaudal glands | Perioral cheek glands | Foot glands | Saliva | Ventral glands | Token urine | Token faeces | Faecal latrine |
|---|---|---|---|---|---|---|---|---|---|---|---|---|
| Canidae | — | + | ? | — | + | + | + | + | — | + | + | + |
| Hyaenidae | + | — |  | — | — |  | + |  | — |  | — | — |
| Viverridae | + | + |  | + | — | + |  | + | —² | + | + | + |
| Felidae | — | + | + | — | ? | ? | ? | + | — | + | + | + |
| Ursidae | — | +/? |  | — | — |  | ? | ? |  | + |  | + |
| Mustelidae | —² | + | + |  | — |  | + |  | + | + | + | + |
| Procyonidae | —² | + |  | — |  | ² |  |  | + |  | + | ? |

There are different sources of social odours amongst the families of the Carnivora. This table summarizes those odour sources that at least some members of each family definitely use (+), those that they do not use (−), and others for which there is equivocal information or only circumstantial evidence (?). A superscript 2 refers to Table 15.2, indicating that one species has a gland that is atypical of its family. The table also introduces the terminology used throughout this chapter:

*Anal sacs*: paired reservoirs, one either side of the anus, which drain into ducts that may open either internally or externally to the anal orifice. The sacs act as reservoirs for the secretions of apocrine and, in some cases, sebaceous cells. Ursids apart, only four of the 231 species of carnivore are thought to lack anal sacs.

*Anal pouch*: a depression within which the anus is sunk in the Hyaenidae, two subfamilies of Viverridae, and three unorthodox members of other families. In these species the ducts of the anal sacs, if present, exit into the chamber comprising the anal pouch. In hyaenas the anus is situated to the ventral edge of the pouch; in the mongooses it is more central.

*Auxiliary glands*: the walls of the anal pouch may be equipped with sebaceous tissue whose secretions can be smeared on the substrate when the pouch is opened or everted. These glands are referred to as auxiliary glands. The terms anal pouch and auxiliary glands describe morphologically similar structures in different families without inferring homology between them.

*Anal glands*: all glandular tissues in the anal region are collectively called anal glands. In addition to anal sacs, pouches and auxiliary glands, this term embraces the many diffuse skin glands on the anal skin and those around the rectum. This general term is useful since so many observations concern glandular odours originating from the anal region but from an unspecified or unknown source.

*Perineal gland*: a secretory depression or chamber in the perineum including glands confined entirely between the anal and genital orifices and also those that extend around and anterior to the genitalia.

*Ventral gland*: glands discharging through the abdominal skin and lying anterior to the genital orifice. Typical of some mustelid subfamilies.

*Supracuadal gland*: an elliptical mass of largely sebaceous tissue on the dorsal surface of the tail, near to the root.

*Subcaudal gland*: secretory cells immediately ventral to the root of the tail. In some badgers (Mellinae) these discharge into a deep pocket.

*Foot glands*: include glandular tissue in the pads of the feet and interdigital cavities.

*Perioral and cheek glands*: glandular tissue of the chin, lips, and region of the mystacial vibrissae and, posteriorly, of the cheeks and genal vibrissae.

[*Notes continued on opposite page*]

Some species are atypical either in having an unusual gland or in not having one otherwise widespread in their family; these are summarized in Table 15.2. Schaffer (1940) and Ortmann (1960) found many small concentrations of secretory tissue about the integument of carnivores; I have tried to limit the number of names used here. Flood (Chapter 1) lists several non-glandular sources of social odours; of these, amongst the Carnivora, there is evidence concerning saliva, urine, and faeces. Since urine and faeces may sometimes be purely eliminative, it is convenient to have a term to describe occasions when they are thought to be sources of social odour (cf. Kleiman 1966). For this purpose I use the adjective 'token', in the sense of a sign, symbol, or evidence of communicative function. Thus the expression 'token defecation' is shorthand for 'defecation that meets the criteria for a scent mark'. In this table, token defecations are split into those which involve accumulations of faeces at latrines or middens, and those which do not. Although they are doubtless a source of social odours, so little is known of the accessory sex glands of carnivores that they are excluded from this table.

Table 15.2 Atypical glands amongst Carnivora

| Species | Gland present | Gland absent | Ref. |
|---|---|---|---|
| **Canidae** | | | |
| *Nyctereutes procyonoides* | | Anal sac | 10 |
| *Lycaon pictus* | | Supracaudal gland | 11 |
| **Viverridae** | | | |
| *Galidia elegans* | Throat gland | | 5 |
| *Fossa fossa* | Throat gland | | 6 |
| *Cryptoprocta ferox* | Sternal gland | | 7, 8 |
| *Prionodon* spp. | | Perineal gland | 9 |
| *Viverra zibetha* | Anal pouch | | 7, 14 |
| **Mustelidae** | | | |
| *Meles meles* | Subcaudal pouch | | 1 |
| *Arctonyx collaris* | Subcaudal pouch | | 1 |
| *Mellivora capensis* | Eversible anal pouch | | 13 |
| *Taxidea taxus* | Abdominal gland | | 1, 12 |
| *Enhydra lutris* | | Anal sacs | 2 |
| **Procyonidae** | | | |
| *Potos flavus* | Abdominal, sternal, mandibular | Anal sacs | 3, 4 |
| *Nasua narica* | Subcaudal slits | Anal sacs | 3 |
| *Ailurus fulgens* | Anal pouch | | 3 |

| | | |
|---|---|---|
| 1. Pocock (1920*a*) | 6. Albignac (1970) | 11. Schaffer (1940) |
| 2. Jacobs (1938) | 7. Wemmer (1977*b*) | 12. Pocock (1925) |
| 3. Pocock (1921*a*) | 8. Vosseler (1929*b*) | 13. Pocock (1920*b*) |
| 4. Poglayen-Neuwall (1962) | 9. Pocock (1933) | 14. Pocock (1915*c*) |
| 5. Albignac (1969) | 10. Seitz (1955) | |

The imbalance of information on different odour sources and behaviour between species and families of the Carnivora thwarted any attempt to structure this chapter uniformly. However, the Canidae are exceptional in that several species have been the subject of intense study and so, following an introductory synopsis, there are major sections on each of red foxes, *Vulpes vulpes*; coyotes, *Canis latrans*; wolves, *Canis lupus*, and their descendant dogs, *Canis familiaris*. These species are considered in order of generally assumed increasing sociability although, in fact, they all have complex social systems. There is potential for confusion because many of the articles cited here are old and employ archaic species names; I have opted for the contemporary names used by Corbet and Hill (1980) and have cited in brackets the superseded synonym.

**Family Canidae**

There are more comprehensive data on the role of odours in the social behaviour of the Canidae than of any other family of the Carnivora. The early anatomical work by Mivart (1882*a, b*) and Pocock (1914*a, b*) was greatly expanded by the meticulous histological studies in the 1930s by Schaffer (1940) and, subsequently, Spannhof (1969) on the same skin glands which attracted the interest of chemists in the 1970s (Albone 1977). All these studies included work on red foxes, *Vulpes vulpes*, and so much is known about the skin glands of foxes as about those of any other carnivore; this does not diminish the caveat that very little is known of the biological function of even red foxes' glands. Added to the knowledge about the skin glands of Canidae, are the fruits of several intensive studies of the communicative significance of their urine and faeces. Since these studies developed from investigations of canid behavioural ecology, at least some patterns of marking behaviour can be interpreted in the context of species' social behaviour.

All wild canids approximate a relatively long-legged cursorial build, with the stocky bush dog, *Speothos venaticus*, departing most from this pattern. All are omnivorous. Their social organizations are varied and flexible and range from monogamy to single and/or multi-male groups, which may be loosely knit or highly integrated cooperative hunters. Most species have the following cutaneous glands: paired anal sacs, a supracaudal gland, interdigital glands, preputial glands, and probably, maxillary glands. In addition, most, and probably all, species of canid deploy both their urine and faeces at sites and in patterns that indicate communicative significance.

*Anal glands*

The anal sacs of the Canidae are paired reservoirs, one lying either side of the anus and each opening through a single duct. Either or both of tubular or sebaceous glands secrete into these reservoirs. In gross terms there are few differences between species, except slight variation in the location of the apertures

Plate 12.1. Adult male rabbit (Dutch-belted strain) chin-marking.

Plate 15.1. Part of a group of over twenty golden jackals on a border patrol which followed a route punctuated by piles of faeces (see Fig. 15.1) (Photograph: D. W. Macdonald.)

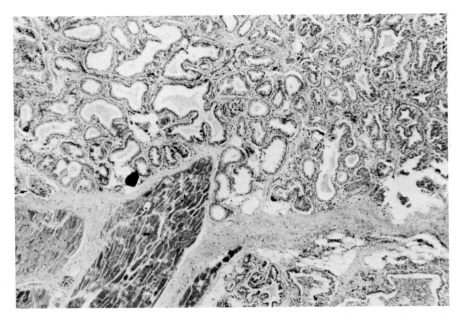

Plate 15.2. Anal sac of red fox: showing tubular apocrine cells clustered around the walls of the sac. (Photograph: D. W. Macdonald)

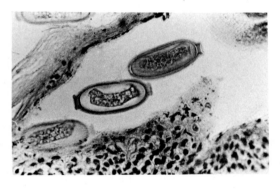

Plate 15.3. The anal sacs of over one third of adult red foxes examined in Oxfordshire were infested with these oval eggs. An association with attractive odorous secretions which may be closely sniffed and licked could play an important role in parasitic transmission. (Photograph: D. W. Macdonald)

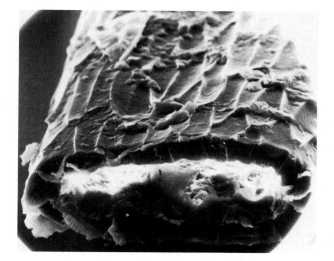

Plate 15.4. SEM of broken shaft of bristle growing from supracaudal gland of a
red fox. The scales on the bristle form a roughened, sculptured surface which
may aid the adhesion of secretion and dissemination of odour. (Photograph
courtesy of K. Kranz.)

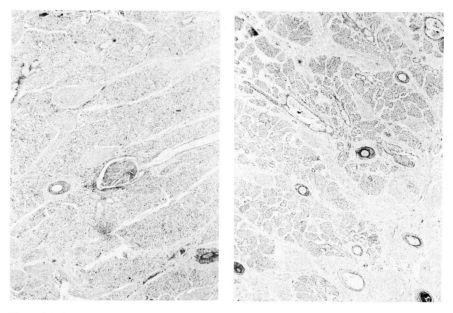

Plate 15.5. Supracaudal gland of red fox: the sebaceous development of these
glands varies between individuals, irrespective of season (although male glands
tend to be more active during the mating season). In active glands (*left*)
abundant vesicles containing secretion are evident, whereas (*right*) in less active
glands these vesicles are scarce and the sebaceous acini embedded in abundant
connective tissue. (Photograph: D. W. Macdonald)

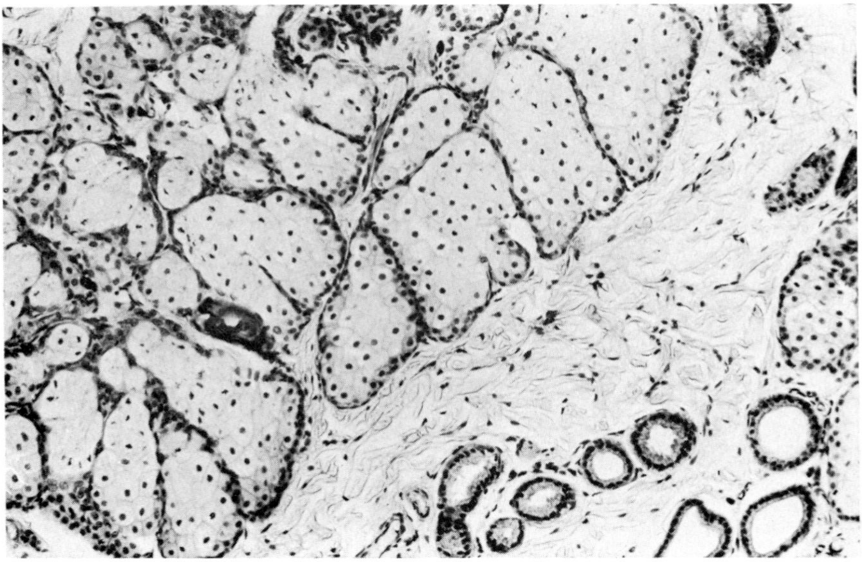

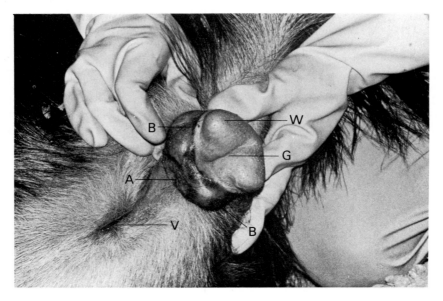

Plate 15.6. Abundant sebaceous tissue and some tubular glands from the skin at the angle of the jaw of a female red fox. (Photograph D. W. Macdonald.)

Plate 15.7. Everted anal pouch of female brown hyaena. The pouch lies dorsal to the anus (A) and vulva (V). When everted a groove (G) bisects the central, sebaceous part of the pouch which secretes a white paste. Nearer to the anus and mouth of the pouch the walls (B) are coated with black secretions of apocrine glands. (Photograph courtesy of M. G. L. Mills and M. L. Gorman.)

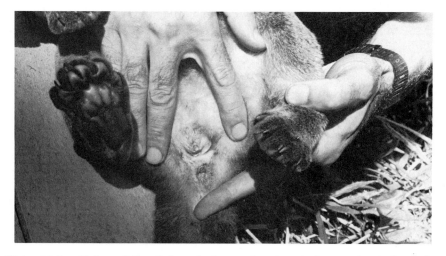

Plate 15.8. Males of the Sulawesi giant palm civet lack a perineal gland. In contrast, the female, shown here, has a 9 mm deep, semi-lunar perineal depression lying posterior to the vulva and which is typical of those found in other palm civets (Paradoxurinae). (Photograph courtesy of C. Wemmer.)

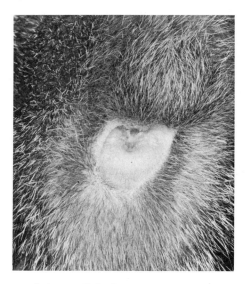

Plate 15.9. The anus of the small Indian mongoose lies within an antechamber or circum-anal depression into which drain a variety of secretions. On both sides of the anus there is an anal sac into each of which two ducts discharge, one from apocrine and one from sebaceous cells. Above the anus there is a raised crescent of skin underlain by abundant sebaceous tissue. In the terminology followed here all these glands are termed anal glands, the sebaceous crescent is called the auxiliary gland and it, together with the anal sacs, all lie within the anal pouch.
(Photograph courtesy of M. L. Gorman.)

Plate 15.10. Dwarf mongooses mark upright objects with the secretions of their anal pouch from a handstand position (*above*), from which the pouch is dragged downward (*below*), so smearing the secretions on to the vertical surfaces. (Photographs couresy of O. A. E. Rasa.)

Plate 15.11. Territorial male cheetah sprays urine on to an isolated sprig of vegetation conspicuously located at the crest of a hillock (Photograph courtesy of G. W. Frame.)

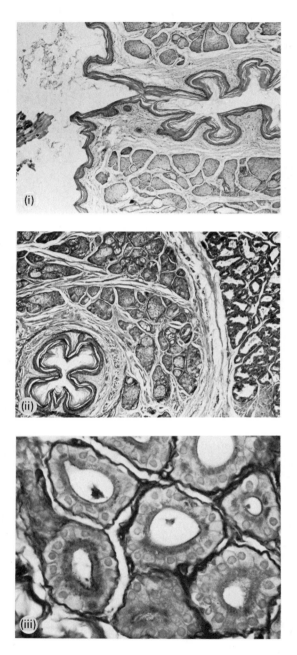

Plate 15.12. Two views of the anal sac of a mink: (i) a longitudinal section along the duct, showing the convoluted passageway leading to the surface of the anal skin and coated with an inner layer of sebaceous tissue; (ii) a transverse section across the same duct, illustrating how the coat of sebaceous acini is itself encased by a layer of apocrine tubular glands; (iii) glycoprotein granules in apical portions of the apocrine cells seem to be associated with the input of sulphur into the secretions of mink anal sac. (Photographs courtesy P. F. Flood.)

Plate 15.13. Three aspects of scent marking by the giant river otter: (a) A fresh spraint aloft a riverside log is covered in copious secretions from the anal sacs; the dark coloured secretions can be seen trickling away from the spraint. (b) Parties of giant otters visit traditional riverside marking sites where they clear a wide circle of vegetation and trample the ground into a muddy quagmire mixed with glandular secretions, urine, and faeces. (c) Giant otter marking vegetation that overhangs the bank beside a marking site. In the course of marking the vegetation is broken down. (Photographs courtesy of N. Duplaix.)

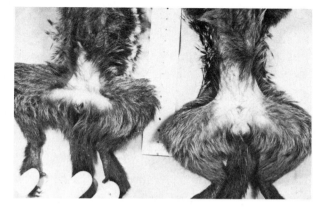

Plate 19.1. Circumgential-suprapubic scent glands of one-year-old twin brother *Saguinus fuscicollis* castrated and sham-castrated at 5½ months of age.

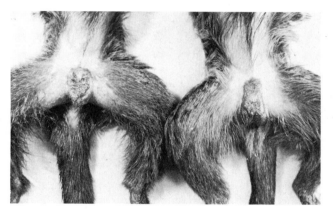

Plate 19.2. Cicumgenital–suprapublic glands of an intact male and a male castrated in adulthood.

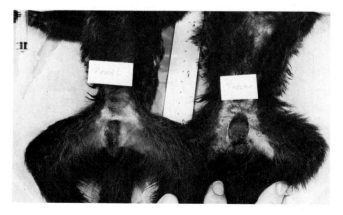

Plate 19.3. Cicumgenital–suprapubic glands of an intact female and a female ovariectomized in adulthood.

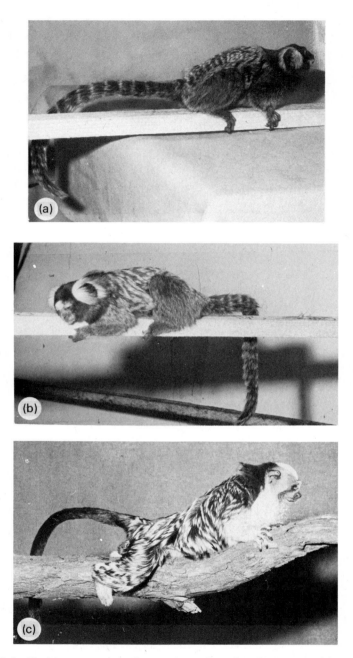

Plate 19.4. Marking patterns of several species of Callitrichidae (a) *Callithrix jacchus* (male) 'sit rubbing'; (b) *Callithrix jacchus* (male) sternal marking and 'sit rubbing'; (c) *Saguinus oedipus geoffroyi* (male) 'pull rubbing'.

(d) *Saguinus oedipus geoffroyi* (female) rubbing the whole ventral surface on substrate; (e) *Leontopithecus rosalia* (female) sternal marking; (f) *Callithrix jacchus* 'partner marking': female sit rubs against the male's body, while he sniffs her scent mark on the perch.

Plate 19.5. Variability in marking patterns of a *Saguinus f. fuscicollis*. (a) sit rubbing; (b) squatting on substrate; (c) straddling small protuberance; (d) pull rubbing; (e) partner marking while male sniffs marked spot on branch.

with respect to the anal opening; in domestic dogs they lie just on the rim of the anus, whereas in red foxes the two duct openings are clearly visible on the cir-cumanal skin, lying about 2 mm to either side of the anus itself. In spite of the anatomical similarities in the sacs, the odours (and appearance) of their contents for at least some species are quite distinct to the human nose. The domestic dog's sac contents vary in colour from creamy to dark brown and often contain large flocculent globules. The evacuate from the red fox's sac is straw coloured with a characteristically sharp smell that catches the back of the nose.

The anal sac acts as a fermentation chamber in which aerobic and anaerobic bacteria convert the original secretions of the glandular tissue into aliphatic acids and other odorous compounds (e.g. Albone and Fox 1971; Albone *et al.* 1974; Bullard 1982). However, there is no study that sheds convincing light on the function of these odours, nor on the roles and importance of primary secre-tions as distinct from secondary products of fermentation.

*Supracaudal glands*

With the exceptions of hunting dogs, *Lyacon pictus*, and most domestic dogs, *Canis familiaris* (Retzius 1849; cited in Schaffer 1940) canids are generally thought to have a supracaudal gland. Despite general agreement with Retzius's findings, veterinarians acknowledge a tendency to seborrhoeic swellings of what are apparently the supracaudal glands of some pet dogs (Anon. 1982). Kleiman (1974) could not find these glands on captive maned wolves, *Chrysocyon brachyurus*, or bush dogs, *Speothos venaticus* (although in the latter species Langguth (1969) reported their presence). The supracaudal gland is generally an elliptical patch of heavily glandular (largely sebaceous) tissue amid thickened skin, on the upper surface of the tail, 5-10 cm below the root. However, as Schaffer pointed out in 1940, and little has changed since then, the evidence is poor for the assumption that this gland is generally present. It has been reported for several members of the genus *Vulpes* (Schaffer 1940), for the Arctic and grey foxes, *Alopex lagopus* and *Vulpes (Urocyon) cinereoargenteus* (Weber 1927; Fox 1971). Pocock (1914*b*) noted that the supracaudal gland of a female Arctic fox was larger than those he had seen on any other species of canid, measuring 25 mm in length. He noted that this gland was devoid of underfur and that (like the feet) it gave out a strong 'foxy' smell. However, in the grey fox, *V. cinereoargenteus*, it is longer (125-200 mm long, 6.5-8.00 mm wide), stretch-ing one-third of the tail's length (Hildebrand 1952). Grinnel, Dizon, and Linsdale (1937) state that there are subspecific differences in the width of the grey fox's tail gland. Weber (1927) mentioned that in the wolf, *Canis lupus*, the gland was rather undeveloped. In coyotes, *C. latrans*, the gland is elliptical as in the red fox (Hildebrand 1952). It is clear from photographs that most members of the genera *Canis* and *Vulpes* have a patch of dark guard hairs where the supracaudal gland should be, and this probably indicates that it is there. Amongst the foxes, for instance, such hairs are clearly visible on photographs of the Fennec fox,

*Vulpes (Fennecus) zerda* (for example, Koenig (1970) fig 6a, page 223) even on a 33-day-old cub. Of the South American Candidae, it was present as an elongated ellipse on the tail of a crab-eating fox, *Cerdocyon thous*, that I examined, and in *Dusicyon* spp. (Cabrera and Yepes 1940; Kuhlhorn 1940).

There is no clearcut evidence of a function for this gland in any species. Fox (1971) thought it smelt more strongly after play among young Arctic foxes. There are indications of increased secretory activity in the supracaudal gland of male red foxes during spermatogenesis (see below), which perhaps ties in with its reported steroid metabolizing properties (Albone 1975; Albone and Flood 1976). However, as with anal sacs, these glands are clearly active throughout the year and so do not function solely in terms of rutting behaviour.

## Foot glands

The pads of canid feet are separated by deep hairy cavities; it is widely assumed in the tracker-dog and fox-hunting literature that foot glands of one sort or another contribute to an animal's odour trail (e.g. Budgett 1933). Both the footpads and the pockets between them are potential sources of odour. Pocock (1914b) describes the anatomy of canid feet and writes of 'scent pockets' between the pads of Arctic foxes. These cavities may be larger amongst vulpine foxes than amongst *Canis* species, but the evidence is equivocal. Certainly the interdigital cavities of red foxes are deep, have a reddish tinge to the skin, and a strong, characteristic smell; these cavities are heavily glandular (pers. obs.). Pocock (1914b) distinguished the feet of Arctic fox, *Alopex lagopus*, from those of other species of Canidae, because they had a distinct, thickish ridge of integument passing forward from the median lobe of the plantar pad reaching to the point of junction between the third and fourth digits. This ridge divided the glandular depression between the digital and plantar pads into distinct left and right portions. In another paper Pocock (1914a) draws little naked pockets in the angles between the digits of dogs and these are presumably present in wild canids too. Much of Pocock's voluminous work on the Carnivora is concerned with the detailed anatomy of the feet. He writes on this topic about all families of carnivores and yet only for the Canidae does he mention foot glands, so by implication he did not find them in other families. Amongst canids, accounts of pack members following each other's line are legion, for example of wild dogs (Estes and Goddard 1967; Frame and Frame 1981) and dholes (Jonsingh 1982).

Prompted by the observation that huskies sometimes suffer injury from iced up sweat on their feet after long and arduous journeys, Sands, Coppinger, and Phillips (1977) compared the histology and response to heating up to almost 50°C of foot pads of huskies, wolves, and both eastern and western subspecies of coyotes. As the pads of anaesthetized subjects were warmed in an experimental box they were tested for signs of sweat. Neither at the hottest temperatures nor spontaneously at any other time was sweat detected on the wolves' feet. In contrast, the huskies sweated copiously, as did the western, but not the eastern, coyotes. Histologically the pads differed between these species and populations of canids. Eccrine sweat glands were superficial and abundant in the dogs and the western coyotes; wolves and eastern coyotes had few such glands, and they were deep within adipose tissue rather than within the connective tissue bordering the epithelium. What is more, only the dogs and the western coyotes (and not the eastern ones nor wolves) appeared to have any apocrine sweat glands.

## Facial glands

Although they have barely been investigated, so much of canid social interaction involves sniffing at muzzle, cheek, and ear that it seems likely that odours from these areas are socially important. At least in the red fox (see p. 636) the skin of the angle of the jaw and the mandible is richly endowed with sebaceous glands.

## Urine and faeces

Both urine and faeces are important sources of social odours amongst the Canidae. Almost all canids have been reported, at least anecdotally, sometimes to cock their legs when urinating and occcasionally to leave their faeces singly or in accumulations at specific types of site.

The following generalizations have emerged regarding the voiding of small volumes of urine in circumstances which behaviourally and spatially are not random and are thus interpreted as scent marking (and which will be referred to as 'token' urinations as partly distinct from eliminative ones (see Table 15.1)).

1. Dominant members of both sexes are more likely to engage in token urination than are subordinates.

2. The frequency of token urinations increases during periods of territorial establishment, and during courtship when they are also associated with over- or double-marking.

3. Victorious individuals mark after defeating a rival, and often do so in the immediate vicinity of the vanquished animal. (However, Lamprecht (1979) specifically mentions that bat-eared foxes, *Otocyon megalotis*, do not urine mark after agonistic encounters.) This emphasizes that the act of urination has a visual component whose signal value is doubtless important.

4. Urine marks are often made at sites where other social odours are present, including faeces, glandular secretions, and scratch marks. They are also made at visually conspicuous sites, which may be repeatedly marked over long periods.

5. Urine marks may be made at places where there are residual odours of food from empty caches or escaped prey, and in this context may serve to minimize time wasted subsequently investigating these places. Food items are sometimes also marked.

6. Urine may be used during allomarking of other individuals.

7. Apart from the information transmitted by each individual urine mark, their overall pattern of deployment probably has communicative significance.

Hints at each of these phenomena can be found in any general account of a canid's behavioural ecology, and do not seem to differ in nature much between those species living in more or less close-knit social units. For example, in Frame and Frame's (1981) account of wild dogs, *Lycaon pictus*, arguably the most obligatorially sociable canid, almost every interaction either within or between packs is prolifically punctuated by urination. Within home-ranges of over 2500 km², patrilinear packs of wild dogs hunt cóllaboratively. The pack is led in almost every activity by an alpha-pair which have the prerogative of reproduction and of urine-marking (Frame and Frame 1976; van Lawick 1974). Double-marking is commonplace, generally involving the alpha-male over-marking the urine of the alpha-female.

Wild dog society differs from that of other canids in one important respect, being typified by female emigration. Frame and Frame (1981) describe the adventures of two females who emigrate together from their natal pack. Almost as soon as they had left the aegis of their parents one of them began to urine-mark for the first time in her life. These female wild dogs encountered the scent trail of a bachelor pack, which they promptly followed. On this journey the dominant of the two urine-marked frequently and stopped to urinate upon, and roll in, a spotted hyaena, *Crocuta crocuta*, latrine. Frame and Frame speculated that, by rolling in the same odours as the males she was following, the female was making herself smell more familiar to them. When the females caught up with the males the meeting involved token urinations: male and female wild dogs forming a pair urine mark in tandem within 30 min of their first encounter.

Generally, canids token urinate throughout their home-ranges. Kleiman (1972) undertook behavioural studies of captive canids to extend Pocock's (1927a) structural comparison of the extremes of variation on the canid anatomical theme—the pack-living bush dog and the solitary, monogamous maned wolf. Urination by the bush dogs, a forest dwelling species, was unusual in several respects: (i) both dominant and subordinate males scent-marked with urine and did so with an unorthodox action—extruding the penis slightly and waggling it laterally to create a spray of urine; (ii) females urinated from a handstand posture (see Kleiman 1966), which left their marks aloft elevated sites and some 15 cm above those of males. More conventionally, the male bush dogs increased their average rate of urinations (one per 10–15 min) when in a new cage (one per 5 min), when introduced to an oestrous female (one per 3 min), and when encountering a strange male (one per 3 min). Also, when members of a pair were reunited after a four-day separation, both marked at just over once per 3 min during the first half-hour of their reunion. Female bush dogs in oestrus urinate once per 6 min, but when one such female was introduced to a male she urinated 81 times in an hour and specifically double-marked all the sites marked by the male during that time. Amongst canids over-marking is more often a male behaviour, e.g. Lamprecht (1979) described how 66.6 per cent of the urine marks of paired male bat-eared foxes were over-marks on females' urinations (or faeces), whereas 10.1 per cent of females' urinations were on the sites marked immediately beforehand by males. Over-marking is generally more frequent prior to the bredding season (see also Golani and Keller (1975) for jackals, *C. aureus*, and Ghosh (1981) for domestic dogs). Montgomery and Lubin (1978) note that pairs of crab-eating foxes, *Cerdocyon thous*, frequently overmark (generally male followed by female) as they forage. Three pairs of these foxes occupied the same area, so the authors dismiss the possibility that the urine-marks served a territorial function.

The female bush dog is exceptional in routinely urinating from a handstand position, but not unique. Keller (1973, cited in Jonsingh 1980) noted that dholes, *Cuon alpinus*, may allomark from a handstand. Male wild dogs and female red foxes have also been seen to urinate in this posture (van Lawick and van Lawick-Goodall 1971; Macdonald 1979*a*).

By comparison, the maned wolves behaved more like other canids (Kleiman 1966, 1972). Males urinated about once per 2 min, normally with a raised leg, and sometimes over a scratch mark they had previously dug. They were also inclined to rub their sides along the sites of urine marks. Bush dogs rubbed with chin, cheek, and neck at spots on the ground of their pen. Finally, Kleiman (1972) noted that, whereas bush dogs scattered their faeces apparently at random, maned wolves defecated at specific sites. This finding is elaborated by Deitz (1981), who has studied maned wolves in an area of the Brazilian Sierra del Canastra where these omnivorous, 23 kg canids live in pairs, the members of which foraged separately, predominantly in the cerrado habitat where their home-ranges average 27 km². Deitz found faeces in two circumstances: either they were in loosely defined latrines in the vicinity of a resting site, or they were scattered singly aloft conspicuous objects (averaging 39 cm high) along main trails. Commonly one, but up to three, latrines were found within 20 m of a resting site and comprised 10–20 faeces on a low flat rock, probably used by both members of a pair. Deitz noted that when he collected the faeces from trails for diet analysis, a fresh replacement was left at the same site within a few days.

Whether as latrines (e.g. wolves, *Canis lupus*; dholes, *Cuon alpinus*; raccoon dogs, *Nyctereutes procyonoides*) or singly (e.g. red foxes, *Vulpes vulpes*; Arctic foxes, *Alopex lagopus*) canid faeces are often left on particular sites, such as tussocks of vegetation, and/or at particular places, such as trail junctions or good feeding sites. Of those species studied in the field, bat-eared foxes, *Otocyon magalotis*, have given least indication of using faeces communicatively (Lamprecht 1979). In general, single or accumulated faeces are scattered by canids throughout their home-ranges, but golden jackals, *Canis aureus*, may ring their territory with middens comprised of accumulations of 100 or more faeces, whereas single faeces are left aloft tussocks and boulders throughout the interior of the territory (Macdonald 1979*b*) (Fig. 15.1). This situation arose where jackals lived in packs of 20 or more animals feeding from a site where offal and garbage were provided routinely. Each evening the members of the pack joined to patrol the border, engaging in much social excitement while urinating and defecating at the middens (Plate 15.1).

The golden jackal's latrines were very precisely defined heaps of faeces, but most canid latrines seem to be more diffuse. For example, wolves may spread faeces along a particular section of path, or around a junction, but not in a single heap. However, the latrine sites described by Jonsingh (1980; 1982) for dholes, or Asiatic red dogs, are conspicuous, often sited at the junction of two paths. In an area where dholes live and hunt in groups of 7-18 animals, occupying territories of 20-40 km$^2$, Jonsingh mentions that several members of a pack

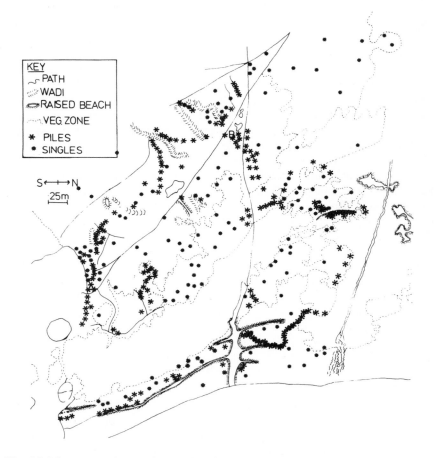

Fig. 15.1 In one study area in Israel, golden jackals left their faeces either in piles or singly. Single faeces were left aloft visually conspicuous objects and scattered within the territory, whereas piles were mainly found around the territorial border and were largest in those perimeter zones most frequently contested with neighbours. (From Macdonald 1979*b*.)

visiting a latrine may defecate more or less simultaneously; indeed, on occasions when packs of 15–16 dholes were seen at latrines between 1 and 9 of them defecated. Dholes apparently do not scrape with their hind-feet after defecating, as do wolves and many other *Canis* species (although *Vulpes* rarely do (personal observation)).

The raccoon dog, *Nyctereutes procyonoides*, is a rather unorthodox canid which, on one small Japanese island, occupies widely overlapping home ranges of 8–51 ha (Ikeda, Eguchi, and Ono 1979; Ikeda 1982). Raccoon dogs defecate and urinate at latrine sites some of which at least are used by all or some of the individuals sharing the area (Ikeda 1982). Yamamoto and Hidaka (1983) showed experimentally that raccoon dogs could distinguish their own faeces from those of conspecifics.

With more diffuse collections of faeces, it is not easy to distinguish true latrines from aggregations of faeces which have accumulated owing to an unrelated factor. For example, trails leading from their day-time rest places are heavily scattered with faeces by red foxes. Casually, this is the simple consequence of the fox's tendency to defecate soon after awakening; functionally, the concentration of faeces may nevertheless convey information. Similarly, Trapp (1978) found 220 faeces within the shadow of a fruit tree beneath which grey foxes had been foraging (see also Trapp and Hallberg 1975).

The tendency to scent mark on conspicuous sites is widespread within the Canidae and throughout the Carnivora (Table 15.3).

### *The red fox*, Vulpes vulpes

#### *Anal sacs*

The anal sacs of the red fox are paired reservoirs lying on either side of the anus and each opening onto the circumanal skin via a duct of about 2 mm diameter. Each sac is approximately spherical and measures about 10 mm in external diameter. These reservoirs each have a capacity of about 0.5 cm$^3$ and contain a mixture of the secretions from the apocrine glands of the sac wall together with sloughed epidermal cells and a microflora dominated by the aerobes *Streptococcus* and *Proteus* (Albone, Gosden, Ware, Macdonald, and Hough 1978) and six genera of anaerobes (Ware and Gosden 1980).

*Histology*   Spannhof (1969) examined the anal sacs of 14 foxes (8 ♀♀, 6 ♂♂), most of which were collected from zoos in Germany. The following description is based on her work, together with my own description of anal sacs excised from 29 wild foxes (17 ♀♀, 12 ♂♂) killed in Oxfordshire, England.

The sac walls are richly supplied with apocrine glands (Plate 15.2) whose ducts, visible to the naked eye, feed into the sac's lumen. A single layer of columnar epithelium lines each glandular tubule. Striated muscle is interspersed amongst the tubules and presumably serves to evacuate the contents of the sac.

Table 15.3 Conspicuous visual features associated with scent-marking in the Carnivora

| Species | Elevated/ conspicuous site | Manu- facture site | Trail junctions | Special posture |
|---|---|---|---|---|
| **Canidae** | | | | |
| *Canis familiaris* | U1, 5, F75 | FU65 | | |
| *Canis lupus* | FU4,6 | FU64 | FU2,3 | |
| *Canis latrans* | FU7,8,77 | U82 | | |
| *Canis aureus* | FU18,19,20,21 | | | U78 |
| *Cuon alpinus* | FU32 | | F32 | |
| *Lycaon pictus* | U18,23 | | | U23 |
| *Speothos venaticus* | U14 | | | U14 |
| *Chrysocyon brachyurus* | FU14,15,16 | | | |
| *Nyctereutes procyonoides* | FU22 | | | |
| *Otocyon megalotis* | U17 | | | |
| *Alopex lagopus* | FU13 | | FU13 | |
| *Vulpes vulpes* | FU9,10,11 | | F9,11,U11 | FU11 |
| *Vulpes cinereoargenteus* | FU12 | | | |
| *Vulpes macrotis* | U76 | | | |
| *Cerdocyon thous* | FU24,25 | | | |
| **Hyaenidae** | | | | |
| *Crocuta crocuta* | G26,27 | | FG26,27 | |
| *Hyaena hyaena* | G28,29 | | | |
| *Hyaena brunnea* | G30 | | | |
| *Proteles cristatus* | G31 | | | |
| **Viverridae** | | | | |
| *Genetta genetta* | G33 | | | G33 |
| *Viverra civetta* | FG27, 39, 81 | | FG27 | U33 |
| *Paguma larvata* | | | | U33 |
| *Nandinia binotata* | | | | U33 |
| *Arctictis binturong* | | | | U80 |
| *Herpestes auropunctatus* | G35 | | | G35 |
| *Herpestes ichneumon* | G36 | | | |
| *Helogale parvula* | G34 | | | G34 |
| *Cynictis penicillata* | G38 | | | G38 |
| *Suricata suricatta* | G37,25 | | | G37,25 |
| *Cryptoprocta ferox* | G33 | | | G33 |
| *Mungotictis decemlineata* | G40 | | | |
| *Galidiea elegans* | G33,41 | | | G33 |
| *Fossa fossa* | G33,42 | | | |
| **Felidae** | | | | |
| *Panthera leo* | U43 | UG43 | | G43, U43,83 |
| *Panthera tigris* | U43,44,45 | UG43 | | |
| *Panthera pardus* | U46, 47 | | | |
| *Panthera uncia* | U52 | | | |
| *Acinonyx jubatus* | FU48,49 | | | |
| *Felis onea* | | G50 | | |
| *Felis concolor* | U51 | U51 | | |
| *Felis lybica* | F55 | | | |
| *Felis sylvestris* | FU56 | | | |
| *Felis catus* | FGU56,57,58,59,79 | | | G57 |
| *Lynx lynx* | FU54 | | | |
| *Lynx rufus* | FU53 | | | |
| *Prionailurus bengalensis* | U52 | | | |

| Species | Elevated/ conspicuous site | Manufacture site | Trail junctions | Special posture |
|---|---|---|---|---|
| **Ursidae** | | | | |
| *Ursus arctos* | GU60 | G60 | | GU60 |
| *Ursus americanus* | GU61 | G61 | | G61 |
| **Mustelidae** | | | | |
| *Martes martes* | FG62 | | | |
| *Martes pannanti* | FUG63 | | | |
| *Mustela erminea* | FG64 | | | |
| *Mustela vison* | FG65,66 | | | |
| *Lutra lutra* | FG67,68 | | | |
| *Aonyx capensis* | FUG70,71 | FUG70,71 | | |
| *Pteraneura brasiliensis* | FUG70 | FUG70 | | |
| *Meles meles* | FG72,73 | FG72,73 | | |
| **Procyonidae** | | | | |
| *Ailurus fulgens* | G69 | | | |
| *Ailurus melanoleuca* | GU84 | G84 | | G84 |
| *Bassariscus astutus* | F74 | | F74 | |

One diagnostic quality of a scent mark is that it is deposited in a way which increases the likelihood of attracting the attention of the desired recipient of the message. In practice this often means marking on an elevated landmark, which, as Peters and Mech (1975) summarize, increases the conspicuousness, evaporative surface, and durability of the mark. This table lists part of the overwhelming body of evidence that the use of such sites is widespread amongst the Carnivora. The first column records evidence of social odours being deposited on or beside visually conspicuous landmarks (but excluding 'olfactory landmarks' such as foetid meat). Some species manufacture landmarks; in the second column I have admitted only cases where the environment is scratched, torn, dug, or scraped and arbitrarily omitted most latrines which are manufactured only in the sense that the accumulating mass of faeces becomes conspicuous as it is added to. Thus, the latrines of *Meles meles* are included because the ground is dug around them, whereas those of *Lutra lutra* are not, because although conspicuous in themselves, they are not manufactured by behaviour directed at the environment. A tendency to mark the junctions of trails is noted, but marking on trails is not, because, although this doubtless does increase the likelihood of a mark being encountered, this could arise because animals travel on trails, rather than because they select trails as marking sites irrespective of where they travel (but see converse, Rothman and Mech 1979). Special postures to elevate odours exclude the leg-lifting and genital aiming stances which are almost universal. This table is not complete (e.g. almost every canid that has ever been seen, has been seen to cock its leg) but serves to emphasize the widespread incidence of marking upon conspicuous sites. References are selected as those with the fullest accounts and most comprehensive bibliographies. Scent marks are categorized as glandular secretions (G), urine (U) and faeces (F). Scratches and scrapes are scored as 'G', as they may combine secretions from foot glands with their visual component. The location of the scent mark (e.g. at territorial border) as distinct from its site at that location may also influence the probability of it being encountered; variation in location is summarized on Table 15.7.

1. Uexküll and Sarris (1931)
2. Seton (1909)
3. Mech (1970)
4. Peters and Mech (1975)
5. Scott and Fuller (1975)
6. Schenkel (1947)
7. Lehner (1978)
8. Wells and Bekoff (1981)
9. Murie (1936)
10. Tinbergen (1965)
11. Macdonald (1979a)
12. Trapp and Hallberg (1975)
13. Hersteinsson and Macdonald (1982)
14. Kleiman (1966)
15. Altman (1972)
16. Dietz (1981)
17. Lamprecht (1979)
18. van Lawick and van Lawick-Goodall (1970)

[*Notes continued on p. 632*]

Abundant sloughed epithelium was present in the lumen of every sac examined.

Variation in the dimensions of the epithelial cells of the glandular tissue and of their nuclei generally correspond to different phases of secretory activity. Spannhof (1969, personal communication) noted that these epithelial cells were noticeably fatter and the nuclei larger in her sample during the winter (rutting) months. She also noted that the epithelial cells' nuclei moved from a predominantly basal to a medioapical or distal position as secretion began. In contrast, cells in the anal glands of some foxes of both sexes in my sample always looked active and contained distended, distal nuclei throughout the year.

*Bacteriology and microbiology* External body cativies, such as the anal sac, which are warm and fluid-filled have a limited supply of oxygen and are ideal for anaerobic bacteria. Common products of anaerobic microbes include short-chain aliphatic acids and other compounds which have strong odours. Several of these characteristic products of microbial action recur among the compounds identified from mammalian scent glands, and this suggests that microbes may be important in the production of some mammalian odours (Gorman 1976).

The anal sacs of the red fox contain an abundant microflora (Gosden and Ware 1977; Ware and Gosden 1980). This means that the compounds voided by the fox when its anal sacs are evacuated are probably not those originally

19. Macdonald (1979*b*)
20. Golani and Keller (1975)
21. Moehlman (1983)
22. Ikeda (1982)
23. Frame and Frame (1981)
24. Montgomery and Lubin (1978)
25. Pers. obs.
26. Kruuk (1972)
27. Bearder and Randall (1978)
28. Kruuk (1976)
29. Macdonald (1978)
30. Mills *et al.* (1980)
31. Kruuk and Sands (1972)
32. Jonsingh (1982)
33. Wemmer (1977*b*)
34. Rasa (1973)
35. Gorman (1976, 1980)
36. Hefetz *et al.* (1983)
37. Ewer (1963)
38. Earlé (1981)
39. Ewer and Wemmer (1974)
40. Albignac (1976)
41. Albignac (1969)
42. Albignac (1970)
43. Schaller (1967, 1972)
44. Whittle (1981)
45. Sunquist (1979)
46. Eisenberg and Lockhart (1972)
47. Hamilton (1976)
48. Eaton (1970)
49. Frame and Frame (1981)
50. Darwin (1845)
51. Hornocker (1969)
52. Wemmer and Scow (1977)
53. Bailey (1974)
54. Saunders (1963)
55. Stuart (1977)
56. Corbett (1979)
57. Leyhausen (1971, 1979)
58. Liberg (1981)
59. Panaman (1981)
60. Tschanz *et al.* (1970)
61. Lloyd (1978)
62. Lockie (1966)
63. Powell (1982)
64. Erlinge *et al.* (1982)
65. Gerell (1970)
66. Birks (1981)
67. Erlinge (1968)
68. Kruuk and Hewson (1978)
69. Roberts (1981)
70. Duplaix (1980)
71. Arden-Clark (pers. comm.)
72. Neal (1947, 1977)
73. Kruuk (1978)
74. Trapp (1978)
75. Sprague and Anisko (1973)
76. Egoscue (1962)
77. Linhart and Knowlton (1975)
78. Golani and Mendelssohn (1971)
79. Fiedler (1957)
80. Kleiman (1974)
81. Randall (1977)
82. Bekoff and Diamond (1976)
83. Bertram (1978)
84. Kleiman (1983)

secreted by the glands around the sac, but rather the products of their microbial degradation (Albone and Perry 1976). The hypothesis that the bacteria of the anal sac are merely contaminant invaders from the nearby anus can be dismissed because *Streptococci* and *Proteus* spp. predominate amongst the aerobes, while coliforms and gut lactobacilli are rare or absent, as are common *Staphylococci* associated with mammalian skin. Furthermore, when Gosden and Ware (1977) attempted to establish *E. coli* from fox faeces in the anal sac community (by inserting it *in vivo*) the initial inoculum of $8 \times 10^9$ organisms perished, reduced to 3 per cent of this number within four days.

An analysis of the anaerobes in the sac revealed populations of obligate anaerobes (predominantly *Clostridia* species) as abundant as those of aerobes. Facultative anaerobes, including *Streptococcus* and *Proteus*, create and maintain an anaerobic environment in the sac in which strict anaerobes can flourish. Chemical studies of the contents of fox anal sac evacuate confirm that the odour derives largely or exclusively from microbial action, and that under field conditions these same products are extremely attractive to red foxes (see below). Ware and Gosden (1980) examined 103 anal sac samples collected during two years from 24 red foxes. These samples yield 239 obligate anaerobic isolates of six genera. They conclude, 'No correlation could be found between particular foxes and their anal sac flora. Different foxes sometimes yielded identical strains, whereas the same fox also frequently yielded different organisms'. Albone *et al.* (1978) reviewed studies of the red fox anal gland, pointing out that the major constituents of anal sac secretion are volatile fatty acids. Indeed anaerobically collected red fox anal sac secretion incubated for 48 hours on cooked meat medium produced volatile fatty acids in concentrations comparable to (or in excess of) those found in the anal sacs. Conversely, when the sacs were washed with antibacterial agents no volatile fatty acids were produced for three days.

These findings provide a tantalizing link between the fox's interest in the anal sac secretions of its conspecifics and in rotten meat—since at least some of the odours of both are the same products of microbial action on mammalian flesh. However, there are experimental indications that foxes spend longer investigating the bacterial metabolites of fox tissue than of other tissues. Albone *et al.* (1978) compared the attractiveness to wild foxes of the products of anaerobic incubations of fox anal sac inocula on media such as fox tissue extract, egg yolk, and meat broth and found that the incubated fox tissue extract had the greatest allure. Blizzard and Perry (1979) confirmed this result by comparing (i) the number of visits and (ii) the time spent sniffing by captive foxes at each of two ports through which these same odours were blown. As a result of these paired comparisons, Blizzard and Perry recorded almost twice as many visits to incubated fox extract (FEI) than to any other odour, and these visits were, on average, of more than twice the duration.

One experiment using the scent-station technique of Linhart and Knowlton (1975) and summarized in Albone *et al.* (1978) took place in a woodland where at least eight adult red foxes lived (two males and six females), and which was criss-crossed by a network of fox trails. Along these paths, vegetation was cleared from two 60 cm diameter circles on opposite sides of the path. Six pairs of these circles were prepared at 5 m intervals along each of four paths. The soil was sifted until it was of a consistency to ensure that individual padmarks would show up. Rectangular sifted areas were also prepared, spanning the path between each of a pair of circles. These were intended to detect the passage of any foxes which travelled the path but did not investigate the sifted circles. Late in the afternoon of each day a clean 15 cm cane bearing a 4.25 cm diameter filter paper was placed in the centre of each earthen circle; 1 cm$^3$ of test odour solution was dropped on to the filter paper. Odours were presented randomly in pairs along each path. Overall the incubated fox extract (FEI) was visited most often and accumulated most footprints per visit. In contrast, the results for the unfermented control fox tissue extract (FEC) did not differ from those for water. The differences between the fox's response to the odours were highly significant, and could be ranked as FEI > urine > FEC = water.

*A parasite on anal-sac marking*    Over one-third of the 29 pairs of anal sacs from Oxfordshire red foxes were infected with nematode eggs (Macdonald 1977). These eggs (Plate 15.3) were present in 33.3 per cent of the sacs from dog foxes and 35.3 per cent of those from vixens, and were oval, with a vase-shaped operculum at the apex. They resembled the eggs of the whipworm, *Trichuris trichuria*. The eggs either lay in the lumen of the anal sac, often close to duct openings, or else they appeared to be encapsulated in the epithelium itself.

The high proportion of infested sacs may suggest that they are a regular discharge point in the parasite's life cycle. Other foxes certainly sniff closely at discharged anal sac secretion and faeces, and may often lick them. One might speculate that an odour that was such a focus of attention (see experiments with FEI above) would make an ideal transfer agent for a parasite.

### The supracaudal, or violet gland

Although it is occasionally sniffed during amicable social interactions, the function of the red fox's supracaudal gland is unknown. Toldt (1907) compared the supracaudal glands of over 190 foxes and noted great variation in their sizes, but found no correlates with either sex or season. The most detailed study is found in Schaffer's (1940) remarkable monograph, where he uses the early nineteenth-century name—violet gland, an epithet derived from the similarity between the odour of the secretion and the fragrance of violets. Schaffer's histological examination of the supracaudal glands of three male and one female red fox led him to speculate that the function of the gland was associated with reproduction, since the tissues of two males shot in February (towards the end of the fox's courting season) were more active than those of other specimens. My own study (below) partly confirms that hypothesis.

*Histology*  Especially on those red foxes with relatively dark pelage the site of the supracaudal gland is pinpointed by an elliptical patch of black guard hairs on the tail. In some individuals the site is visible due to a slight parting of the fur. The gland is a pinkish, horizontal sliver of tissue immediately beneath the skin; it is roughly elliptical in shape, more rounded at the anterior end, and about 3 cm long, 1 cm at its widest, and 0.3 cm deep.

Externally the supracaudal gland is typified by the reduction or absence of underfur and the modification of the guard hairs into yellowish bristles. The bristles and skin are smeared in a yellowish sebum. Müller-Schwarze, Volkman, and Zemanek (1977) found that on the metatarsal gland of the deer the hair scales have a characteristic structure that may promote the release of odours. A scanning electron micrograph of a bristle collected from the supracaudal gland of a male red fox is shown on Plate 15.4. The surface of these hairs is heavily sculptured, which may increase the adhesion of secretions.

I dissected the supracaudal glands from 186 foxes killed in Oxfordshire, of which 67 were fresh enough for detailed histological work. Formalin-fixed, paraffin-embedded serial sections (approx 5 μm) were made of each gland, cutting transversely across the long axis of the gland. Successive sections were stained with haematoxylin and eosin, Schiff's reagent, and the third of each set was left blank, for examination with ultraviolet light. Ultraviolet examination of the unstained sections revealed that droplets of highly fluorescent material were scattered throughout the gland (Macdonald 1977). The fluorescent material was presumably the secretion (see Albone and Flood's (1976) results). An attempt to count these droplets as an index of secretory activity was thwarted since the fluorescence was highly photolabile, decaying within as little as 20 s. However, the fluorescent material was highly PAS positive (scarlet colour) when stained with Schiff-periodic acid reagent. The material secreted by the gland and coating the external bristles fluoresced strongly and so did droplets within the glandular tissue. Once stained with Schiff's reagent, these secretions could be seen within and between sebaceous cells, massed within lacunae amid the glandular tissue, running up hair follicles within the dermis, and smeared across the surface of the skin and emergent hairs.

The appearance of the tissue varied markedly between specimens (Plate 15.5). In some, almost the entire section was comprised of sebaceous cells amongst which were a few apocrine sweat glands; in others, the sebaceous cells were, by comparison, greatly reduced, densely packed and interspersed amongst connective tissue and arrector pili muscles. Where there was abundant sebaceous activity the glandular acini were separated by hair shafts and narrow tracts of secretory material, but connective tissue was hardly visible; in such cases two types of sebaceous cell were distinguishable: in one type the nucleus was large and the cytoplasm highly eosinophilic, in the other type the nucleus was degenerate and sometimes displaced by secretory droplets. Schaffer (1940) noted that vacuolization was greater in cells nearer to the ducts, where nuclei decreased in size until they disappeared, being replaced by oxyphil granules. These secretory vesicles measured up to 70 $\mu$m in diameter. In the most active glands abundant PAS-positive material could be seen around hair follicles and was extruded out onto the skin. Many follicular canals were empty of hair but packed with secretion.

As an index of secretory activity, counts of vesicles containing PAS-positive material were made in 12 different microscope fields of view ($\times$10), from widely separated areas in the centre of each gland (giving an adequate vesicle count, Southwood (1966) p. 19). Although the mean vesicle count for the pooled sample of dog-foxes is greater than that for vixens (7.8 $\pm$ SD 5.4 versus 5.9 $\pm$ 6.8; $t$-test on means, $t = 6.9$, d.f. 65; $p < 0.001$), there was considerable variation in the vesicle count for both sexes. The samples were biased towards the winter months, when more foxes are killed in Oxfordshire. Nevertheless, neither season nor sex explained all the variation in vesicle count between individuals. Consequently, I compared vesicle count with age, weight, condition (kidney-fat weight index), and reproductive status (testis weight). Many foxes arrived as tangled, incomplete corpses, but it was possible to make all five measures on 22 dog foxes and to make at least some of them on a total of 39.

For the 39 dog foxes for which there was a vesicle count, this was regressed with each of these five variables for which there were data. These analyses revealed only a weak positive correlation between mean vesicle count and testis weight ($r = 0.37$, d.f. 34; $p < 0.05$). For the 22 dog foxes for which all five measures were available, a multiple regression analysis of vesicle count against all four other variables slightly strengthened the correlation ($r = 0.45$; $r^2 = 0.2$; $p < 0.05$). A step-wise introduction of each of these variables confirmed that the only significant relationship was of secretory count with testis weight ($r = 0.39$, d.f. 21; $p = 0.038$). In conclusion, the supracaudal gland of male red foxes is generally more active when they are sexually active. One might speculate that the additional variation arose because of differences in status between the males, irrespective of their breeding condition. However, a partial correlation analysis of vesicle counts versus each of body weight, age and fat index whilst controlling for testis weight provided no significant results.

*Maxillary gland*

Female red foxes have been seen to rub their face and neck against vegetation and even to drag sticks and leafy vegetation through the angles of their jaws, chewing as they do so (Fig. 15.2). This 'mouthing' behaviour is positively correlated with the frequency of token urination among vixens, and mouthing and rubbing are most common at heavily marked token sites (Macdonald 1977). In general, rubbing and mouthing on vegetation was only frequent during December and January, the mating season. However, rubbing on food items occurred throughout the year. Histological examination of the perioral region of one vixen (Plate 15.6) revealed an abundance of sebaceous tissue and some tubular glands.

*Subcaudal gland*

Schaffer (1940) described a subcaudal gland on the red fox's tail as similar in structure to the violet gland. However, in contrast to the violet gland, he found that the abundant sebaceous cells of this gland are underlain by large tubular apocrine glands. He likened this gland to a similar one in goats.

*Urine and faeces*

*Pattern of urine-marking*   Red (and Arctic) foxes mark frequently with token urinations and leave their faeces aloft visually conspicuous objects along trails (see, e.g. Murie 1936; Macdonald 1979a; Hersteinsson and Macdonald 1982). Both sexes urine-mark and both may cock a hind-leg to varying extents while doing so (Fig. 15.2). Jorgenson, Novotny, Carmack, Copeland, Wilson, Katona, and Whitten (1978) are wrong in saying that the sexes can be distinguished because males invariably cock their legs and that females do not as a rule urinate on raised objects. The distinction between those urinations which conform to Kleiman's criteria for scent-marking and those which do not lies not so much in the posture of the fox, as in the amount of urine voided. Until they are 5 months old (and thereafter if they are socially subordinate) vixen red foxes squat to urinate, with no apparent heed to their immediate surroundings and generally without sniffing the site first. During such squats they urinate for an average of 9 s and are thereafter unlikely to urinate on the same site on a subsequent occasion. In obvious contrast, older and socially dominant vixens urinate on visually and/or olfactorally conspicuous sites, generally after careful sniffing and at sites on which they are likely to mark again on a subsequent visit. On such occasions urine is voided only momentarily. There is no obvious difference between the sites selected for marking in this way by vixens or dog foxes, and the posture and extent of hind-leg elevation of both sexes is clearly determined to a great extent by the shape and location of the site at which they direct their urine. However, perhaps due to their pelvic anatomy, females rarely perform a maximally elevated raised-leg urination, whereas males frequently do so against a vertical surface. Consequently male urine may be squired up or down but it invariably falls forward of the hind-legs, whereas female's urine invariably falls between or behind the hind-legs. This distinction can be important in reading snow signs (cf. Henry 1979). Here, irrespective of posture and sex, the term token urination embraces all those which share the common denominators of (i) small volume and either or both of (ii) preliminary sniffing, and (iii) being directed at a recognizably conspicuous site. The postures and locations typical of token urination are illustrated for red foxes in Fig. 15.2. Both males and females may token urinate frequently: Macdonald (1979c) reported watching radio-tagged red foxes, of which a male marked at 1.0 per min and a vixen from the same group marked at 1.2 per min (both during February), and Arctic foxes of both sexes have been watched marking at similar rates on comparable sites

Fig. 15.2. To judge by the frequency with which they are deployed, scents play an extremely important role in communication among red foxes. During social encounters foxes sniff each other's mouths, cheeks, and ears; although they are rarely seen to sniff at each other's supracaudal gland, subordinate foxes (a) lash their tails when approached by a dominant, whereupon wafts of odour may complement their visual display. The behaviour patterns of mouthing (b) and rubbing (c) may occur at urine-marking sites before, during, and/or after

(Hersteinsson and Macdonald 1982). Henry (1979) found that along sample tracks of approx 1 km, red foxes of both sexes marked at mean frequencies varying between 0.46-0.99 tokens per 100 m. Whitten, Wilson, Wilson, Jorgenson, Novotny, and Carmack (1980) found urine marks in winter in Maine, USA, every 15-130 m of snow tracking. The smell of fox urine repeatedly at the same site is well known in the countryside (Tinbergen 1965) and the characteristic odour apparently derives from $\Delta^3$-isopentenyl methyl sulphide and 2-phenylethyl methyl sulphide (Wilson, Carmack, Novotny, Jorgenson, and Whitten 1978).

In an area where wild red foxes maintained territories averaging 40.4 ha, each occupied by one male and a mean of 3.4 adult, related females (Macdonald 1981), I studied the token marking behaviour of hand-tame foxes which were allowed to walk more or less where they pleased through the countryside, on long leashes (Macdonald 1979*a*). In particular, one yearling vixen was studied during 112 walks, each of approximately one hour's duration, and during which she left 1283 token urinations and 135 faeces on 615 separate sites. Each site was distinguished (at a distance) by a numbered cane and reflective tag (most of the work was done at night). During each walk the reference numbers of each site visited were spoken into a tape-recorder. The vixen generally confined her travels to an area of about 20 ha (larger than the smallest wild fox territory locally) and the greatest number of token urinations she made in this area during a typical walk was 115. After the 112 walks, taken between November and March, it was possible to analyse the cumulative spatial and temporal patterns of her marking. Figure 15.3 shows how her daily total of token urinations rose throughout the winter, but then declined sharply around the day when she

---

bouts of token urination or, less commonly, in isolation. When mouthing (b) the fox draws vegetation through the angle of its jaw, smearing it with saliva and sometimes biting and shaking the plant. The lips and nearby skin are rich in sebaceous tissue and this may also add to the odour deposited during mouthing and rubbing. Rubbing (c) (and rolling) also occurs on strong or foetid smells and on prey (d). The idea that the fox is trying to mask its own odour by rubbing on prey seems incompatible with the observation that prior to rolling the fox may urinate and defecate on prey. Socially dominant adult red foxes of both sexes token urinate. Urine from males (e) is more forwardly directed, but both sexes may 'cock' their legs. Vixens raise their legs to varying extents ((f)–(i), drawn precisely from photographs) depending on the elevation of the site to be marked. Urine (i) and, rarely, faeces may even be voided from a handstand posture. Similar sites are used for both urine and faecal marking (j). One to three faeces, but seldom more, commonly accumulate at such sites, generally on well travelled trails. Often, but probably not invariably, faeces are anointed with drops of anal sac secretion. Sometimes this can be seen clearly to drop after the faeces, following a distinct contraction of the anal region. Anal sac secretions can be voided without defecation when the fox is frightened and probably also following territorial clashes.

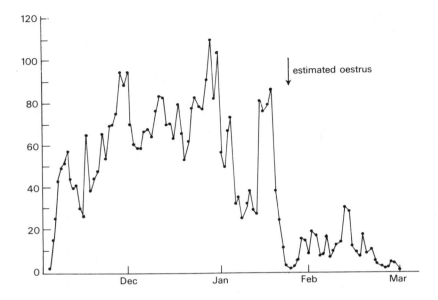

Fig. 15.3. The mean number of urinations per walk each day from early November, when a then ten-month-old vixen was first walked within the study area, to early March when observations ceased. Calculating back (52 days) from the birth of her cubs indicates that she conceived on or around 26 January, the date which sees her marking frequency plummet to a minimum.

mated. She reached (and then maintained) peak marking rates by early December, almost two months prior to oestrus. Coincident with the declining number of token urinations immediately post-oestrus came a contraction of the area in which she marked until it was limited to the vicinity of the enclosure which the vixen shared with her mate. Prior to this contraction, there was a very marked spatial pattern to the vixen's token urination. She led me through ramifying paths of the woodland, pausing every few metres to mark a tussock, fallen bough, or molehill. She would then move towards the open fields and make her way down the hedgerows, continuing to mark similar sites, together with clods of soil thrown up by the plough. However, there was a border at which the vixen would generally turn back, marking at about 30 per cent of her previous rate. If she chose to walk further afield, then she would not continue to token mark. Excluding the area within 100 m of her outermost token site, the vixen cumulatively marked more sites, and more frequently at each of them, in direct proportion to the number of walks she had made down each path. So, within her marking range, more marks accumulated (both in terms of number of sites and number of tokens per site) where the fox went most frequently. When walking off paths she marked about 3 per cent of the number of sites per distance walked in comparison to her behaviour on paths. On paths which were poorly endowed with conspicuous sites she marked each site more often than she marked each of those which were along paths where many sites were available.

Wild foxes trespassing in a neighbouring territory do not token urinate (Fig. 15.4) (Macdonald 1979a). Wild and captive foxes carried surplus food stolen from a neighbour's territory back into their own marking range in order to cache it. When a tame vixen was presented with alien vixen urine within her marking range, she invariably overmarked it (sometimes on several successive days). She was presented with the same urine outside her marking range and never overmarked it. Generally when she found the alien urine outside her marking range

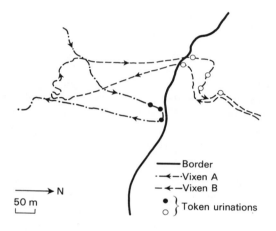

Fig. 15.4. Female red foxes who are regularly seen to token mark within their territory do not do so while trespassing. This figure shows the route of two neighbouring vixens as they approached a track which acted as a clear-cut territorial border. Both vixens marked at the border, but Vixen 'B' did not urinate while she was trespassing to the south of the border. (After Macdonald 1979a.)

she turned for home. Apart from demonstrating that alien odour alone could at least sometimes alter her behaviour, she also thus demonstrated the ability to distinguish her own urine from others, since she was never prompted to turn back on the occasions when she encountered her own (transplanted) urine outside her marking range. Blizzard and Perry (1979) showed that male red foxes under experimental conditions responded differently to the urine of male cage mates and that of unfamiliar males.

Many sites are marked repeatedly and those ones that were marked only once were generally off the fox's normal tracks (Macdonald 1979a). Henry (1979) found that about 30 per cent of fox token marks made on winter snow overlay previous marks. Of course, such a figure rests on the longevity of the biologist's ability to detect them, just as its biological significance rests on the longevity of the fox's ability to detect them. Henry's (1979) results also show that there is no difference between the fox's overall tendency to mark on top of other marks during and after the courtship season (in contrast to the suspicions of Tembrock (1957)). During the winter snows, 18 per cent of the token urinations that Henry found were on empty caches or other signs of prey whose further investigation would have been fruitless (88 per cent had been on food remains in his summertime study). These particular marks probably had an additional function, 'book-keeping', which he had postulated (Henry 1977) for tokens made on empty caches in leaf litter during the summer months.

*Book-keeping hypothesis*    The 'book-keeping' hypothesis was formulated by Henry (1977) regarding red foxes (see Harrington (1981) re wolves) and is that urine marks are made on sites where the odour of food persists, but where there is not an appreciable amount of food. By not investigating these alluring but fruitless sites a foraging fox saves time. In a wooded area of Canada, Henry recorded that about 88 per cent of urine marks in summer fitted the book-keeping hypothesis, being deposited at sites where the fox had scavenged for food. Henry showed experimentally that wild red foxes spent less time investigating sites where he had introduced the odours of food and urine together than where he had introduced only the odour of food.

*Urine chemistry*    Jorgenson *et al.* (1978) analysed the chemistry of wild red fox urine collected from fresh snow, and identified with GC–MS an apparently uniquely male factor, 2-methylquinoline (quinaldine), but Bailey *et al.* (1979) have subsequently found 2-methylquinoline in the urine of vixens. One of the observations of Jorgenson *et al.* was that the pungency of the fox urine increased during the mating period and Henry (1979) tested this suggestion by measuring the distance from which he could detect marks in the field. Henry found that he was able to smell urine marks from significantly ($p < 0.05$) further away during the period December–February than he could during March and April, and that this difference applied equally to marks thought to be made by dog foxes and by vixens. In fact, Bailey *et al.* (1979) looked at seasonal variations in the chemistry of red fox urine and found that each of two males sampled over three years showed a peak in the production of both 4-heptatone and 3-methylbutyl methyl sulphide in February (Fig. 15.5) (the same compound is present at about one-fifth the concentration in vixen urine). The correspondence in the pattern of production of these chemicals with that of males' spermatogenic activity and the mating season all lead to the suspicion that 3-methylbutyl methyl sulphide is involved in signalling in a sexual context, and/or is an excreted by-product of a sex-linked physiological change (see Bailey *et al.* 1979). Jorgenson *et al.* (1978) and Bailey *et al.* (1979) both speculate on the possible biochemical pathways that may give rise to some of the compounds they found.

Some of Jorgenson's co-workers have synthesized fox urine from a mixture of the compounds they discovered therein, amongst which $\Delta^3$-isopentenyl methyl sulphide may be species specific to red foxes (Wilson *et al.* 1978). Wilson *et al.* (1979) placed samples of the synthetic urine aloft small snow mounds, alternating the synthetic urine with a control of 20 mg citronellal per litre of water (similar report in Whitten *et al.* (1980)). Their results showed that the synthetic urine was about 20 times more likely to be token marked than was the control.

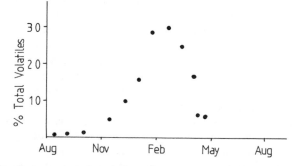

Fig. 15.5. Two compounds identified from the urine of male foxes have been found to increase (as a percentage of total volatiles present) around the mating period in January–February. The more abundant of these two is 3-methylbutyl methyl sulphide (shown here). The other compound, 4-heptanone, that follows the same pattern rarely exceeds 4 per cent of the volatiles present. (After Bailey *et al.* (1979).)

Their 'urine-marked' snow mounds were placed at 40 to 80 m intervals along paths, and in some ways it is surprising that of those which foxes passed closely by (*n* = 119) only 22 per cent were marked. My leash-walking dominant adult vixens invariably overmark alien urine, and in caged conditions Blizzard and Perry's (1979) males token marked on 63 per cent of the alien male samples with which they were presented. From other preliminary results, Wilson *et al.* (1979) suspect that the principal attractant to foxes from the smell of urine derives from the sulphur compounds therein.

*Faeces*   Fox faeces are generally left on conspicuous sites along trails (Fig. 15.2). Typical sites are on tussocks of grass, clods of soil, and molehills. Sometimes they are found aloft precarious positions and I have very occasionally seen foxes defecate from a handstand position to achieve this. Mostly only one faeces is left at a site but two, or even three faeces sometimes accumulate at one place. Seventy per cent of the faeces left by one leash-walked vixen were at sites she had previously urine-marked (Macdonald 1979*a*). Faeces accumulate along trails in the vicinity of a red fox's day-time lair, perhaps because they often defecate soon after waking up. Foxes defecate throughout their home-range (Macdonald 1980). The biggest accumulations of faeces that I have seen in the wild have been on and around the carcasses of large mammals, such as sheep.

*Coyote,* **Canis latrans**

*Urine and faeces*

*Field studies*   Three studies have been published recently on coyote scent-marking (Bowen and McTaggart Cowan 1980; Barrette and Messier 1980; Wells and Bekoff 1981; and see Lehner 1978). Each of these studies was undertaken in a different habitat, and so in comparing their findings it is important to remember that coyotes are amongst those canids on which there have been sufficient studies to confirm intraspecific variation in social behaviour. Nevertheless, all three studies report marking with urine and faeces, alone and in combination, at least sometimes by almost all the animals they studied, and sometimes at high frequencies. Gier's (1975) observation that coyote traps baited with the urine of coyotes of one or other sex killed more of the opposite sex, gives a good indication that coyote urine yields social odours.

Barrette and Messier (1980) examined 949 scent marks found in the snow during one winter in a woodland area of southern Quebec, Canada. There, coyotes live as territorial pairs, occupying ranges of about 22 km², sometimes accompanied by their juvenile offspring. Solitary adult coyotes occupied larger (49.4 km²) ranges which overlapped each other's and those of the territorial adults. These authors defined a scent mark such that it could comprise some or all of urine, faeces, and scratches. Of their total sample of marks, 90 per cent included urine, 19.3 per cent included faeces, 80.6 and 9.7 per cent were urine alone or faeces alone respectively, and 9.5 per cent urine and faeces. Scratching was associated with 30 per cent of urine marks irrespective of the presence of faeces, and with only 14 per cent of faeces, if urine was absent. By comparison with the average figure, irrespective of group size, of 2.17 per km per coyote, Barrette and Messier discovered that coyotes travelling in pairs token marked more than single animals (*average* rates: 1.25 vs 2.8 per km per coyote). When juvenile members of a territorial group were temporarily separated from their parents and travelled alone, they marked at a similar high rate to their parents, and more than did lone non-territorial adults. Similarly, there was a tendency for a greater proportion of the total marks of territorial coyotes than of solitary coyotes to be urine, and the marks of territorial animals were more commonly associated with scratching, and urine was more often added to their faeces.

Bowen and McTaggart Cowan's (1980) study was also in a wooded area of Canada, but there coyotes lived in packs of 2–8 which defended territories of 8–20 km² and fed mainly on scavenged moose and white-tailed deer for which they may have hunted co-operatively. Owing to the limited availability of snow Bowen and McTaggart Cowan's study only spanned November to March, but during that period they found 1047 snow signs which conformed to Kleiman's definition of marking. Of these, 66.3 per cent were urine alone, and most of the remainder were urine and either faeces or scratching. They found that all pack members scent-marked, although dominant males marked more often than dominant females (12.7 versus 6.6 times per hour). The average coyote marked at 1.8 marks per kilometre, and marked more often (and more often with multiple marks at each site) within a 0.25 km wide strip near territory boundaries. Coyotes would venture into a neighbouring territory to feed from a carcass, but unlike the territory holders they would not then mark around that carcass. However, in the absence of such an attraction neighbours rarely trespassed. In general, coyotes are inclined to urine-mark on carcasses (Bowen and McTaggart Cowan 1980; Camenzind 1978). Bowen and McTaggart Cowan saw coyotes scent mark at average rates as high as once every 150 m, so that only three or four minutes would elapse between coming upon fresh scent as a coyote followed a typical trail.

Both of the foregoing studies were hampered in that they relied solely on information from the snow which, apart from other limitations, is only available in winter. The work of Wells and Bekoff (1981) is unique amongst studies of scent-marking of the Canidae in that they were able to watch undisturbed wild coyotes. From a vantage point called Blacktail Butte they could readily measure the rate of marking in terms of marks per hour of coyote activity, as quite distinct from the measure of marks per distance travelled used in the snow-tracking studies cited above. These data revealed that the rate of raised leg urinations (RLUs) per hour of observation of coyotes varied seasonally. From November to February, which is the courtship season for coyotes, RLUs were maintained at an overall high rate of about 0.8 per hour (remembering that during these months coyotes are only active for just under half the time). By the time pups were born in April, the rate of RLU marking was falling and by May it had reached a minimum of 0.2 RLUs per hour. Although the overall rate of RLU marking fell just prior to the birth season, the rate per active hour peaked then. That is to say, the coyotes were increasingly lethargic as the birth of pups approached, but during this period the more inactive they became the more they marked during the diminishing periods when they marked at all. This paradox, like much in Wells and Bekoff's study, illustrates the horrendous complexity of resolving patterns of scent-marking. Indeed, they paid particular heed to the multivariate interactions of criteria used to diagnose marking as distinct from elimination. Considering four possible candidates for scent marks (RLU, squat, forward lean urination, and defecation) they analysed each in terms of whether it was preceded by sniffing at the site, followed by scratching, and/or whether it was directed at a site where a coyote had previously been seen to urinate. On the basis of these criteria they could not dismiss any of the four as a potential scent-marking behaviour, and most notably, they could not dismiss squats. Certainly, more squats (12.0 per cent) occurred in the absence of any criterion than did any other type of elimination, and more (11.6 per cent) RLUs were associated with all three than was any other, and a discriminant analysis showed that RLUs differed from squats in that they were more often directed at previous urinations and were more often performed in the company of other coyotes. In contrast, squats more frequently followed sniffing than did RLUs. This led Wells and Bekoff to conclude that if any or all of their criteria were robust, then for these coyotes any interpretation of RLU versus squat as parallel with mark versus eliminate was not tenable. There was a clear-cut sex difference: adult males adopted the RLU posture at 78 per cent of urinations and never squatted, whereas adult females almost invariably (97.5 per cent) squat-urinated. In contrast to Barrette and Messier's (1980) findings concerning the numbers of marks per unit distance travelled, Wells and Bekoff found no difference in the marking rate per coyote per hour for solitary coyotes as compared to pairs, but they did find a highly significant increase in this rate for members of trios in comparison with members of pairs and singletons. Indeed, there were more marks per coyote per hour for trios than for members of quartets and larger packs, whereas Barrette and Messier recorded more marks per coyote per km for members of pairs than of trios. One cannot judge whether these differences reflect chance variation (perhaps in the sociological composition and hence marking behaviour in groups of different sizes) or whether the differences are biologically relevant. It is possible, for example, that ecological differences between study areas could cause consistent differences in the composition and structure of groups between the areas (e.g. in the ratio of breeding to non-breeding adults, cf. Macdonald and Moehlman (1982)). This uncertainty clouds any interpretation of these average marking rates since there is no such thing as an average coyote. Amongst coyotes the frequency and type of marking varies with the individual's sex and age, with the time of the year and with its activity of the moment (Wells and Bekoff (1981) show that RLUs were more often associated with travelling then were squats, which were in turn more often associated with hunting). In addition to these variables, there is every reason to assume that the subtleties of an individual's status and social relationships with each of its companions also affect its inclination to mark in a given way under given circumstances. Hence, subsumed within the categories of different group sizes lies the product of many interacting variables that make the mean figures alone for marks per member of packs of different sizes uninterpretable. Indeed, it is daunting that after the admirable detail that these authors incorporated into their analyses of a uniquely large sample of observations, their conclusions regarding the functions of the behaviour they had watched remained few and speculative. Nevertheless, they were able to show that, although members of one pack spent roughly equivalent amounts of time in each area, their corporate marking rate was much higher (0.72 marks per h) in a part of their territory where there were many intrusions, than in the vicinity of their den (0.09 marks per h) or an area where there were few intrusions (0.21 marks per h).

*Urine chemistry and lures*   Concern with predator control has stimulated several investigations of the chemical constituents of coyote urine and their relative attractiveness (Murphy, Flath, Black, Mon, Teranishi, Timm, and Howard 1978; Teranishi *et al.* 1981; Fagre *et al.* 1981). Teranishi *et al.* (1981) present lists of the compounds they have identified and Fagre *et al.* (1981) have investigated the responses of penned coyotes to some of these (mainly grouped as acid, base, or neutral fractions, each of which contained at least 50 compounds). Their carefully conceived methodology for presentation of odours involved measuring the time coyotes spent sniffing a capsule containing a trial odour and affixed to a 0.3 m² stainless steel plate within a 100 m² test area. In general the coyotes spent more time investigating acid and neutral fractions of the urine than they did the base fractions, and they spent more time on fractions which were artificially concentrated than they did investigating those presented at their natural concentration in urine. Under the same circumstances, even the most concentrated of these mixtures had only a fraction (40 per cent) of the allure of the best commercial trap lure (Carmen's distant call lure). At a practical level these authors are concerned only with the relative efficacy of each substance they test, and hence their results are deceptively straightforward. At a biological level the interpretation of such differences in attractancy of various mixtures is difficult, if not impossible.

*Faeces*  Coyote faeces may be associated with urine and/or scratching. For the most part they are deposited singly. However, Wells and Bekoff (1981) found that defecation was recorded proportionally more around carrion than was any other of their categories of marking. It was also interspersed with 'lying down' amongst the elements of their behavioural inventory. The two may be linked to the extent that coyotes lie down to rest near a carcass, and defecate soon after waking up. The consequence has been described by both Camenzind (1978), and Bowen and McTaggart Cowan (1980) as an accumulation of faeces around carrion.

## *Wolf,* Canis lupus

### *Urine and faeces*

*Field studies*  Biological interest in scent marking by wolves began in 1909 with Seton's description of their tendency to leave both urine and faeces at trail junctions. Young and Goldman (1944) described scent posts, noting 'as wolves pass over their runways, they stop at these posts, invariably voiding urine, and often faeces as well'. Thompson (1952) described just such concentrations of wolf droppings along fire lanes in Wisconsin forests, and noted that he never found coyote faeces similarly distributed. However, Thompson did find piles of 4-10 coyote faeces on the wooden bridges that spanned small logging roads, where in turn he never found wolf faeces. Subsequently, Peters and Mech (1975) have shown that 40 per cent of what have come to be known as RLUs (raised leg urinations) and many faeces are located at trail junctions. Peters and Mech's paper on wolf scent-marking is itself a landmark in the study of carnivores' social odours: in an area of Minnesota where the spatial organization of the wolves was already known (Mech 1977) they used aircraft to plot the movements of 13 radio-marked packs whose paths they subsequently tracked in the snow along 240 km of trails. From observations in the field and captivity, they learnt to distinguish the marks left in the snow by two sorts of urinations, firstly those delivered with a cocked leg (the RLUs) and secondly those delivered while squatting (where the wolf's hind-legs bend, lowering its rear without either hind-leg leaving the ground). Using their tracking skill, Peters and Mech found an average of one RLU per 450 m of pack trail, with variation between 20 RLUs per km of trail and none in seven kilometres. In the context of their data on pack territoriality they mapped out the 583 RLUs which they found, and so demonstrated that the mean frequency of marking per km of trail was much higher near borders than it was elsewhere within territories (Table 15.7) and furthermore that, irrespective of location within the territory, the mean frequency of RLUs along a pack's trail was greater on trails than off them. The significance of these findings is not diminished (nor their interpretation seriously altered) by Bekoff's (1980) important methodological points concerning the accuracy (about 70 per cent) that is attainable in distinguishing from snow signs the RLUs of domestic dogs from their squats, nor the under-representation of squats that snow sign gives for coyotes (Wells and Bekoff 1981).

Among members of a wolf pack, the frequency of RLUs per km of trail did not increase with pack size over two. Combining this finding with observations on captive packs led Peters and Mech to the conclusion that raised leg urinations, which were almost invariably directed at conspicuous sites, were the prerogative of the alpha-pair. They noted that the elevated positioning of the urine not only facilitated its dispersal on the wind, and increased the evaporative surface as the urine trickled downwards, it also minimized the chance of the mark being covered by fresh snow and created a visually conspicuous beacon. Unfortunately, their observations in the wild were confined to the snowy months, but even within the winter they found a variation in RLU frequency from a peak of 3.5 per km in February to a minimum of 1 per km in March. Importantly (viz. Hediger's *pars pro toto* model), there was evidence of wolves turning back when trespassing if they encountered the recent marks of a neighbouring pack.

In a similar study, Rothman and Mech (1979) clocked up a further 72.2 km of tracking, this time of lone wolves. In all that distance they found only one series of RLUs, and no evidence of scratching. Moreover, the RLUs they did find were left far from the normal travel routes of local packs. Similarly, lone wolves did not overmark the scents of strangers they encountered, which packs invariably did, nor did they mark in the vicinity of kills. Finally, loners consistently defecated away from the trails, not because they travelled less on trails, but because they deliberately left them in order to defecate. These differences were observed in spite of the fact that lone wolves were just as inclined to discover and investigate scent marks as were the newly formed pairs which Rothman and Mech also tracked.

*Captivity studies* By comparing the investigation times of captive wolves sniffing the urine of individuals from another pack Brown and Johnston (1982) were able to show that the urine of different individuals could be discriminated by odour. They achieved a comparable result for beagle dogs.

In experiments on two captive packs of wolves Harrington (1981) confirmed for this species the essence of Henry's book-keeping hypothesis (see pp. 642 and 702). The wolves never marked cache sites while making them and rarely marked them while they contained food, but usually marked the emptied cache within one minute of removing the food. Harrington dug artificial caches which he provisioned with chicks, and he also dug identical, but empty control holes. Most emptied caches were marked, but only 10 per cent of control holes prompted urination although all were thoroughly investigated. Subsequently when wolves revisited these sites they spent an average of 19.6 s sniffing at control holes and unmarked, emptied caches, in contrast to 12.7 s at emptied and marked sites. Harrington concluded that the residual odour of food at the cache site stimulated the urination and that it functioned to minimize subsequent wasted time on fruitless investigation of emptied sites.

### Domestic dog, Canis familiaris

#### Anal sacs

Canid anal sacs are best known for their pathology (Ashdown 1968) and hence are described in textbooks of veterinary anatomy (e.g. Miller *et al*. 1964; Montagna and Parks 1948). The terminal part of the rectal canal of domestic dogs is normally completely closed by powerful sphincter muscles which obliterate the lumen of the passage. These striated muscles lie just inside the dog's perianal skin and form a sleeve around the anal sacs such that contraction of the sphincter would seem almost inevitably to put pressure on the sac. Ashdown (1968) replaced the 1–1.5 ml contents of greyhound anal sacs with coloured fluid and found that at the next defecation the dogs ejected the entire contents of the sac, squirting it over a distance of 60 cm. Repeating the experiment, but filling the sacs with only 0.5 ml, he found that the dyed contents were expelled intermittently over about 80 h. The stain was generally on the first or last element of the stool, and rarely the middle. Sometimes the dye was ejected before defecation, sometimes afterwards and rarely in the absence of defecation. Grau (1935) reported that the odour of the anal sac secretions of both dogs and bitches varied with sexual activity and Donovan (1967, 1969) suggests that the secretion of oestrous bitches attracts male dogs. Donovan gives few details, but describes how he evacuated the anal sacs of oestrous and anoestrous bitches in his veterinary surgery, gathering each sample of evacuate on surgical gloves. In some cases he then smeared this material on the rump of other dogs. He found that male dogs mounted others on to which the anal sac secretion of oestrous bitches had been smeared, but showed little interest in the secretions from anoestrous bitches. He also collected samples from dogs that had recently been frightened and wrote that other dogs markedly avoided these samples. Donovan concluded that these glands produced odours which had amongst their functions sexual attractance and a warning effect; however, in the absence of details these conclusions are unsubstantiated.

Indeed Doty and Dunbar (1974) in a carefully controlled set of preference tests found no difference in the time that male dogs spent investigating the anal sac secretions of dioestrous and oestrous bitches. In the same tests they found that both dogs and bitches spent more time investigating the odour of urine of either sex than they did investigating the odour of anal sac secretions. Dunbar (1977) presented dogs with paired odour samples from the anal sacs of males and oestrous and anoestrous females. He found that there was no difference in response by any of the three classes of subject to the anal sac secretions of any other, irrespective of how the samples were paired in the trials. However, although neither oestrous nor anoestrous bitches showed any difference in response to the faeces of either of the three classes, males did show a preference for female faeces over those of males (see below).

At a chemical level, the dog's lack of differential response to oestrous and anoestrous anal sac secretions is reinforced by Preti, Meutterties, Furman, Kenelly, and Johns' (1976) analysis. They found that neither the contents of the anal sacs of beagles nor the ratio of volatiles varied consistently between dogs and bitches or between bitches at different stages of the oestrous cycle. Indeed, the constituents did not vary in composition between dogs and coyotes, although dogs secreted greater quantities of volatiles than did coyotes.

The compounds identified from dog (and coyote) anal sacs by Preti *et al.* (1976), in common with the secretions of the anal sacs of other species, include short chain ($C_2$–$C_5$) aliphatic acids, trimethylamine, and acetone, which could be the products of bacterial action on proteins, carbohydrates, and lipids secreted into the sac (Albone *et al.* 1978).

*Urine and faeces*

Doubtless a dog cocking its leg against a lamp-post is the most familiar of all examples of mammalian scent-marking, and it is the one which prompted Lorenz (1954) to make an impish jab at established sensibilities by suggesting that it was the functional equivalent to the nightingale's melodious song. In fact, although the analogy was heuristic it is partly incorrect since free-ranging, but owned, domestic dogs do not seem to use urine to maintain territories (Scott and Fuller 1965; Bekoff 1979*b*), nor even, generally, to be territorial (Beck 1973). Nevertheless, study of the social odours of dogs offers opportunities for observation unparalleled by any wild canid and has considerable potential relevance to public health—Beck (1973) estimated that New York City dogs deposited some four million litres of urine and 18 million kg of faeces on the streets annually, and it is clear (e.g. Ghosh 1981) that the pattern in which at least some of these excreta are distributed relate to the dogs' social behaviour.

Sprague and Anisko (1973) compiled a directory of the postures used by laboratory beagles during elimination, recognizing 12 postures associated with urination. Of Sprague and Anisko's terms, 'squat' is commonly associated with urinations that do not meet Kleiman's criteria, whereas various positions in which a hind-leg is lifted from the ground in the course of directing urine (such as raise, lean-raise, squat-raise, and elevate) are generally regarded as scent-marking. Strictly speaking, Bekoff (1979*b*) has pointed out that the raised leg urinations (RLU) of Peters and Mech (1975) incorporate Sprague and Anisko's elevate and raise. At least amongst the Canidae there is such variation in postures during urination that otherwise meet Kleiman's criteria, that posture seems to be an inadequate indicator of any communicative function of urination. Perhaps the volume voided is a more useful criterion.

*Park studies*    Bekoff (1979*b*, 1980) watched 51 free-ranging domestic dogs around university campuses and confirmed that, although males generally (97.5 per cent of occasions) cock their legs (RLU) and females generally (67.6 per cent) squat, both sexes at least sometimes used each of the four postures he noted. In spite of considerable individual variation, males marked more frequently (once per 1.6 min) than females (once per 4.7 min) and of all the urinations he saw, a greater proportion of those by males (71.1 per cent) than by females (18.0 per cent) could readily be defined as scent-marks (although Bekoff himself notes how tricky this distinction is, since it does not involve any information on the effect of the urine on another animal smelling it). Bekoff's (1979*b*) data showed that male dogs were more likely to urine mark within a given 5-s interval if they had just met another dog (of either sex) or had just seen another dog urinate, and that of these effects, the most significant was the threefold increase in likelihood of a male urine-marking after he had seen another male do so. Bitches exhibited similar, but much less significant, trends. Most interestingly, Bekoff coined the term raised leg display (RLD) to describe the motion without the event of raised leg urinations (called pseudo-urination by Sprague and Anisko (1973) and dry marking by Rothman and Mech (1979)) and he demonstrated that RLD was significantly ($p < 0.001$) associated with situations where another dog was in sight of the actor (see also Ghosh 1981). Harrington (1982*b*) argues that for coyotes RLDs may simply indicate an empty bladder, rather than a separate type of signal.

Males scratched after 14.3 per cent of urine marks and females after 9.5 per cent (a significant difference, $p < 0.05$), and this behaviour was not particularly associated with any specific posture of urination for either sex. However, as with RLDs, scratching was associated (76.9 per cent) with other dogs in view (Bekoff 1979*a*). Chantry (1982) and Reid (1982) have studied dogs under somewhat similar conditions to Bekoff, recording their behaviour in parks and urban areas, with and without accompanying owners. Chantry compared behaviours when the subject dog had another dog in sight with those when it did not. Both dogs and bitches were generally less active when another dog was in view, since they stood still to look at the other, but there was no difference in RLU, scraping, or defecation behaviour between the two situations. On the contrary, bitches squat-urinated for longer when another dog was in view and males 'squat-raised' more often and for longer. The differences between Bekoff's and Chantry's findings very probably relate to the different social structure of the two dog communities. Bekoff's dogs were probably much more familiar with each other than were those in an English urban setting which Chantry studied and where dogs were probably uncertain of the identity, far less status, of many other dogs they met. In these circumstances squat-raise urination may symbolize the social uncertainty of the signalling dog.

Working on the same population as Chantry, Reid (1982) examined aspects of eliminative behaviour of dogs in the light of whether or not they were accompanied by their owners. In parts of the study area faeces were scattered at a density of 2.67 per 100 m$^2$, and dogs running beside their owners but off their leads were more likely ($p = 0.02$) to defecate than those on a lead. Similarly, dogs of either sex off the lead were more likely to urinate than those on leads, and still more likely to do so if they were in the park entirely without their owner ($p = 0.006$). Males in every situation were more likely to urinate than females in the same situation and, intriguingly, mongrel males were more likely to urinate when off the lead than were pedigree males.

Considering the sanitation aspects of elimination and/or scent marking by dogs, Reid (1982) concluded that walking on a leash reduced the excretory material left in an area. Whatever the reason for the increased probability of defecation by dogs running loose in the company of their owner, there is no such increase in comparison to dogs on a leash with respect to dogs running free in the absence of their owners. In contrast to most populations of loose dogs, the pariah dogs studied by Ghosh (1981) in India were territorial. Ghosh concluded that urine marking (and RLDs) played a role in territorial defence. Male dogs marked during encounters with interlopers and after expelling them (whereupon they sometimes defecated). Males urine marked 0.45 times per minute, whereas females did so 0.15–0.22 times per minute (but 0.5 times per minute when in oestrus). Over-marking was common and males urinated at a higher rate in unfamiliar parts of their home range outside the core area.

*Laboratory studies*    In the laboratory Beach and Gilmore (1949) produced experimental confirmation of the layman's conviction that male dogs could discern the odour of an oestrous bitch from her urine—in their experiments sexually active dogs spent longer investigating the urine of oestrous bitches than they did that of anoestrous ones (an effect which one sexually inactive dog did not show (see also Hart 1974; Hart and Haugen 1971). Doty and Dunbar (1974) followed this by a series of experiments in which three male dogs with previous sexual experience and three with none, together with six ovario-hysterectomized bitches (under various hormone treatments) were confronted with the odours of each other's urine, anal sac, and vaginal secretions through one of two odour ports, each being paired with water behind the second odour port as a control. During 2.5-min tests Doty and Dunbar scored the time spent sniffing the test odour and subtracted the time spent sniffing the water. Figure 15.6 shows the response of the experienced and inexperienced males to the three types of odour from the bitches under normal conditions and when treated with hormones. An analysis of variance of these data indicated that all three factors (experience, odour source, and hormonal state of the bitch) influenced the sniffing times of the males. Males and females spent more time investigating urine than they did either vaginal or anal sac secretions. However, despite individual differences,

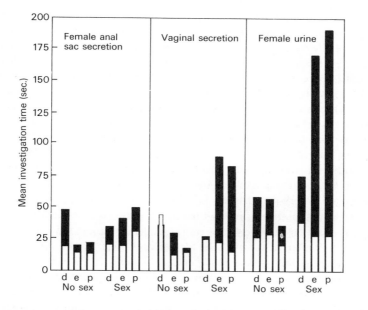

Fig. 15.6. Doty and Dunbar (1974) demonstrated how a complex array of factors may interact to affect the time male dogs spend sniffing at the odours of bitches. They presented female odours together with a control of water from each of two ports and recorded the time (during 2.5 min trials) spent investigating the test odour (*black bars*) *versus* the water (*white bars*). The males differed in whether they had previous sexual experience (sex *v*. no sex). Trials were repeated over 18 days, over which the originally dioestrous (d) bitches were successively treated with oestrogen (e) and then progesterone (p) to manipulate their hormonal state. The results show that the type of secretion, experience of the male and hormonal state of the female all interact in producing the male's response. (After Doty and Dunbar 1974.)

only sexually experienced males spent more time on oestrous urine (almost 2 min) than dioestrous urine (about 1 min) and, equally, only sexually experienced males spent longer on oestrous vaginal secretions (see also Beach and Merari 1970). However, when the roles of experimental dog and donor were reversed, oestrous and dioestrous bitches spent the same time on average investigating male urine, which was also relatively uninteresting to other males. Doty and Dunbar discuss several methodologically important aspects of 'preference' experiments of this sort: first they point out that there is no reason to assume that the volumes of each odour placed in the test apparatus were similar in any way other than quantity from the dog's viewpoint; secondly, the interpretation of investigation times is fraught with difficulty, although they did find that their odour investigation times for given sexes and/or hormonal states correlated well with times that Le Boeuf (1967) had found dogs spent investigating other individuals of similar categories. However, the social significance of even Le Boeuf's investigation times for dog-to-dog interactions is difficult to interpret.

Dunbar (1977) staked-out test dogs within an arena and noted the number, nature, and duration of visits to them by other dogs. Male dogs were very interested in bitches even when they were not on heat and their direct and enthusiastic approach to them indicated to Dunbar that males could distinguish bitches from a distance. For example, males spent more than twice as long (92 s versus 39 s) with bitches as they spent investigating males, averaging 10.6 visits to anoestrous bitches per test and 1.6 visits to males. These visits were of a different nature, too. Dunbar showed that males sniffed five times more often at the anogenital region of anoestrous females than at the anogenital region of males, nearly four times as often to the muzzle and face, and over 20 times as often both towards the ears and to the rest of the body.

In a second experiment, Dunbar (1977) found not only that males spent more time investigating oestrous bitches than non-oestrous ones (and both more than males) but also that males urinated more frequently against the holding cages of animals they 'preferred to visit'. In 85 per cent of tests they urinated at least once against cages holding oestrous females, but only in 27 per cent of tests on other males. Anoestrous females showed no especial tendency to prefer other anoestrous bitches or males, but oestrous females visited males more often and for longer than they visited other bitches. Finally, Dunbar altered his experiments to test the response of dogs to urine, faeces, vaginal and anal sac secretions, saliva, and ear wax. He found that males' preference for, and tendency to urinate upon urine and vaginal secreta mirrored their preference for the caged animals which donated the sample. Males spent longer investigating the faeces of females and this was not affected by the hormonal state of the female. When urine samples from non-oestrous bitches and males were paired the non-oestrous females spent more time investigating the bitch odour (and urinated upon that sample more often), but when the subject bitches were in oestrus these results were reversed.

Dunbar's observation that faeces themselves have communicative potential is relevant to Sprague and Anisko's (1973) note that a quarter of male defecations were directed at elevated surfaces.

## Vaginal secretions

Goodwin, Gooding, and Regnier (1979) followed the cytological changes associated with the oestrus of three laboratory bitches, during which time they observed each of the bitches in the company of dogs and documented changes in behaviour throughout the oestrous cycle. During the bitches' oestruses the males urinated more frequently, overmarked with urine more frequently, and increased anogenital sniffing. Throughout these observations the authors also collected vaginal smears each day from each bitch and analysed them with gas chromatographic and mass spectrometric techniques. This enabled them to identify a substance which was in vaginal secreta only during oestrus (25 or more other compounds were there all the time). When they synthesized this compound, methyl-*p*-hydroxybenzoate, and smeared it on the vulva of anoestrous bitches, males reacted to them as if they were oestrous, trying to mount in 18/21 trials in spite of the female's rebuffs. It seems that this single chemical was sufficient to elicit all the male behaviour patterns associated with females on heat. This same compound is used as a preservative in many foods and cosmetics, shampoos and handcreams.

## Family Hyaenidae

The four species of the family Hyaenidae show a fascinating inter- and intra-specific variation in social organization (Kruuk 1976; Mills 1978).

Where prey are abundant and stable, the spotted hyaena, *Crocuta crocuta*, lives in matriarchal clans averaging 55 members and defending territories of 10–50 km² whose borders are demarcated by latrines littered with different odours (Kruuk 1972). Brown hyaenas, *Hyaena brunnea*, live in small groups of related males and females who forage solitarily through territories 250–540 km², within which are dotted latrines and pastings from the anal glands (Mills 1978; Owens and Owens 1979). The striped hyaena, *Hyaena hyaena*, is distributed in Africa northerly to the brown hyaena and in the Serengeti ecosystem it lives solitarily in home-ranges of 44–72 km² throughout which its faeces are scattered singly (Kruuk 1976). The aardwolf, *Proteles cristatus*, weighs only 12 kg in comparison to spotted hyaenas of up to 80 kg, and with a reduced dentition forages for termites in 1.5-km² territories within which faecal middens are scattered (Kruuk and Sands 1972; Bothma and Nel 1980). Studies in contrasting habitats have revealed variations on each of these social themes, and in the patterns of scent-marking.

*Anal glands*

All the Hyaenidae have anal scent pouches and in the sequence of aardwolf, spotted, striped, and brown hyaenas there is also progressive differentiation of subcaudally situated sebaceous glands which open into a pouch above the anus. Matthews (1939) described these subcaudal sebaceous glands (which Ewer (1973) calls auxiliary glands) as lying between the anal sacs and opening directly into the upper portion of the subcaudal pouch, in the same way that Flower (1869) described for the aardwolf. Kruuk and Sands (1972) state that the aardwolf uses its anal glands in a similar way to those of spotted hyaenas. At the other end of the spectrum, Mills, Gorman, and Mills (1980) describe the extraordinary glandular armaments of the brown hyaena's anus in which the eversible subcaudal pocket consists of an upper and (when everted) outer bilobed section on which the whitish secretion of large sebacious glands gathers, and to each side of which lies an area where the black secretions of apocrine glands accumulate (Fig. 15.7 and Plate 15.8) (an arrangement first described by Murie (1871)). The pasting behaviour and associated 'anal bulging', as Kruuk (1972) calls it, by which these secretions are applied to individual grass stalks, involves the brown hyaena straddling the stalk, squatting slightly and everting its anal pouch (Fig. 15.7). Once everted, the two lobes of sebaceous gland masses are divided by a central groove into which, by deft and deliberate wigglings, the brown hyaena slots the grass stalk. The hyaena then moves forward, retracting the everted pouch as it does so, thus firstly drawing the stalk through the central crease and so coating it in a blob of white secretion, and then, as the pouch inverts, drawing it through the black secretion. The result is that, about 70 cm above the ground, a centimetre long blob of white sebaceous secretion adheres to the stem, and above this, separated by just over 1 cm, lies a slightly shorter smear of black secretion. On average, brown hyaenas of both sexes repeat this performance once every 2.6 km that they travel.

The anal pouch of the striped hyaena, as described by Pocock (1916b) shows an intermediate stage in the development of the sebaceous auxiliary glands, and Kruuk (1972) states that the pasting action of striped and spotted hyaenas is the same. The secretions of striped hyaena anal pouches have been analysed and contain an unusual thioester (Wheeler, von Endt, and Wemmer 1975).

Rieger (1977, 1979) notes that, during social encounters, striped hyaenas sniff at each others' everted anal pouches. In addition, they will rub on the pasting sites of other individuals and in so doing smear their shoulders with the secretions.

Kruuk's (1972) classic field-study of spotted hyaenas revealed something of the social significance of 'pasting' by that species. The secretion is creamy white, and is deposited on a grass stalk when the spotted hyaena drags its anal pouch over the stalk. The hyaena bulges out its pouch to a distance of about 3 cm and so exposes the everted inner walls, which leave a 2–5-cm long smear on the grass. This process may be repeated 4–5 times in succession, and might either be followed by, or alternating with, scratching. The resulting smell is detectable by humans

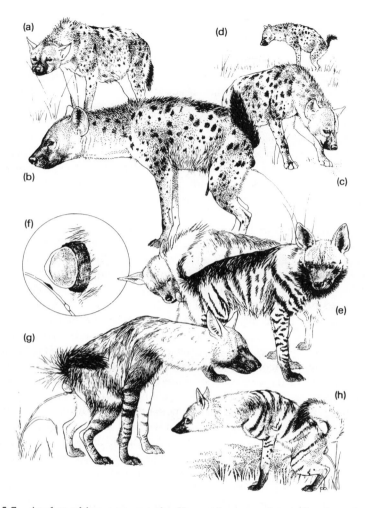

Fig. 15.7. Anal marking amongst the Hyaenidae: members of a clan of spotted hyaenas (a)–(d) in the Ngorongoro Crater visit pasting sites where they inspect (a) their secretions on long grass stalks and add more pastings from their anal pouches (b). While at pasting sites spotted hyaenas can be highly aroused, with females (b) exhibiting clitoral erections. Pawing at the ground (c) occurs frequently at pasting sites and at latrines (d), which also occur at the territorial borders, but generally separate from pasting sites. Marking behaviour and the pattern of pastings and faeces differ between populations (Kruuk 1972). Despite interspecific differences in their social organization, sniffing of the anal pouch is an important component of social greeting for all hyaenids, as shown here for two striped hyaenas (e). When everted, the anal pouch of the brown hyaena (f)–(g) has a central crevice through which a grass stalk is drawn and coated with white secretion. As the sac inverts a further smear of black secretion from its walls is left higher up the grass stalk (see Plate 15.7) (Mills *et al.* 1980). The smallest hyaenid, the insectivorous aardwolf (h), pastes throughout its territory, but at a higher rate at the borders (Kruuk and Sands 1972).

from several metres downwind. Kruuk described 'pasting places', many of them along territory borders, where spotted hyaenas, often in parties, would return repeatedly to add new paste marks. In addition, pastings were made elsewhere in clan territories, for example around kills or when a party of hyaenas were confronted by a lion. They were also found at faecal latrines near territorial borders (see below). Wherever they were, spotted hyaenas were stimulated to paste when they encountered other pastings.

The analysis of the distribution of some 5144 brown hyaena pastings (Mills *et al.* 1980) showed that all members of a family group pasted wherever they travelled within their territory. Displaying these data on a three-dimensional plot showed clearly that not only were some parts of the range marked more often than others, but that in general over the year the border areas accumulated fewer marks than did the interior. This was in spite of the fact that when brown hyaenas travelled in the border zones of their territories (which did overlap somewhat with those of neighbouring groups) they marked more frequently per km travelled. So, brown hyaenas mark at a higher frequency at the borders of their territories, but they travel less in these areas and so the net effect is that fewer marks accumulate around the perimeter than in the interior. Within the territory, males marked more frequently than females, but at the border the two sexes marked at the same rate during one year, but females marked significantly more frequently per distance travelled during the following year. One factor which may have contributed to the higher frequency of marking in border zones was the greater likelihood of encountering the pastings of neighbours. Mills *et al.* (1980) showed by a combination of opportunistic observations and pasting transplant experiments, that brown hyaenas invariably pasted on top of alien marks (and so demonstrated that they can distinguish members of their own group from neighbours on the basis of paste odour). This recognition may be based on consistent individual differences in the relative concentrations of compounds within the secretion. Mills *et al.* show, for example, that two male brown hyaenas differed consistently (over 72 hours) in the amounts of each of a dozen or more chromatographic peaks in both the white and black pastes of their marks (collected from grass stalks in the wild).

Kruuk and Sands (1972) describe the pasting behaviour of the smallest hyaenid, the insectivorous aardwolf, *Proteles cristatus*. As aardwolves criss-crossed their home-ranges in the Serengeti they pasted on grass stems throughout, but sometimes they undertook bouts of intensive pasting, lasting up to 22 min and at a rate of about one pasting per min. These bursts of pasting occurred exclusively in the vicinity of what Kruuk and Sands interpreted as territorial borders.

*Urine and faeces*

Although pasting sites of spotted hyaenas living in the stable matriarchal clans of the Ngorongoro Crater were often at territory borders, they were not often used as latrines. However, the spotted hyaenas did defecate together repeatedly (often simultaneously, Bigalke 1953) at latrines which comprised 10-m long stretches of ground littered with visually conspicious white faeces (Matthews 1939). Kruuk (1972) found that parties of spotted hyaenas would travel together to these latrines, often making substantial detours to reach them and there amidst much social sniffing they would defecate, paste, and scratch the ground. Such latrines were concentrated along territory borders at Ngorongoro Crater, but out on the Serengeti plains where hyaena social units were ephemeral and territories unstable or non-existent, latrine sites were instead located along travel routes. For example, Kruuk found eight latrines on 270 km of trail across the Serengeti, and none across 230 km of open country. The use of border latrines in Ngorongoro crater was clearly associated with territorial defence. Some latrines were visited by members of both adjoining clans and territorial conflicts were punctuated by visits to latrine sites and frenzied marking. Kruuk (1972, p. 257) describes the rout of the Lakeside Clan by 12 members of the Scratching Rocks clan, and the subsequent intensive burst of marking at the nearest border latrine. He speculates that the aggressive circumstances of latrine use and the presence of anal bulging behaviour (without marking) during everyday agonistic encounters between hyaenas indicates a visual element in the displays at pasting places and latrines, and perhaps suggests that the olfactory display evolved from a ritualization of the visual one. The border latrines and pasting places were not at sites distinguished by any topographical feature that Kruuk could recognize, perhaps other than an abundance of long grass stalks.

In addition to defecating at latrines, spotted hyaenas will also defecate without ceremony, nor apparent marking function, wherever they happen to be, and similarly, Kruuk concluded that their urination was purely eliminative. The random distribution of faeces also seemed to be the rule for striped hyaena in the Serengeti, but around a clumped and stable source of abundant food near the Dead Sea, Israel, Macdonald (1978) found a small (probably family) party of striped hyaenas using latrines which averaged 5.5 faeces per site. Single faeces were also found in the same area. Elsewhere in the Dead Sea study area striped hyaena faeces were found only solitarily, although concentrated in the vicinity of a lair, so it seems that the middens occurred only around the feeding site. Mills *et al.* (1980) state that brown hyaenas use latrines that are scattered within their territories.

Aardwolves defecate at latrine sites, whose distribution differs from that of their glandular pasting sites which are especially concentrated at range borders (Kruuk and Sands 1972). Each aardwolf home range, of approximately 1.5 km$^2$, may have up to four active latrines, often on top of a wildebeast dung pile (see Gosling 1981). Kruuk and Sands (1972) describe these latrines, which are scattered throughout the animal's range as being 'a thick carpet of aardwolf droppings'. Aardwolves walk for several hundred metres to the nearest latrine when they want to defecate and on arrival they dig a small hole in the carpet of existing faeces in which to deposit the fresh faeces, which they then cover

over. Kruuk and Sands speculate that these middens may simply be sites where faeces are disposed of. Aardwolf faeces smell of their termite prey, and this odour might confuse the foraging aardwolf in its hunt for prey. The practice of burying the fresh faeces might hasten their breakdown (largely by *Macrotermes vitrialatus*, one of the aardwolf's prey species, but one which is never eaten in the immediate vicinity of latrine sites). Bothma and Nel (1980) favour the same hypothesis and note that aardwolf latrines are larger in the Namib desert than in the Serengeti plains. This and other differences in scent marking behaviour may stem from the possible absence of territoriality in the Namib population.

In addition to the intraspecific variation in the pattern of scent-marking by spotted hyaenas described by Kruuk (1972), Bearder and Randall (1978) describe further variations. They studied spotted hyaenas in southern Africa where they preyed largely on giraffe, *Giraffa camelopardalis*. They distinguished temporary and long-term latrine sites (and noted that the hyaenas also urinated and defecated wherever they happened to be). Temporary latrine sites developed around a giraffe kill where the hyaenas might camp for several days. The latrine commenced where a hyaena urinated and scraped. Subsequently other urinations and then defecations accumulated also. In contrast to temporary latrines, the 65 long-term latrines studied by Bearder and Randall were all on roads or paths, and 61 per cent of them were at junctions. Indeed, only 12 per cent of long-term latrines were not associated with a landmark which was obvious to the authors. Latrines were permanent over several years, the biggest contained 144 faeces within 800 m$^2$. Bearder and Randall record that spotted hyaenas in their area used border latrines in a different way from those within their range: in a given night, marking at border latrines was equally often performed by one hyaena or by several, but elsewhere it was more than twice as common for only one animal to mark at a latrine in a given night. There was also a temporal pattern: in the dry season scraping was less frequent than were defecations, but in the wet season, when grass was lush and the ground soft, they found almost twice as many signs of scrapes as of faeces. Each latrine was generally visited at least once every 10 days, but longer intervals were not uncommon. Latrines near the borders of territories were visited more frequently than those elsewhere.

## Family Viverridae

Scent of viverrids is notorious. In *As You Like It* Shakespeare wrote that 'Civet is of baser birth than tar, the very uncleanly flux of a cat', but his disgust at this product is rather unexpectedly not shared by perfumiers, who prize it. The gland from which civet musk emanates is called the perfume or perineal gland, and the principal component, civetone, was in 1926 amongst the first chemicals to be isolated from a mammalian scent gland (Dorp *et. al.* 1973).

Added to the fact that little is known of their behaviour, the civets and mongooses that presently comprise the family Viverridae are horrendously complex in the structure of their scent glands and, partly in consequence, in their taxonomy. Indeed, Wozencraft (personal communication) believes that they should be split into two families, the Viverridae and Herpestidae.

Pocock (1915*a*, *b*, *c*) uses cutaneous features, including the perfume glands, to support Mivart's hunch that the viverrine and paradoxurine genera should be relegated to distinct subfamilies within the Viverridae. The differences between the perineal glands of these two subfamilies can be summarized (see Table 15.4) as follows:

1. Viverrinae: the labia are hairy both without and within, and form a pair of tumid masses closely juxtaposed or (as *Viverra zibetha*) confluent anteriorly. In no case does the gland extend in front of the prepuce or vulva, nor does it ever form a flattened naked area covered all over with secretory orifices. The most complex perineal glands are found among the Viverrinae.

2. Paradoxurinae: the labia form upstanding ridges with naked internal surfaces. They are not closely apposed and so can be spread, exposing the secretory area as a flat surface, studded all over with pores.

In addition to the perineal gland, many viverrids have been noted to rub their cheeks and chins on specific sites (Wemmer 1977*b*). Kleinman (1974) describes how male and female binturong regularly pressed their flanks against two places in their cage and also often sniffed protractedly at these same places. In particular they rubbed with their foreheads and cheeks at these sites. Cheek glands may be widespread among herpestine mongooses; Rasa (1973) describes their function among dwarf mongooses, *Helogale parvula*, and Rensch and Dücker (1959) note a sweet secretion from the vibrissae of the Indian grey mongoose, *Herpestes edwardsi*. Ewer (1973) saw the kusimanse, *Crossarchus obscurus*, and the marsh mongoose, *Atilax paludinosus*, rub their cheeks on objects.

Rubbing the neck and chin against the substrate, especially against strong scents and dead prey, seems to be widespread amongst civets and genets. It has been recorded for *Genetta, Viverra*, (*Civettictis*), *Nandinia, Paguma* (Wemmer 1977*a*, *b*; Dücker 1965, 1971; Ewer 1973). In addition the fossa, *Cryptoprocta ferox*, rubs its underside against vertical objects while adopting a hugging position. Vosseler (1929*b*) noted fatty material on the fur of the underside, and reasonably interpreted this as an indication of an underlying sternal gland. Both sexes of *Cryptoprocta* engage in this upright ventral rubbing but Vosseler noted that the gland is largest among males and secretes most copiously during the mating season.

## Subfamily Viverrinae

A well-documented arrangement of the perfume or perineal glands is exemplified by the members of the genus *Genetta*. Citing Chattin's publication of 1874, Pocock (1915*c*) describes the lips of the genet's perineal pouch as two elongated eminences covered with hair both internally and externally (Fig. 15.8). These lips normally abut against each other, forming a longitudinal sulcus which is Y-

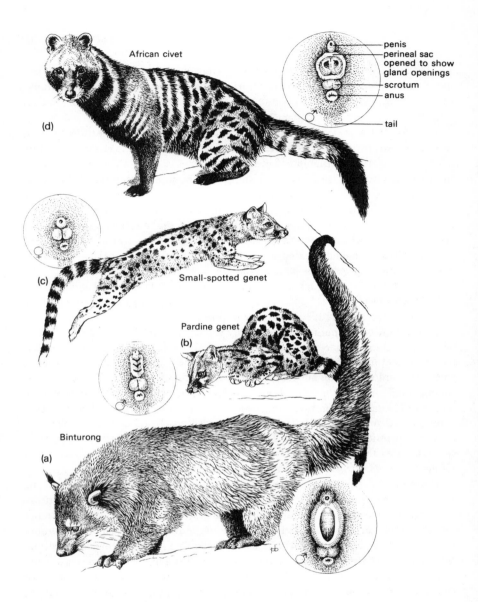

Fig. 15.8. The perineal or perfume glands of the civets and genets are confusingly varied in the detail of their structure. Those of the paradoxurine palm civets are naked pockets within loosely opposed labia, as illustrated by the male binturong (a). In some Viverrinae, such as the forest genet (b), both sexes have tripartite chambered perineal pockets closed by hairy labia. This arrangement also typifies the males of some other species, including the European genet, whose females (c) have a simpler, one chambered perineal pocket. The perineal gland is most complex in the African civet (d), where the inner face of each labium contains a hair-lined cavity which extends inside anteriorly.

shaped anteriorly where it divides just behind the vulva or prepuce. When these lips are pulled apart the space between them can be seen to be imperfectly divided into three compartments, separated by two transverse ridges of integument. At the bottom of each compartment the secretion of the glands can be squeezed from a pair of laterally placed clusters of minute orifices. Thus, in total, there are six centres from which liquid exudes.

This tripartite arrangement seems to be typical of all male genets but of the females only in some species. Summarizing the variation, it seems that the species of *Genetta* can be divided as follows (Pocock 1915c):

1. The large spotted genet, *G. tigrina*, forest genet, *G. pardina*, (or *G. rubiginosa*, no longer recognized): The tripartite chambered pouch is similar in both sexes.

2. European genet, *G. genetta*, and Servaline genet, *G. servelina*: the male has the tripartite chambered pouch, the female has a simpler unpartitioned structure.

Wemmer (1977b) has studied the marking behaviour of the genet, *Genetta tigrina*, in captivity, where he found evidence that genets of both sexes adopted various postures, combined with rhythmic pelvic oscillations, in order to daub secretions from their perineal glands on marking sites. These postures include squats (generally females) and handstands (generally males). Within the labia of the genet's pouch Wemmer found a coating of oily emulsion, whereas the exudate deep within the glandular crypts had denser consistency. Captive female genets showed a greater frequency of marking around the time of (largely after) oestrus, and Wemmer points out that the contiguity of perineal and vulval labia provide the potential for perineal glandular secretion to be mixed with both urine and vaginal fluid. Genets also scuffed with their metatarsi, and rubbed with their heads and necks and with arched backs. Perineal marks were made from strictly terrestrial positions by genets of both sexes. Anal sacs were evacuated by the genets under different circumstances, and invariably during intense, largely agonistic, interactions with another genet. Wemmer (1977b) described the fluid as a thin translucent coffee-coloured stream ejected in squirts which travel about 20 cm. He pointed out that the spray also splashed onto the genet's fur around the anus and tail.

The perfume gland of the large Indian civet, *Viverra zibetha*, lacks partitions, but is otherwise a more complex structure. The pouch made by the glandular labia is much wider in its deeper recesses than at the orifice, and at the anterior the labia are fused to form a roof covering a large storage pouch. Secretion is exuded from five areas on each side of the pouch (Pocock 1915c). Furthermore, in this species the scrotum is divided and the broad area between the two lobes is naked and glandular. This glandular bald patch extends from the rear of the perfume gland, between the scrotal sacs to the anus. It does not quite encircle the anus, but almost does so. Thus the anus is cupped in an upstanding rim, rather as in some mongooses and especially the fossa, *Cryptoprocta ferox* (Pocock 1916a).

Table 15.4. The glands of the Viverridae are particularly confusing; this table summarizes the distribution of perineal glands, anal sacs, and anal pouches (including auxiliary glands) among some species of each subfamily (see also Fig. 15.8).

| Species | Anal pouch | Anal sacs | Perineal |
|---|---|---|---|
| **Viverrinae** | | | |
| Genetta tigrina | − | + | + |
| Genetta pardina | − | + | ♂+♀ Tripartite, open chamber |
| Genetta genetta | − | + | ♂ Tripartite |
| Genetta servelina | | + | ♀ Single, open chamber |
| Viverra zibetha | ? | + | ♂+♀ Single chamber, labia fused anteriorally |
| Viverra civetta | − | + | ♂+♀ Pockets within labia |
| Prionodon spp. | − | + | — |
| **Paradoxurinae** | | | |
| Paguma larvatus | − | + | ♂+♀ Shallow chamber, naked labia; ♂ Penile collar |
| Paradoxurus hermaphroditus | − | + | ♂+♀ Single depression, naked labia, surrounds genitalia |
| Arctictis binturong | − | + | ♂+♀ Single chamber posterior to genital orifice |
| Nandinia binotata | − | + | ♂ Single pouch extends forward of penis; ♀ Entirely anterior of vulva; ♂+♀ No separation of gland into rt/lft halves |
| Arctogalidea trivergata | − | + | ♀ Naked glandular skin largely anterior of clitoris. ♂ No gland |
| Macrogalidia musschenbroekii | − | + | ♂ Shallow chamber, naked labia; ♂ No gland |

| Taxon | | | |
|---|---|---|---|
| **Hemigalinae** | | | |
| *Fossa fossa* | − | + | +/− |
| *Euperes goudoti* | − | + | − |
| *Eupleres major* | − | + | − |
| *Hemigalus hosei* | − | + | Single chamber |
| *Hemigalus derbyanus* | − | + | Shallow depression, no labia |
| *Cynogale bennetti* | − | + | +? |
| **Galidiinae** | | | |
| *Galidictis* | − | + | ♀ Pocket wholly perineal with closely opposing lips. ♂ Unknown |
| *Galidia elegans* | − | + | ♀ Open pocket extending in front of vulva, with hairy interior ♂ Perhaps absent |
| | | | |
| **Herpestinae** | + | + | + |
| *Herpestes auropunctatus* | + Deep sebaceous crescent above anus | Open beside anus | − |
| *Herpestes ichneumon* | ⎱ Uniformly dispersed sebaceous auxiliaries | | − |
| *Atilax paludinosis* | + | + | − |
| *Cynictis penicillata* | + | Open dorsal to anus | − |
| *Mungos mungos* | ⎱ Pouch folded to form pockets | Open beside and close to anus | − |
| *Helogale parvula* | | | |
| *Suricata suricatta* | | | |
| **Cryptoproctinae** | | | |
| *Cryptoprocta ferox* | Pouch mainly dorsal to anus | ? Open ventral to anus | − |

A yet more modified viverrine perineal pouch is found in the African civet, *Viverra (Civettictis) civetta*. In the male of this species the two halves of the scrotum are not widely separated and there is no integumental collar around the anus. The two glandular labia lie in close apposition, separated by an undilated slit, but when they are pulled apart a moderately large orifice can be seen on the inner face of each labium. Each orifice leads into a large, hair-lined sac or pouch, which extends upwards within the glandular lobe. Secretion pours into these cavities and then drains through the orifices into the space between the lobes. The female's gland is similar: in this case a pair of deep pockets, visible when the labia are parted, extend forwards alongside the vagina.

The linsangs, *Prionodon*, lack a perineal pouch (although they have anal sacs) and for this reason Pocock (1915c, d) favoured removing them from the Viverrinae.

### Subfamily Paradoxurinae

The perineal gland of the masked palm civet, *Paguma (Paradoxurus) larvatus*, was described by Pocock (1915a). In the male the main portion of the gland consists of a pair of thickened ridges of skin extending between the scrotum and the penis, and forming the side walls of a longitudinal fossa which is everywhere perfectly smooth. The glandular labia do not extend as far forward as the penis, but a rim of thickened skin, naked internally, does so and circles the penis like a collar. The narrow space resulting between penis and collar is highly glandular, producing a secretion which Pocock found repulsive and quite different to the 'mousy' odour of the labia themselves. The female's perineal gland is similar, as is that of the male of the common palm civet (toddy cat), *Paradoxurus hermaphroditus*, except that there is no penile collar and the labia are less pronounced.

The perfume gland of the male binturong, *Arctictis binturong*, is like that of *P. larvatus* excepting the penile collar (although scantily furred labial skin does surround the penis) (Fig. 15.8). The labia converge both fore and aft, and can be opened very widely. As with the Viverrinae, the male binturong's gland is wholly perineal, lying behind the vulva or prepuce, and hence in contrast to that of *Paradoxurus* where it extends both anterior and posterior of the genitalia (Pocock 1915a). However in the female the glandular skin extends anterior to the vulva (Story 1945). Wemmer and Murtaugh (1981) note that female binturong in oestrus secreted copious vaginal mucus, which mixed with the secretions of the perineal gland which encircles the vulva (Pocock 1939). During oestrus, the labia of the female's perineal gland become firmer.

The binturong is a rather specialized paradoxurine, being a shaggy-coated prehensile-tailed, arboreal creature, with a largely frugivorous diet. Kleiman (1974) studied a pair in captivity, describing their perfume glands as large, oval shaped, with a smooth, naked interior, normally covered by paired, hairy labia. The gland is smaller in females, in which it lies more anteriorly. In addition to the accidental smears that must be left as this gland strikes passing vegetation, Kleiman noted that both members of her pair routinely marked on four sites within their cage.

Some sites were marked in an upright posture, another in an inverted quadrupedal stance (hanging below the branch, rather like a sloth) which amounted to an upside-down perineal drag. Although common among the Herpestinae (mongooses) these postures seem to be unique amongst the Viverrinae, save for *Cryptoprocta's* upright marking, which is, however, with its sternal gland (Wemmer 1977*b*). Kleiman describes how perineal gland marking by binturong is often associated with so-called metatarsal scuffing, itself an element of urine marking.

The perfume gland of the female spotted palm civet, *Nandinia binotata*, extends into a pouch anterior to the vulva, which was once thought to be the functional homologue of the marsupial pouch. Similarly, in the male, a naked glandular area extends from the scrotum a long way in advance of the penis. When relaxed the labia meet, but if they are pulled apart they expose a subcircular depression with tumid margins. The secretory cells are not differentiated into right and left portions, as they are in the glands of *Paguma*, *Paradoxurus*, *Arctictis*, and the viverrines.

Pocock (1915*a*) described a different arrangement for the female of *Arctogalidia trivergata* (*leucotis*), the small-toothed palm civet from South East Asia. In this species the vulva, preceded by a large clitoris, lies at the posterior of a small but distinct glandular tract. The most active glandular tissue apparently nestles just in front of the clitoris.

Wemmer (1977*b*) made comparative studies of several civets. He noted that the perineal glands of members of the genera *Nandinia* and *Paradoxurus* (palm civets) were thin walled and hairless and devoid of accumulated secretions, in marked contrast to those of the genus *Genetta* and the African civet, *Viverra* (*Civettictis*) *Civetta*. The Perineal gland of the African civet is similar to that of the palm civet, *Nandinia binotata*, in that the labia are sealed anteriorly to make a pocket in which material accumulates. In female *Nandinia* the gland lies anterior to the vulva; in the male it surrounds the penis (Pocock 1915*a*). The secretions of these species are thus likely to spend longer in the pouch prior to marking (as indicated by their darkening colour) than do the secretions of species with completely open pouches.

Recently, Wemmer, West, Watling, Collins, and Lang (1983) found that the giant palm civet of Sulawesi, *Macrogalidia musschenbroekii*, shares with *Arctogalidia* the unusual trait that the males lack a perineal gland (although they do have a naked patch of skin). The female giant civet's perineal gland was similar to that of female *Paguma larvata*: a shallow semicircular depression behind the vulva and about one centimetre anterior to the anus (Plate 15.8). One pocket measured 16 mm in length and 9 mm in depth, and its interior was hairless and pink. Wemmer and Watling (in press) report that *Macrogalidia* inhabits primary rain forest, feeding on small mammals and fruit, and apparently drops its scats at random, in contrast to the sympatric, similar, but less frugivorous Malay civet, *V. tangalunga*, which uses latrines (Macdonald and Wise 1981). *Macrogalidia* makes claw-marks on smooth barked trees, but their significance is unknown.

## Subfamily Hemigalinae

Pocock (1915*d*) reports that the Madagascan fanaloka, *Fossa fossa*, and the banded civets, *Eupleres* spp., are entirely without perineal glands. He supposes that they represent divergent types of a primitive group of viverrids, antedating the ancestor of the groups now characterized by the possession of such glands. However, other members of this subfamily, such as *Hemigalus*, do possess rather genet-like perfume glands (Pocock 1915*e*). The otter-civet, *Cynogale bennetti*, has a simplified perineal pouch, but Pocock (1915*b*) prefers to explain this as an adaptation to an amphibious habit.

## Subfamily Herpestinae

*Anal glands*    The anatomy and chemistry of anal glands amongst the subfamily Herpestinae is best known for the small Indian mongoose, *Herpestes auropunctatus* (Gorman, Nedwell, and Smith 1974). When this mongoose is at rest, its tail hangs down and folds of skin around the genital area close together leaving a transverse orifice. When the tail is raised the anus is seen to be seated within a depression or antechamber which is referred to here as the anal pouch. Below the mongoose's tail there is a crescent of raised hairless skin which contains abundant sebaceous glands and which is especially developed in males (Plate 15.9). These sebaceous tissues will be referred to as auxiliary glands, as is comparable tissue within the anal pouches of Hyaenidea. On either side of, and exterior to, the anus lies an opening of an anal sac, and the lumen of each sac contains dark brown secretion. Two ducts drain into each sac: one from a cluster of sebaceous cells and the other from apocrine cells. Each sac is encased in a layer of striated muscle and when this contracts the contents are extruded into the circumanal depression (Pocock's anal sac). Chemical analysis of the secretion of these sacs revealed a series of six short-chain odorous carboxylic acids (acetic, propionic, isobutyric, butyric, isovaleric, and valeric). However, these carboxylic acids were absent from pockets treated with penicillin and this led Gorman *et al.* (1974) to suggest that they were compounds of secondary origin, the products of bacterial action on substances, such as triglycerides, secreted in the first instance. The sacs did indeed contain a rich bacterial flora. Subsequently, mounting evidence has accumulated in favour of this hypothesis (Albone *et al.* 1978). Gorman (1976) trained Indian mongooses to distinguish between the anal sac secretions of different individuals. His experiments supported the chemical evidence—namely that individual differences in the proportions of each carboxylic acid present in the anal pocket secretions did enable mongooses to identify each other individually. Gorman based the procedure for his training experiments on that used by Rasa (1973) in her pioneering study on social odours of a related species *Helogale parvula* (*undulata*).

The findings of Gorman *et al.* (1974) concerning the histology of the anal pouches of the Indian mongoose are largely mirrored in the study of Hefetz *et al.* (in press) on the African mongoose, *Herpestes ichneumon*. However, the chemical composition of the glandular secretions were quite different for these two species. The anal pouches of the African mongoose contain apocrine anal glands and a row of sebaceous glands. The light brown fluid of the anal sacs mixes with the whitish paste of the sebaceous cells to produce the brown mixture smeared on marking posts. The African mongooses in Israel (Hefetz *et al.*, in press) lived in female kin groups of up to seven individuals, whose home-range is at least partly overlapped by that of a solitary adult male. All the females rubbed their anal pouches on communal scenting posts. These secretions contained saturated carboxylic acids of much longer chain length ($C_{10}$ to $C_{22}$) than those in the secretions of Indian mongoose and with varying degrees of methyl branching. They also contained a compound specific to males (2, 4, 6, 10-tetramethylundecanoic acid). The proportion of components varied more between males than females, perhaps due to cross-infection among the females of a group at common scent posts.

Hefetz *et al.* (in press) mention that the African mongooses smear their own fur with anal pouch secretions which may in turn taint vegetation as the mongooses brush past.

Pocock (1916*c*) believed that the anuses of all mongooses are seated within a more or less developed saccular depression. He called this depression the anal sac, but I am going to resist that nomenclature since it causes such confusion with the different structure given the same name for all other carnivores (i.e. the paired reservoirs lying on either side of the anus and most typically illustrated by those of the Canidae). The depression described by Pocock amounts to an anal antechamber or pouch, and derives from a hairy involution of the circumanal integument. Outside the anal sphincter, but within this antechamber the ducts from the paired anal sacs emerge. Pocock states that these ducts emerge close to the anus of *Mungos*, further away in *Herpestes* and *Suricata*; in *Ichneumia* they emerge somewhat higher up the antechamber, and considerably higher in *Cynictis*. Pocock (1916*c*) dismisses earlier reports that the anal antechamber is absent from the Marsh mongoose, *Atilax paludinosus*. On the contrary he describes it as exceptionally developed in that species.

Pocock (1916*c*) makes especial mention of *Mungos mungo* (*Ariela fasciata*) which has a relatively large anal chamber partitioned by three pairs of integumental folds or depressions. Two pairs of these lie above the anus, one lies at the same level. Each such depression contains a glandular pit and gathers secretion from enlarged sebaceous glands. These structures are quite distinct from the normal anal sacs. Nevertheless, in the male banded mongoose the integumental folds are more heavily involuted and so give more of the impression of real anal sacs, an impression which enabled Pocock to take a characteristic swipe at his rival Mivart, 'This secondarily acquired similarity . . . seems to have misled Mivart into thinking that the two normal anal glands present in all Aeluroid Carnivores had become broken up . . . in the male *Ariela fasciata* into five pairs of glands described by Chattin (1874).'

Pocock (1916c) describes a different arrangement in the grey meerkat, *Suricata suricatta*, where in both sexes the anal antechamber is pitted with conspicuous hair follicles, from the pores of which a liquid can be squeezed. These pores are concentrated in an oblique slit-like depression on either side of the anus, with the anal sacs being situated at the deepest extremity of these depressions. Ewer (1973) points out that in *Suricata, Helogale* and *Mungos* the pouch wall is folded into pockets which she supposes serve to store secretions. Lynch (1980) weighed the excised anal glands of grey meerkats; the mean mass of those of adult males was greater than that of glands from adult females (males 2.65g; females 1.07g). Although the differences were not significant, pregnant females tended to have the heaviest anal glands. In contrast, Lynch found that oestrous female yellow mongooses, *Cynictis penicillata*, had significantly heavier anal glands than did other adult females. Overall, male *Cynictis* glands were heavier than those of females (males 4.74g; females 2.47g). There were distinct seasonal differences for gland mass of male meerkats (coinciding with breeding season) but not for male *Cynictis*.

Finally, although the text is hard to relate to the accompanying figure, Pocock (1916c) describes the anal pouch of the dwarf mongoose, *Helogale parvula* (*undulata*). This species is of especial interest here because of the experimental work that has been done on it (see below, Rasa 1973). The anus of the dwarf mongoose lies in a depression, in the walls of which are two pairs of supplementary pouches and one midline, larger and unpaired supplementary pouch. Just below and external to the openings of the lower, larger and outermost pair of supplementary pouches lie the openings of the ducts to the anal sacs.

### Subfamily Galidiinae

Pocock (1915e) felt that the perineal glands of the Galidiinae resembled those of civets in some respects and of mongooses in others (Table 15.4) but he concluded that they should be classified with neither. In this account Pocock is very explicit in that he regards scent glands as more robust taxonomic indicators than feet and teeth. To the extent that scent glands function in social behaviour, which is doubtless under considerable selective pressure, it is not clear that they should be as resistant to change as Pocock supposed.

Albignac (1976) noted that the narrow-striped mongoose, *Mungotictis decemlineata*, anal drags on the ground and on vegetation. Higher rates of anal dragging were associated with agonistic encounters.

*Behavioural experiments on gland function*    Rasa (1973) drew attention to the interesting fact that many mammals, but in particular dwarf mongooses, *Helogale parvula* (*undulata*), mark the same site with odour from more than one source (anal and cheek glands, urine and faeces, in this case). The glands opening into the dwarf mongoose's anal pouch produce a thick, creamy, grey fluid with a strong musky scent, which dries to form a waxy solid. The cheek glands lie between the eyes and ears on each side of the head—an upper gland lies at the level of the jaw angle, a lower one slightly forward and at the level of the lower jaw. The cheek glands produce an adhesive, opaque secretion (which is odourless to humans).

Marking with the anal glands involves everting the pouch and dragging it across the surface to be marked (Rasa 1973; Zannier 1965). The dwarf mongoose marks upright surfaces from a handstand, during which it walks away from the upright with its front feet, so dragging its anal region downwards (Plate 15.10). The stance adopted by the mongoose depends on the object to be marked—low flat objects are anointed during an 'anal drag', whereas intermediately promi- nent sites (such as the bodies of other mongooses) are marked with a 'leg lift' posture. Cheek-marking involves grasping the object between the front paws and stroking the cheek against it. A typical marking session involves sniffing for an average of 3–4s, four strokes with the cheek gland (two on each side), a 2s mark with the anal glands, a further burst of sniffing and then, sometimes, the whole cycle may be repeated (up to 37 times). The average marking episode lasts 3–4 min, and marking sites are distributed throughout the animal's cage.

Within her captive groups of dwarf mongoose, Rasa found that dominant males marked more often than dominant females with both cheek and anal glands (anal: ♂ = 5.9, ♀ = 3.1; cheek: ♂ = 33.8, ♀ = 15.6 marks per 5 min). Sub- adult males marked less than adult males with their cheek glands, but not with their anal glands. There was an overall peak in marking frequency by the entire group when there were babies in the nest. Dwarf mongooses 'babysit' infants produced by the α-female of the group, and, in captivity, Rasa (1973) noted that each individual taking over from another for a stint of babysitting marked immedi- ately before doing so, whereas the animal that was being relieved marked as soon as it emerged from the nest box. This led Rasa to speculate that being out of the social milieu while babysitting diminished the potency of that individual's social odour within the group, and that their insistence on marking at the last and first opportunities before and after isolating themselves while babysitting was to minimize the risk of getting socially out of 'touch', in olfactory terms. This hypothesis presupposed that mongooses could distinguish fresh and old marks, and that they could recognize individuals through their marks. Rasa's demonstra- tion of both these abilities remains one of the few convincing studies of its ilk.

Rasa (1973) trained three dwarf mongooses to sniff an odour smeared on a glass slide, then to inspect a series of other odours placed in separate Petri dishes and run back to the experimenter for a reward on finding the Petri dish contain- ing an odour matching that on the original smear. The mongooses could readily distinguish between the age of anal sac secretions that were fresh, one hour old and two hours old, and could also distinguish the anal sac secretions of different individuals. Since the secretion of the apocrine anal sacs and of the nearby sebaceous cells are mixed in the anal pouch it is not possible to say whether one or both contribute information on identity and age of the sample. In contrast, the mongooses failed to distinguish between the secretions of cheek glands from different individuals. Rasa pointed out that since identity was signalled by anal sac odours and since cheek glands were never rubbed in the absence of anal marking, the information on identity in cheek gland odours would be redundant. However, the frequency and vigour with which mongooses marked with their cheeks increased both when they were confronted with strange conspecifics and when they were presented with an object marked by strangers. Rasa hypothesized that at sites where cheek marking had been especially vigorous this might convey heightened excitement. This hypothesis requires mongooses to be able to disting- uish between thicker and thinner smears of cheek gland secretion. A further series of choice experiments revealed that they could do so. The longevity of the signals from the anal and cheek glands differed: dwarf mongooses could detect anal pouch smears after 10 days (but not after 20), but they could not detect cheek gland smears for more than one day.

Odours from the anal pouch are deposited in the absence of cheek-gland marking during allomarking, of which the majority (64.8 per cent) is by the alpha-male on the alpha-female, and solely because of the alpha-male this increases to a peak on the first day of the female's oestrus. The frequency of object marking by the alpha-male did not increase with the onset of oestrus in the alpha-female. Second to the alpha-female, the alpha-male was the most frequently allomarked member of the group, and 66.6 per cent of the marks on him were made by juveniles. Kingdon (1978; personal communication) describes anal-gland allomarking among groups of banded mongoose, *Mungos mungos*, which, after slight separation or a mild scare, will pile on top of one another, uttering a warbling chitter and rubbing their anal pouches on one another's backs. A similar melee of scent-marking, including rubbing their bodies under their companions' tails, typifies the behaviour of *Helogale* prior to a battle with another group (Rasa, personal communication).

### Field studies of viverrid glandular marking

There are only a few field studies concerned with scent-marking among wild viverrids, notably those of Charles-Dominique (1978) on the African palm civet, *Nandinia binotata*, and of Bearder and Randall (1978) on the African civet, *Viverra (Civettictis) civetta*.

Charles-Dominique radio-tracked 28 palm civets in the tropical jungles of Gabon where females maintained exclusive 45-ha territories over-lapped by the larger (100-ha) territories of males. The palm civet is a largely frugivorous omni- vore and Charles-Dominique's observations suggested that the perineal gland was involved in territorial marking. Glandular marks were deposited in the vicinity of territorial borders, and around fruiting trees within each territory.

Ewer and Wemmer (1974) describe the use of the double-pocketed perineal gland of *V. civetta* in captivity (the structure having been described by Pocock 1915c). Bearder and Randall (1978) describe the fresh secretion as whitish-yellow and grease-like in consistency. This secretion was deposited along the civets' trails, but was most abundant at civetries (latrines), at which they also defecated (see below). Some civetries were on territorial borders, but Bearder and Randall could not establish if they were especially concentrated at such borders.

Gorman (1980) presents a map of the anal-marking sites of a tame Indian mongoose, which indicates that individuals of this species deploy these marks throughout their home-ranges. Earlé (1981) notes that male yellow mongooses, *Cynictis*, mark with anal glands throughout their territories. On 'marking trips' these mongooses anal marked every 5–10 m. In contrast, Earlé notes that cheek gland marks were made by all colony members of both sexes and were largely confined to the vicinity of burrows. Yellow mongooses apparently rubbed with their backs against traditional sites on their territorial borders and dominant males anal marked on their subordinates.

### Urine and faeces

Most viverrids squat to urinate (*Viverra, Nandinia, Paradoxurus, Paguma, Hemigalus, Fossa*, and *Genetta*). The toddy cat, *Paradoxurus hermaphroditus*, defecates along branches (Bartels 1964) or, in captivity, on top of nest boxes (Wemmer 1977b), as does *Paguma*. Wemmer reported that males of the genet, the African civet, and the palm civet, *Paguma*, scuffed their hind-legs in their urine and, in the case of *V. civetta,* males actually urinated onto their hind-legs during social encounters with a female. Kleiman (1974) describes the same phenomenon for binturong, *Arctictis binturong*, which make alternate shuffling movements of their hind-legs while supporting most of their weight on the fore-limbs. The binturong's toes are turned up during this procedure, so the naked heel is mostly in contact with the ground. Scuffing may occur in isolation or with urination and perineal marking. In the latter case, the binturong scuffs before, during, and after urinating and then also drags its tail through the urine pool. Males scuff more than females (10.8 vs 6.0 per 30 min). Both sexes increased marking when the female was treated with sex hormones.

The African civet, *Viverra (Civettictis) civetta*, studied in the wild by Bearder and Randall (1978) and Randall (1977) defecates almost exclusively at civetries which comprised compact accumulations of faeces, often in open clearings or in some sort of depression. The Malay civet, *V. tangalunga*, similarly favours depressions, defecating into the crevices between rocks (Macdonald and Wise 1981), whereas *Paradoxurus hermaphroditus* latrines were in clearings (Bartels 1964). The pattern of use of civetries remains mysterious: Bearder and Randall discovered that African civets do not mark each civetry they visit, and that irrespective of the numbers of civets visiting a given civetry only a rather constant number of them (1–3) will defecate there. At each civetry, active periods of defecation were commonly interspersed with breaks of 2–3 months duration. When a civetry was translocated to a new site 5 m from the original one, the civets continued to use the original site. African civets did not urinate at civetries, but males did spray urine in a posteriorly directed stream onto conspicuous sites along their tracks.

Rasa (1973) noted that members of a group of dwarf mongooses defecated and urinated at a communal latrine site. In the wild these sites are almost exclusively associated with termite mounds in which the animals sleep, and are usually close to a communal marking post. In one experiment Rasa moved one such site and over the course of 24 hours the mongooses progressively and permanently shifted further eliminative activity to the new site. Rasa thus concluded that the elimination products themselves, more than their particular location, drew the mongooses' attention. She achieved the same result when the new site was made up of excreta from strange mongooses, with the difference that the alien faeces prompted frenzied and prolonged (>1 hour) anal pouch dragging and general investigation. Rasa supposes that the olfactory significance of faeces is bestowed upon them by traces of anal pouch secretions.

Both yellow mongooses and grey meerkats live in groups of approximately ten members. Grey meerkats forage in a tight knit band whereas yellow mongooses often forage alone. Both species defecate at communal latrines and at least the latrines of yellow mongooses are concentrated at territorial borders (Earlé 1981; Ewer 1963; Macdonald, unpublished observations).

## Family Felidae

*Anal glands*   In addition to paired anal sacs whose ducts open just inside the rectum, cats have enlarged anal glands in the anal skin below the tail (Schaffer 1940). The secretions of the anal glands of both lions (Albone, Eglington, Walker, and Ware 1974) and domestic cats (Preti *et al.* 1976) differ from those of all canids studied to date in lacking trimethylamine. Verberne and de Boer (1976) note that anal sac secretions can evoke flehmen amongst cats. The walls of the anal sacs of at least domestic cats, lions, and tigers have more sebaceous tissue than do those of canids and produce correspondingly lipid rich secretions (Albone and Grönneberg 1977; Greer and Calhoun 1966).

*Facial glands*   Rieger and Walzkoenig (1979) compared the histology of the skin from the inner left thigh, the left cheek, and ear region of three male cats. In each case the sebaceous glands were small and connected with hair follicles. Consequently the authors dismissed the idea that cats were marking when they cheek rub. However, they noticed that such rubbing was common on odorous objects and so they concluded that it was a form of scent-rubbing (Rieger 1979).

This result is irreconcilable with the findings of Verberne and de Boer (1976) and Verberne and Leyhausen (1976), who concluded from laboratory experiments that secretions from cheek glands (in addition to those in urine) inform tom cats of the hormonal state of females. These authors found rubbing with the cheeks on objects (especially poles placed in their experimental arena) was a behaviour common to both sexes of domestic cat, but one which varied in frequency between individuals.

The response of cats to cheek-gland secretions differed from that to urine in that flehmen was rarely involved. Places where a cat had cheek rubbed usually prompted sniffing and further rubbing. Experimental pegs which had previously been rubbed upon evoked longer rubbing times from experimental cats than did clean pegs, and females on average sniffed for longer than did males. There was clearly a visual aspect to rubbing, as it was sometimes done on to the pegs as part of a greeting by the cats to the approaching experimenters. Verberne and de Boer (1976) also showed that the interest of male cats in the rubbing sites of a female followed a cyclic course, in phase with their interest in the female's urine and in phase with her hormonal condition. These authors dismiss the idea that what is being rubbed onto an object is saliva, on the grounds that the behaviour not infrequently occurs when the cat has its mouth shut. Further, although they found that cheek-gland secretions only occasionally evoked flehmen, saliva never did so.

*Catnip*    Although it has long been asserted as common knowledge that catmint or catnip, *Nepeta cataria*, was highly attractive to cats, the phenomenon first came under a biologist's scrutiny when Todd (1962) demonstrated that this response was inherited as a dominant autosomal gene. The active ingredient of catnip is an unsaturated lactone, *trans-cis*-nepetalactone (McElvain *et al.* 1941; Bates and Sigel 1963). Todd noticed that the chin, cheek, and body rubbing, and the rolling shown by cats which reacted to catnip were similar to the behaviour shown by oestrous female house cats in the presence of males. Palen and Goddard (1966) followed up this observation by presenting cats of each sex (both intact and castrated) with either a rat or a cat-sized dummy in the presence or absence of catnip aerosol. They found that gonadectomy did not affect the response, that catnip increased the attention paid by some cats to the stuffed model (and decreased their interest in the prey) and that the rolling reactions of these cats to the model closely resembled an oestrous female's behaviour to a male. Confusingly, some males also showed the catnip response.

*Interdigital glands*    Darwin (1900) described trees which had been worn smooth and scarred with gouges a metre long from the clawings of jaguar, *Felis onca*. Schaller (1972) describes the same behaviour for lions, *Panthera leo*, of which several may gather together to scratch at a particular tree which is visited repeatedly for this purpose. Panaman (1981) saw farm cats, *Felis catus*, scratching logs and trees, but mentions no pattern nor communicative significance to this behaviour. Scratching is also widely thought to be a method of sharpening claws.

*Supracaudal glands*    Schaffer (1940) describes a supracaudal gland (dorsal Schwanzorgan) of domestic cats as a largely sebaceous glandular area sited at the root of the tail, and sometimes reaching downwards on both sides of the tail. He records the gland of a female measuring 0.85 × 0.5 mm and that of a reproductive male as 1.84 × 1 mm. These small organs are distinguished principally by big sebaceous glands with cistern-like cavities running into ducts. Tubular, apocrine glands lie between and at the base of the sebaceous tissue.

*Urine and faeces*   Most of the information on urine and faecal markings by felids consists of assorted snippets (e.g. Fiedler 1957). For example, at least the males of many species have been seen to spray urine backwards on to visually conspicuous objects (e.g. lions, *Panthera leo* (Schaller 1972); leopard, *P. pardus* (Hamilton 1976); cheetah, *Acinonyx jubatus* (Eaton 1973); puma, *Felis concolor* (Hornocker 1969); bobcat, *Lynx rufus* (Bailey 1974); lynx, *Lynx lynx* (Saunders 1963); snow leopard, *Panthera uncia*, and leopard cat, *Prionailurus bengalensis* (Wemmer and Scow 1977); tiger, *Panthera tigris* (Schaller 1967)).

Hornocker (1969) described how pumas will scrape together piles of vegetation prior to urinating on them, so creating their own conspicuous object. Furthermore, tracks revealed that a puma might abruptly change course, or go back the way it had come, having sniffed one of these scent marks. However, Seidensticker *et al.* (1973), concluded that spraying did not serve a territorial function among their population of pumas. In common with canids, token urinating felids may spray frequently: Bailey (1974) found bobcats spraying at a frequency of 7.5 times per km travelled, and Saunders (1963) scored a similarly high rate for lynx (approximately 11 urinations per km travelled). Schaller (1967) tracked one tiger which sprayed urine 11 times in 30 min.

Schaller (1972) saw much in common between the scent-marking behaviour of lions and tigers, save that tigers sometimes defecated onto the sites of their urine spraying and scraping, whereas lions did not. Indeed, Schaller concluded that the seemingly casual way in which lions defecated wherever they happened to be, indicated that faeces were not involved in marking. He believed that urine spraying and scraping were interchangeable in lion olfactory communication. From Schaller's description a lion stops frequently (3–4 per km) at tufts of grass or shrubs, and with eyes closed rubs his face on them with what Schaller termed languorous and apparently pleasurable movements before turning around and with raised tail, squirting a mixture of urine tainted with anal sac secretions back and up at an angle of 30–40°. Tigers also mix urine and anal sac secretion when spraying (Brahmachary 1979). Schaller (1967, 1972) saw male lions and tigers spraying frequently. In contrast, although tigresses also sprayed often, lionesses very rarely did so (but see B. C. R. Bertram's photograph in Macdonald (1980)).

Scraping with the hind-paws visibly scars the soil, and not only do the lion's feet get impregnated with urine-soaked dust in this way, but whilst scraping the lion may urinate, further drenching its hind-legs. It was Schaller's (1972) impression that urine dribbled onto the hind-legs while scraping did not contain anal-sac secretion. Scraping by male lions occurred more often than did spraying (in a ratio of approximately one spray per three scrapes). Both sexes focused their scraping and spraying activities at traditional sites, and Schaller notes instances when members of each sex marked sites previously marked by members of either the same sex, the opposite sex, or both, and he saw both resident and nomadic males marking, in a variety of contexts.

The importance of urine-marking in the social behaviour of cheetah could scarcely be more forcefully emphasized than by its recurrent mention in the narrative of Frame and Frame's (1981) book: although they suspect (p. 17) that adult cheetahs normally use the smell of urine as a means of avoiding each other, males and oestrous females spray very frequently on almost every visually conspicuous site they find, especially rocks, termite mounds, and trees (Plate 15.11). They also defecate at these sites, and may return repeatedly to particular sites to mark again with both urine and faeces. One oestrous female is described as clawing at a tree, then urine spraying onto it and then defecating. The male who came to court her sprayed at a comparable frequency, and scuffed with his back feet as he sprayed. The pair then travelled and sprayed together. These observations contrast with Eaton's (1970), who thought that females only urinated for eliminative reasons, although they took great interest in sprays made by males. Frame and Frame also describe the behaviour of a young male cheetah who commandeered a section of his mother's territory and, having done so, began to spray urine at a high frequency. This male was unusual since in the Serengeti most females remain in their mother's home-range, but young males generally emigrate. Sometimes these males live in coalitions as adults. They do not hunt communally, but co-operate in territorial defence and fighting for mates. Frame and Frame (1981, p. 212) suggest that coalitions of males are at an advantage over singletons in being able to spray urine more frequently and to drench an area in odour more effectively.

In Kenya, Eaton (1970) found cheetah in small groups which avoided contact with each other. He concluded that they adopted a 'time-plan' form of territoriality, after the model of Leyhausen and Wolff (1959). Urine-spraying by male cheetah was fundamental to Eaton's interpretation: he noted 14 occasions where one group followed the path of another after a day had elapsed, and did so warily. In contrast, on nine occasions a group began to follow another's route within one day of the first group's passage, and in each case they changed course after sniffing the relatively fresh spray marks which were left at 30–50 m intervals in dense cover (10–100 m in grass terrain). Eaton shows (1970, Fig. 2) how cheetah kneel down to sniff at spray marks on small shrubs. Eaton (1970) also notes (albeit only for one of his study groups) a behaviour first described by Schaller (1967, p. 254); two males spray marked a habitually sprayed site and then scuffed the ground with their hind legs prior to defecating ('a small amount') and/or urinating into the area scraped clear.

Although lions apparently defecate at random (Schaller 1972), several other felids leave their faeces along trails and on conspicuous objects, e.g. wildcat *F. silvestris* (Corbet 1979), lynx *Lynx lynx* (Lindemann 1955), puma *F. concolor* (Seidensticker *et al.* 1973). Bobcats, *Lynx rufus*, also defecate along trials as well as at latrine sites (Bailey 1974), as may domestic cats (see below). In riverine habitat in the Namib, *Felis lybica*, used latrines of up to 149 faeces (Stuart 1977). Presumably in the desert this habitat represented a rich resource.

### Domestic cat, Felis catus

*Spray marks and time-sharing*    The social behaviour of more or less feral domestic cats, *Felis catus*, has been of increasing interest since, in 1956, Leyhausen proposed a vertically stratified spatial system in which a number of cats might time-share their occupation of the same area. Scent marks, and more particularly urine sprays, played a critical role in Leyhausen's model and in 1971 he likened their function to traffic signals: a fresh mark means 'section closed', and, by analogy, represents a red light at a traffic signal. An older mark indicates proceed with caution whereas an old mark indicates that it is safe to go ahead (see de Boer 1977). Leyhausen (1971, 1979) suggests that an old mark also stimulates over-marking in order to re-calibrate the signal for subsequent animals (see also Leyhausen and Wolff 1959). Several authors have at least limited evidence for such a time-sharing system among other felids (e.g. cheetah, Eaton (1970); puma, Hornocker (1969); and lions, Schaller (1972)). However, Macdonald and Apps 1978) and Panaman (1981) found that the home-ranges of female farm cats from the same colony overlapped widely and while they hunted there was little evidence of avoidance based on scent or visual signals.

*Field studies*    Macdonald, Apps, and Carr (unpublished data) describe the social behaviour among a small kin group of farm cats, including three adult females whose overlapping ranges averaged 13.1 ha. The one adult male from the colony travelled over 83 ha, and spray urinated frequently. Fifty-seven per cent of these sprays were made in the vicinity of the home barn and of those, 10 per cent were made in an agonistic context. Sprays made during an aggressive interaction differed from others in that 60 per cent of them were followed by cheek rubbing on the urine, whereas only 21 per cent of other sprays were followed by this behaviour (see also Leyhausen 1965). The same study indicated that during social interactions there was considerable asymmetry in the frequency with which given pairs of cats rubbed on each other. It seemed likely that being rubbed upon was a correlate of dominance. Amongst the femals of this colony of cats there was a marked difference in their urination and defecation behaviour in the vicinity of the barn and when they were travelling further afield. They were more likely to leave both urine and faeces exposed while away from the barn, and to bury them when at home. Panaman (1981) similarly noted that female farm cats were more likely to cover faeces within their 'common core area'. Panaman's females sprayed most in the context of hunting, with overall rates of one spray per 16.7 min of hunting time or 70 m of distance. During one hour of hunting one female reached a maximum of one spray per 3.3 min or 44 m. These high rates were never seen amongst females in the colony described by Macdonald *et al*.

Liberg (1981) studied the scent-marking of feral tom cats from the rural population in Sweden (described in Liberg (1980)). He found that all the adult male cats sprayed frequently, especially while they were travelling. These male cats were classed either as breeders or as one of three successive non-breeding casts. Overall, the non-breeding males averaged 12.9 urinations per hour of travelling, whereas breeders averaged 22.0 marks per hour. Apps (1981) recorded a high rate of spraying by males on Dassun Island, where feral cats lived in overlapping home-ranges occupied by solitary adults and groups of juveniles. Although Apps saw toms marking at rates as high as 313 sprays in 5 h (62.6/h whereas females never sprayed more than 6/h), he could detect no response to these marks by other cats nor detect any spatial pattern that hinted at their function. A few scrapes were made in the mouths of penguin burrows which the cats investigated and this might have suggested a process akin to Henry's (1977) book-keeping, but the inconsistency of the behaviour led Apps to dismiss that possibility.

**Family Ursidae**

Bears have either very reduced anal sacs, or none at all (Pocock 1914b, 1918). However, grizzly bears, *Ursus americanus*, not only rub their urine-soaked fur against marking trees, they also repeatedly scratch with their fore-paws at such sites. Indeed one of the earliest suggestions for an aversive territorial function of scent marks was made with regard to this behaviour pattern (Bilz 1940). Lloyd (1978) found that grizzly bears especially selected amabilis fir trees (known for their abundant sap which may be a good odour-retention agent) as marking sites. Lloyd studied the pattern of marking by placing a fishing line across the torn bark and noting the frequency with which it was displaced. The average marking tree was 79 years old in Lloyd's study, and was first marked on average at 20 years old. Most were in the vicinity of a water course, where the bears' movements were also focused. In these more intensively used areas marking trees were found on average every 3.4 km of watercourse, and at a density of about 20 per km$^2$. These trees were all along regularly used bear trails, where Lloyd also found stamping marks, where the bears had apparently trampled the soil, but he could neither confirm nor refute his suspicion that these areas had some communicative significance. Tschanz, Meyer-Holzapfel, and Bachmann (1970) describe how the European brown bear, *Ursus arctos*, rubs on trees with its back, sometimes also rubbing with its chest and biting the vegetation. Males are said to be able to distinguish the sex of the animal which rubbed at a given spot, and rubbing reached a peak frequency in the mating season.

*Anal sacs*   In 1918 Pocock wrote that no specialized cutaneous glands had been discovered in any of the Ursidae. However, anal sacs are apparently present in some but not all species. Pocock found them, although greatly reduced, in the American black bear, *Ursus americanus*, but not in the brown bear, *Ursus arctos*.

*Hair structure*    Although biologists studying bears can smell them in the field
(F. Bunnell, personal communication), the Ursidae do not apparently have any
external scent glands, although the findings of Lewin and Robinson (1979) raise
one interesting possibility. These authors discovered that the 'greening' of polar
bears, *Ursus maritimus*, in zoos is due to hollow guard hairs on the outer side of
the legs and rump, which can harbour algae. There is no evidence that these
specialized hairs are involved in harbouring social odours, but it is an intriguing
possibility that they might (see below).

*Urine and faeces*    There is little evidence for faeces having any communicative
function among bears. Indeed, brown bears, *Ursus arctos*, seem to spread their
faeces randomly and, on this assumption, Roth (1980) has produced an index of
bear density based upon the abundance of their faeces. Of course, the random
distribution of faeces with respect to the bears' movements does not in itself
imply that they have no socially significant odour, since, for example, if Roth
can use faeces as an indicator of brown bear abundance then bears themselves
could doubtless do the same thing. Tschanz *et al.* (1970) report that brown bears
mark with urine and that males, at least, can distinguish the sex of another bear
from its urine.

It is clear that at least one subspecies of the brown bear (see Ustinov 1971),
namely the grizzly, *U. a. horribilis*, does not invariably deploy its faeces randomly.
Mealey (1980) describes his work on the diet of this species and mentions that he
had a different procedure for collecting faeces which he found singly rather than
those he found in groups of 10 to 25.

### Family Mustelidae

As with other Carnivora, much of the information available on the Mustelidae
concerns species living in cold regions where snow retains tell-tale signs of odour.
For example, in Haglund's (1966) classic studies of wolverine, *Gulo gulo*, tracks
he noted scent-marking by (i) urine and faeces, (ii) chewing certain tree species
(those with abundant sap), and (iii) anal-gland marking. Pulliainen and Ovaskainen
(1975) confirmed some of Haglund's observations, reporting 25 anal-sac marks
on tussocks, trees, and stones in 7.5 km of wolverine trail and they, like Haglund,
interpreted these marks as territorial. The anal sacs of wolverines, and of all
mustelids, are similar to those of the Canidae, being paired reservoirs, sited on
either side of the anal orifice (Pocock 1920*b*, 1921*a*, *b*; Stubbe 1970). Pocock
(1921*a*) considers that all mustelids share a typical pair of anal sacs which open just
within the anus, but that they can be broadly categorized on the basis of whether
or not these sacs are grossly developed. They are not enlarged in *Martes, Mustela,
Gulo, Tayra, Lutra, Taxidea,* and *Meles.* They are enlarged in *Mellivora, Galictis,
Ictonyx, Mephitis,* and *Conepatus.* Generally the species in the second category
produce foul-smelling evacuate, and this also applies to the polecats, *Putorius*
spp. Pocock felt that all these foul smelling 'aposematic Mustelines' shared the

behavioural trait he characterized as obstinate pertinacity when describing the black-footed ferret, *Mustela nigripes* (Pocock 1911). Subsequently, Pocock (1927*b*) added the striped weasel, *Poecilogale albinucha*, to the list of species with unspectacular anal sacs and this led him to hypothesize that this weasel was a Batesian minic of the zorilla (*Ictonyx striatus*). He had also proposed a similar notion (Pocock 1908) to explain the similarity in colour of two South American jungle forms, the tayra, *Tayra barbara* (Mustelinae), and the bush dog, *Speothos venaticus* (Canidae). Without having seen or smelled a tarya, Pocock thought that their anal sacs might be as offensive as those of the grison, *Galictis vitatta*. Subsequently, this turned out not to be the case so the hypothesis had to be abandoned. However, the tayra is unusual amongst mustelids in having a shallow concavity just above the anus, although the glandular status of this depression is unknown.

### Subfamily Mustelinae

*Urine and faeces*   Although data are few, urine and faeces are probably widely involved in musteline communication; this is the conclusion for fishers, *Martes pennanti*; pine marten, *M. martes*; mink, *Mustela vison*, polecats, *M. putorius*; weasels, *M. nivalis*; and others (Powell 1982; Goethe 1938; Lockie 1966; Birks 1981; Poole 1967). Birks (1981) describes how mink urinate at the sites where they defecate, and how male mink may leave trails of urine 20–40 cm long. Mink scats are coated in strong smelling jelly which probably originates from the proctodaeal glands within the rectum rather than the anal sacs, and the faeces are left on visually prominent features (Gerell 1970). Gerell found faeces concentrated where neighbouring mink met, whereas Birks (1981) found mink scats in accumulations outside their dens. Birks (1981) saw dominant mink produce a blob of proctodaeal jelly following agonistic encounters.

In Finnish Lapland pine martens, *M. martes*, travel home ranges of 100 km². Pulliainen (1982) describes how, within these vast areas both sexes mark with urine and faeces (and anal glands). Females urine marked more than males and, overall, urine marks were left more frequently (0.9/km² rising to 1.6/km²) as winter progressed. All categories of scent mark were deployed through each marten's home range. Scent marks of all kinds were made on elevated sites. Faeces were generally deposited singly an average of 1.3 km apart, but latrines of up to 6 faeces were found near resting places.

*Skin glands*   Sables, *Martes zibellina*, are known to have chin glands (Petskoi and Kolporskii 1970), and fishers (after eating), American martens, long and short tailed weasels, and ferrets rub their chins (Powell 1981, 1982, personal communications). The ventral glands and anal sacs of stoats, *Mustela erminea*, were studied by Erlinge, Sandell, and Brinck (1982). Stoats caught in the wild were introduced singly or in pairs, into an arena of 35–40 m². Their anal drag and body rubbing behaviour, in this arena were recorded when each individual was either alone or in the company of one of a variety of other stoats of known and varied social status. They found that dominant stoats of either sex marked overall about three times more frequently than did subordinates. The dominants maintained a larger differential over the subordinates in body rubbing (2.3 versus 0.5 marks per 10 min) than they did in anal dragging (2.1 and 1.0). When dominant and subordinate stoats were confronted with each other in the arena, the dominant animals immediately increased their rates of both types of marking, whereas the subordinates reduced theirs. Overall, dominant stoats left a rather higher proportion of body rubs to anal drags than did socially inferior stoats. Erlinge *et al.* (1982) noted that the two types of mark were made in different situations: dominant stoats rubbed the ventral gland on the substrate during agonistic interactions with subordinates, after fighting and chasing and after taking over a nest. In contrast the anal sacs were dragged over objects at regularly used sites often where urine and faeces were also voided, and also on strange objects introduced into the arena. When a pair of stoats met in the test arena, they avoided each other initially. Then, when the dominant began to mark, principally with the ventral gland, the subordinate cowered. Indeed, when subordinate stoats were introduced to the arena alone they cowered on encountering a scent mark, whereas dominant stoats over-marked it.

Powell (1982) suspects that fishers, *Martes pennanti*, mark with glands associated with whorls of hair on the pads of their hind-paws, and that they have chin glands with which they rub after eating, drinking, or playing. Birks (1981) noted the ability of mink to follow each other's scent trails, although the source of odour is uncertain. Ventral glands have been described for the American marten, *M. americana*, and the wolverine (Hall 1926), and are probably common to all Mustelinae. Pine marten, *M. martes*, mark throughout their home ranges in winter with anal sac secretions (Pulliainen 1982). Early in the winter the sexes mark at a comparable rate, but later males mark more than females (the opposite of the case for urination). Overall, males and females anal mark 7.5 and 3.6 times per km respectively.

Stoddart (1979) has demonstrated what is, to the weasel, *Mustela nivalis*, at least, presumably an undesirable behavioural consquence of anal-sac gland marking. He showed that short-tailed voles, *Microtis agrestis*, avoided Longworth traps anointed with the extracted odour of weasel anal-sac secretions. In rather surprising contrast, woodmice, *Apodemus sylvaticus*, did not show this cautious response.

*Chemical studies*    Chemical analyses by Erlinge *et al.* (1982) and Brinck, Erlinge, and Sandell (1983) revealed first that the constituents of the stoat's ventral gland secretions differed from those of the anal sac, and secondly that there were individual differences in the presence or absence and in the relative proportions of compounds in the anal-sac secretions of different stoats. There were no such differences in the secretions of the ventral gland from different individuals.

The findings of Erlinge and his colleagues seem to differ from those of Crump (1980*a, b*) who mentions explicitly that four of five peaks on his gas-liquid chromatographic analysis of stoat anal-sac evacuate were invariable in their proportions between different stoats. However, one peak was highly variable. Crump's (1978, 1980*a*) findings were that sulphur compounds (thietanes and dithiolanes) predominated in the stoat's anal sac.

Brinck, Gerell, and Odham (1978) have studied the chemicals within the mink's anal sac secretion and found that when secretions from known individuals are taken throughout a year the complex chromatographic peaks retain individual patterns. Perhaps mink can distinguish each other on the basis of these consistent differences, but this is unknown. The anal sac of the mink has been shown to be both structurally and chemically interesting by Sokolov, Albone, Flood, Heap, Kagen, Vasilieva, Roznov, and Zinkevich (1980). By detailed histological examination these authors found that the glandular tissue around the sacs and their ducts comprise two layers, an inner one of sebaceous tissue, and an outer one of apocrine tissue in which electron microscopy revealed granules of a sulphated compound visible in the lumenal part of the cells (Plate 15.12). The products of these two layers also differ, although both are rich in sulphur—the secretion ($150 \mu$l per sac) is a yellowish mixture of oil (45 per cent by vol.) and an immiscible straw-coloured aqueous solution with abundant suspended material. The oil was largely the product of the sebaceous cells and was made up of wax monoesters, whereas the aqueous fluid contained fatty acids, and other compounds thought to derive ultimately from the apocrine cells, but more immediately probably as products of bacterial action. Indeed, one compound, 2-piperidone, has also been found in the anal sacs of dogs and coyotes (Preti *et al.* 1976) and is closely related to 5-aminopentanoic acid, which is found in the anal sacs of red foxes (Albone, Robins, and Patel 1976), where it is thought to be the product of anaerobic fermentation. So the likelihood is that microbial action gives rise to this odorous substance in all four species. Interest in the similarity between at least some of the chemicals in the mink anal sac and those of other carnivores is enhanced by the finding of Sokolov *et al.* (1980) that a chemical (isopentenyl methyl suphide) previously thought (Jorgenson *et al.* 1978) to be unique to red fox urine is present, albeit in traces, among the head-space volatiles of mink secretions. On the other hand, two of the sulphur-rich compounds of the mink's anal sac were not found in the secretions of eight other mustelids that Sokolov *et al.* (1980) investigated. Crump (1980*a*) makes the interesting point that although both mink and stoat anal sac secretions contain thietanes and dithiolanes, these compounds apparently derive from different precursors in the two species (isoprenoids in the mink, and straight-chain fatty acids in the stoat). Surprisingly, these two related species have similar sulphur-containing compounds in their anal sacs, but apparently they derive from different biosynthetic pathways (see also Schidknecht, Witz, Enzmann, Grund, and Ziegler 1976).

*Subfamily Mephitianae*

The skunks are the most notoriously odorous of the Carnivora, their anal sacs being grossly enlarged (Anderson and Barnstein 1975). Blackman (1911) describes how the musculature around the sacs is well developed and when it contracts the squeezing of the sacs squirts their contents through the paired openings at the anus.

*Urine and faeces*    In his monograph on the spotted skunk, *Spilogale interrupta*, Crabb (1948) stated that they seldom made latrines, and that they simply defecate wherever they happen to be, excluding the nest. Nevertheless, as Crabb points out, they do sometimes use latrines and he pictures one of these in a barn, lying between a hay bale and the barn wall. Crabb also illustrates the tunnels of a spotted skunk den which he excavated. At the side of the main tunnel, and nearer to the entrance than to the skunk's nest, there was one cavity filled with skunk faeces.

There are no data on the distribution of skunk latrines aside from their frequent proximity to a den site. Intriguingly, Chapman (1946) suggests that skunks deliberately defecate in places to which they subsequently return in order to forage for dung beetles.

*Subfamily Lutrinae*

*Anal sacs*    The anus and environs of all otters closely resemble those of typical mustelines, such as *Martes* (Pocock 1921*a*, *b*), excepting the sea otter, *Enhydra lutra*, which lacks anal sacs (Jacobi 1938).

The anal sac secreta of European otters, *Lutra lutra*, are deposited at latrines where they also defecate and urinate. Evacuation of the sacs is normally associated with excretion. Gorman, Jenkins, and Harper (1978) noted that anal-sac secretions were daubed on faecal marking sites by both members of a captive pair of otters in Scotland during synchronized bouts of 11–14 days of intensive marking, interspersed with 30–40 day periods when the animals left excreta on their marking sites, but no anal sac secretion. Periods of anal sac marking coincided with mating activity and on two occasions when the female underwent pseudopregnancies lasting 60 days the entire 'gestation' period elapsed without any anal-sac marking. Subsequently, and using the same animals, Jenkins, Makepeace, and Gorman (1981) attempted to develop immunoelectrophoretic assays to aid recognition of each individual's anal-sac secretions, but the technique failed.

The anal sacs of the giant river otter, *Pteronura brasiliensis*, open just inside the anus and secrete a dark brown viscous fluid with a powerful musky smell. Duplaix (1980) describes the riverside marking sites in Surinam where members of a party of giant otters break down the vegetation, trampling it into the mud, and churning up a bog of anal-gland secretions, urine, and faeces (Plate 15.13). She also describes how the giant otters bend saplings and run them under their bodies. Giant otters deliberately smear faeces on the vegetation at these sites, and then rub material from their bodies over twigs and saplings (see Duplaix 1980,

p. 550, and Laidler and Laidler 1983). Since the giant otters' bodies are besmirched in the scent-impregnated mud of their communal latrines it is not possible to distinguish which particular odour they are rubbing on vegetation. Duplaix surmises that these marking sites are involved in territorial maintenance, but there is no evidence as to their function. Arden-Clark (personal communication) reports similar marking sites on which Cape otters, *Aonyx*, perform a 'shuffle dance' and roll in their faeces.

*Faeces*   More sociable species of otter are reputed to use communal latrines, e.g. *Aonyx*, *Amblonyx*, *Pteromera*, and *Lutrogale* (Duplaix 1980). Even rather solitary species build up latrines. For example, the marking (sprainting) sites of the generally monogamous European otter, *Lutra lutra*, are composed of faeces (called spraints), urine, and, periodically, anal-sac secretions. Their distribution within the home-ranges of these otters varies intraspecifically. For example, Kruuk and Hewson (1978) studied a population occupying a coastal habitat on the west coast of Scotland where the otters principally foraged at sea, and where their ranges overlapped considerably. However, otter holts were distributed along the coastline at rather uniform intervals. Spraining sites were also found along the coast and through protracted use they had developed into grassy mounds known as otter seats. These seats were heavily concentrated near the vicinity of the holt, and not near the home-range borders. This pattern contrasts with Erlinge's (1968) finding for a population of the same species living in a series of lakes linked by streams in southern Sweden. There the otters were territorial and middens were concentrated around border zones. Erlinge found activity at latrines throughout the year (with a slight peak in the mating season) with 30–40 sp 
ainting sites per km of river. He reports several observations of males strategically siting their faecal marks in places where rival males were known to pass. In yet a different habitat, where otters hunt the lochs and connecting streams of NE Scotland, Gorman *et al.* (1978) found no evidence that scent marks of any sort were concentrated at range boundaries.

In Surinam Duplaix (1980, p. 591) describes how *Lutra enudis* scratches up sand to create a mound on which urine and 'scent' (presumably anal sac secretions) were left. Duplaix found the relatively small *Lutra enudis* (*canadensis*) sympatric with the giant river otter, *Pteronura brasiliensis*, and the two species used the same spraining sites, although not simultaneously. Giant river otters (see above) create spectacular trampled areas along river banks where they mark with anal sac secretion, urine, and faeces (see Plate 15.13). Duplaix studied giant otters in two habitats, one typified by slow-flowing forest creeks, and the second an upstream area of falls, wide rivers, and abundant pools. The pattern of faecal marking varied between these two areas: riversides in the forest were punctuated by the communal trampling sites, whereas single faeces were carefully sited aloft conspicuous boulders and promontories in the falls area. Furthermore, upriver, not only were these single faeces not trampled, they were often in sight of each other. So, Duplaix describes a place where the river and pools were almost 1.2 km wide and where a transect of boulders had been marked with faeces across the river, each mark in sight of the next, and a total of a dozen marking sites spanning the river. The different habitats doubtless posed different communicative problems for the otters who may even have adopted a different social system under these two circumstances. Hence not only may the upriver otters have been sending messages in a different environment to that of the otters of the linear habitat of the forest creeks, but they may also have been sending different messages altogether. Certainly, in either case, an alien otter would not have penetrated far into the territory of a resident giant otter before encountering a sprainting site. This intraspecific comparison adds more weight to the general contention that the ways in which social odours are deployed vary with ecological circumstances, as does social organization in general. However, the comparison has an additional interesting feature: Duplaix suspected that some otters undertook seasonal migration between the falls area and the forest creek. This raises the possibility that a given individual might engage in quite different patterns of scent marking at different seasons in different places.

### Subfamily Mellivorinae

The sole representative of the Mellivorinae is the honey badger or ratel, *Mellivora capensis*, which ranges throughout Africa, and east through the Middle East to India.

*Anal glands*    The anal glands of the honey badger were unique amongst the mustelids in being eversible. Pocock (1920*b*) relied on these glands in an attempt to resolve a debate between earlier authors such as Miller and Gray concerning the species' taxonomy. One school held that the honey badger showed affinity with the wolverine, *Gulo gulo*, and hence might be included within the subfamily Mustelinae, whereas the school that Pocock favoured dismissed the idea of any such affinity since the wolverine's anal sacs were conventional whereas those of the ratel were 'of great size, and discharge copiously a suffocating fluid exactly as the Skunks (*Mephitis, Conepatus*), Zorillas (*Ictonyx*), Grison (*Grisonix*) and Teledu (*Mydaus*).' One tantalizing aspect of Pocock's account is that he illustrates it with a drawing of the ratel's anal sac and clearly labels a 'bristle' apparently running through the duct of the gland. However, he makes no further mention of this odd bristle elsewhere. In 1921*a* he noted that in contrast to other Mustelidae, the anus of *Mellivora* was surrounded by a glandular pouch.

### Subfamily Melinae

*Subcaudal pouch and anal sac*    The glands of the anal region of most badgers are simply paired anal sacs, similar to those of the Canidae. Pocock (1920*a*, 1925) describes those of the American badger, *Taxidea taxus*, as lying beside a rather protruberant anus, and producing a colourless, sweetish odour, whose smell was not unpleasant.

The anal-sac secretions of Eurasian badgers *Meles meles* are a yellow-brown colour and gel like in consistency. Kruuk (1978) notes that this material is found on faeces or in the pits associated with latrine sites. The Eurasian badger is unusual amongst the Mustelidae, and even amongst the subfamily Melinae, in having not only anal sacs but a well developed subcaudal pouch. The pouch was first described by Chatin (1884, cited in Pocock 1920a). The pouch lies below the tail and is partially divided into right and left pockets by a vertical partition. Pocock (1920a) noted that *Meles meles* had the habit of rubbing the secretion from this pouch onto objects.

The hog badger, *Arctonyx collaris*, is a little known Asiatic member of the Melinae which also has a subcaudal pocket. This was first, and last, described by Evans (1839) who wrote '(there) is a caudal pouch directly under the origin of the tail . . . but quite distinct from and wholly unconnected with, the anus or genital organs. The sac (has) . . . a lining of naked membrane, secreting a brown unctuous matter, not unlike cerumen, or wax of the ear'. Gairdner (1915) emphasized the potency of the odour from this pocket by noting that '. . . the stench so pervaded the beast that the coolies were unable to eat it'.

The subcaudal pocket of *Meles* has a romantic history. Apparently in the seventeenth century it was accepted amongst huntsmen and countryfolk that badgers survived the winter by sucking the fatty secretions of their subcaudal pockets. By the time Herbst was writing in 1882 he could dismiss this notion as fantasy, but his contemporary version was no less extraordinary. Calling the gland the 'sucking hole' (saugloch), Herbst described how young badgers in captivity suck their own and their siblings' subcaudal pouches. He cites evidence that badgers elsewhere, including Tibet and Japan, behave similarly. Schaffer (1940) took this very seriously and after a long discussion said he could not dismiss the idea. Schaffer stated that adult badgers avoid even their own subcaudal glands, and thus he argues that Herbst's theory is only plausible if there is a developmental change in the nature of the secretions and hence of the gland's structure. Schaffer believed that there is such a change. He pointed out that in the adult badger, large bunches of sebaceous tissue are separated by apocrine glands which run into the narrow ducts of hair follicles, or else discharge into the cisterns of sebaceous glands. However, in two young badgers Schaffer found that the apocrine glands predominated. Furthermore, he described in the lobes of these tubular glands a granular secretion that reminded him of de-fatted milk. Perhaps, he argues, this is what the badger cubs eat, and perhaps the need to probe their anal pouches explains their unusually narrow tongues. My own hunch is that the whole incredible theory developed from observations of the misdirected tendency of many hand-reared animals to suck on the bodies of their siblings.

More recently Kruuk (1978) has described the scent-marking behaviour of a population of badgers in Wytham woods, Oxfordshire, where social groups, called clans, numbering about seven animals (excluding dependant cubs) occupied territories of about 60 ha. Squat-marking, whereby the subcaudal gland was momentarily pressed to the ground, was very frequent. Sometimes these marks were daubed on a visually conspicious site, such as a tussock, and almost every badger passing that site would overmark it. As badgers trundle along their paths their movements are punctuated by these rapid pressing movements. Using a tame badger Östborn (1976) found that some sites are anal marked repeatedly by the same individual. She also showed that badgers could distinguish each other's subcaudal pouch secretions and anal-sac secretions. Kruuk (1978) saw badgers of both sexes pressing throughout their territories, and both marked in this way at a higher rate in the vicinity of territory borders. During one visit to a border latrine a badger might anal-mark many times.

Kruuk describes badgers adopting a 'handstand' posture in order to deposit subcaudal secretions 30–40 cm above the ground, and anal-pressing also occurs off badger trails and away from the border. Eurasian badgers, in lowland Britain at least, principally eat earthworms, *Lumbricus terrestris*, which are caught on the surface while the badger slowly quarters a small (perhaps 25 m$^2$) plot of land for many minutes  (e.g. 20 min or more) (Kruuk 1979). Even away from trails and borders, whilst hunting in this way in the middle of a pasture and away from obvious landmarks, badgers continue to anal-press, at rates of up to 2/min, so that foraging patches are heavily marked (Macdonald 1977).

*Tail glands*    Schaffer (1940) claims that in adult Eurasian badgers, *Meles meles*, there is a gland beneath the tail made up of an extension of the sebaceous tissue of the subcaudal pouch.

*Clitoral glands*    Pocock (1920a) describes small (6 mm deep) glandular pockets lying on each side of and slightly below the vulva in *Taxidea*. It is hard to visualize these pockets from his drawings, but it seems that at the bottom of each there are several setae, each planted in a shallow pit.

*Urine and faeces*    Eurasian badgers generally defecate and/or urinate in communal latrines which comprise a cluster of small pits in 2–4 m$^2$ (Neal 1948). These badgers often travel their ranges along well-worn paths, some of which lead to latrine site. Kruuk (1978) discovered that at least some sections of the border between neighbouring clan territories are demarcated by badger latrines linked by conspicuous paths. He found that in Wytham woods these latrines were common (0.72 per ha) within 50 m of the territorial borders and also in the vicinity of setts (0.81 per ha), but were present at much lower densities elsewhere (0.09 per ha). Each latrine site measured 2–4 m$^2$ within which 1–60 pits were scraped in the ground to a depth of 5–10 cm. Some, but not all, pits contained faeces. These latrines are foci of olfactory activity within the badger community. Badgers make special trips to latrines, and stay there for

20-90 seconds of intensive marking activity. Kruuk found not only faeces and urine at latrines but all or some of anal sac secretions, subcaudal gland secretions and, by inference, interdigital gland secretions. He noted one latrine site that had been active for 12 years, and found that the activity at latrines peaked in February-May, and again, to a lesser extent, in October and November. In Britain, European badgers are inactive for long periods in winter, and the periods of peak latrine use coincide with the spring emergence of the badgers and the period during which they are accumulating fat reserves for winter.

Kruuk (1978) summarized the very convincing evidence that border latrines at least were involved in territorial maintenance. Quite apart from the pattern of their distribution, the density of latrines was significantly higher ($p<0.001$) at sections of border which abutted a neighbouring territory as opposed to those which abutted on no man's land (which were rather nebulously defined). Latrines near borders were more often near a visually conspicuous landmark than those elsewhere; but although border latrines were bigger than others (i.e. had more faeces) latrines everywhere had a similar number (just over one per pit) of faeces.

### Family Procyonidae

Knowledge of social odours among the Procyonidae is even more fragmentary than among other carnivore families. I have found little material to indicate whether or not they mark with their urine or faeces, but Kaufmann (1962) does mention that male coati, *Nasua narica*, indulge in 'penile dragging' and J. Russell (personal communication) tells me that female coatis will occasionally drag the perineum on the ground, especially while foraging.

*Glands* The anal glands of the coati differ from those of any other carnivore. Mivart (1882a, b) described how, when the anus is open, the modified anal glands appear as a series of 4-5 parallel slits. These slits are the apertures of a corresponding number of narrow sacs produced by a folding of the anal epithelium.

Mivart (1882a, b) describes the anal sacs of racoons, *Procyon* spp, as 'canid-like', but relatively small (see Ough 1982), and Pocock (1921b) interprets the earlier writings of Gervais to mean that *Bassariscus* is similarly endowed. Of the subfamily Ailurinae, Pocock (1921c) also describes the anal sacs of the red panda, *Ailurus*, as normal. However, the integument around the anus of red pandas is highly glandular and partially indented, reminiscent of the anal pouch of mongooses. Apparently, adults of both sexes of red panda rub their anal regions on branches within their cages (Roberts 1981). Morris and Morris (1966) describe how the giant panda, *Ailuropoda melanoleuca*, marks with its anal region by rubbing on objects and Kleiman (1983) notes that both male and female giant pandas have a large glandular region in the genital area; in one pair anal rubbing generally increased in pro-oestrous for the female but decreased for the male; even juveniles anal rubbed. Anal rubbing sometimes involved a handstand, and bark was stripped from marking sites on trees.

The kinkajou, *Potos flavus*, stands alone amongst the procyonids with an unusual pattern of cutaneous glands. Mivart (1882*a, b*) had stated that kinkajous have anal glands but Pocock (1921*c*), dogged in his determination to contradict Mivart, pointed out that they had none. However, kinkajous do have two patches of glandular tissue along the mid-ventral line, one abdominal, the other sternal. Each comprises a small patch of skin scantily covered with short hairs and equally developed in both sexes. Pocock wrote 'These glands are conveniently placed for rubbing the secretion along the branches of trees to enable kinkajous to track each other by scent. Although I have never noticed these animals behaving in a way to suggest that this is the function of the glands, I do not doubt that it is so' (Pocock 1921*c*).

Kinkajous also have almost naked skin on the lower jaw and around the lips. Pocock thought that this skin was not especially glandular and so might simply be an adaptation to avoid the stickiness of their food, such as honey. However, Poglayen-Neuwall (1962) describes mandibular glands.

*Urine and faeces*    In Trapp's (1978) field study of the cacomistle, *Bassariscus astutus*, he found faeces scattered along their trails aloft bolders and rocks. They were also left in forks of trees. In one attic, Trapp found a large heap of faeces. Raccoons, *Procyon lotor*, defecate at communal latrines (Stains 1956). Ough (1982) describes anal rubbing by raccoons which resulted in smears of faecal-like material. Fritzell (1978) speculates that raccoon latrines serve a territorial function but Ough notes that the species is not territorial throughout much of its range, so restricting this explanation.

### Discussion

It is clear that as biologists become more attuned to detecting behaviour linked with social odours it is hard to describe any aspect of mammalian social behaviour without the implication of scent. It is also clear that for the Carnivora even this descriptive task is only in its infancy, from both morphological and behavioural standpoints. Nevertheless, these accumulated descriptive snippets amount to a mountain of knowledge in comparison to the handful of cases for which a function can demonstrably be assigned, or even strongly inferred for an odour (Table 15.5). The following paragraphs summarize these functions:

#### Functions

*Identity*    Information regarding aspects of identity, either individually or in terms of sex, or group membership is probably contained in many carnivoran odours. The ability to recognize individuals purely by their odour has been demonstrated unequivocally for dwarf and Indian mongooses (Rasa 1973; Gorman 1976) and for Eurasian badgers (Östborn 1976). Both species of mongoose recognize individuals through their anal-gland secretions (and, at least, in the case of the dwarf mongoose, not their cheek-gland secretions), whereas the badgers could distinguish each other at least on the basis of subcaudal gland secretions. Wolves and dogs can discriminate between the odours of urine from different individuals and hence can probably recognize individuals on the basis of this distinction (Brown and Johnston 1982; von Uexküll and Sarris 1931). Both Macdonald (1979a) and Blizzard and Perry (1979) have shown that red foxes can distinguish their own urine from that of other foxes, but neither study was designed to test the, nonetheless likely, ability of individual recognition through urine. As chemical studies become increasingly linked to behavioural ones it emerges that individual variation in the proportional representation of components in secretions is commonplace. For example, Mills *et al.* (1980) show this clearly for brown hyaena. It will be surprising if these individual differences do not facilitate recognition but it remains generally undemonstrated that they do so. An important caveat is that the most abundant chemicals in a secretion are not necessarily the ones with communicative significance, so chemical analyses without a bioassay are unreliable indicators of function. While there is mounting evidence that information on individual identity is contained in social odours (and that it can be discerned by the receiver), there are no demonstrations of the functional significance of this information, except in so far as it may distinguish individually recognizable group members from strangers (whose individual identity to, say, a territorial male facing an intruder, may be immaterial).

The existence of a distinct group odour has not been proven for any carnivore, as distinct from the possibility that an animal could simply recognize the individual odours of each member of a group. The idea of a shared group odour became increasingly plausible as the role of bacterial action in scent production was established—one could envisage animals catching the group odour in the same way as catching a sore throat. However, since studies of the bacteriology of the red fox's anal sac showed that the microflora varied from week to week within a given sac, the idea of a sustained group odour for that species had to be abandoned. Of course, animals within a group could keep up to date with subtle changes in each other's microbial odours and so continue to recognize each other; moreover, due to repeated cross-infection, their microfloras might change in parallel and so confer a distinct, if ever changing, group odour. Cross infection would be most likely in species whose scent-marking involves several individuals smearing their glands on the same site. Another possibility, namely that members of kin groups share similar odours due to their genetic relatedness has not been investigated. Scent-marking after even a brief separation from the group seems commonplace, as Rasa (1973) described for babysitting dwarf mongooses, and this may indicate that each individual must strive to keep its own odour fresh in the nose of its fellows. Allomarking is found among social viverrids, mustelids, canids, and felids. This, together with the habit of rolling on each other's scent marks,

Table 15.5. Some functions of social odours in the Carnivora

| | Demonstrated discriminations | | | | | Implied functions | |
| --- | --- | --- | --- | --- | --- | --- | --- |
| | Familiar versus alien | Individuals | Status | Receptivity | Sex | Foraging | Territorial |
| **Canidae** | | | | | | | |
| *Canis familiaris* | +U6 | +U12 | | +S1,5,−A1,3,+U4,1 | +S2,+U2,35,+F2 | | |
| *Canis lupus* | | +U12 | | | | +U13 | +U10,11 |
| *Canis latrans* | | | | | +U7 | +U8 | +U9,34 |
| *Canis aureus* | | | | | | | +F14 |
| *Canis mesomelas* | | | | | | | +U15 |
| *Vulpes vulpes* | +U16,17 | | | ?U23,24 | ?U23,25 | +U18 | +U16 |
| *Nyctereutes procyonoides* | +F33 | | | | | | |
| **Hyaenidae** | | | | | | | |
| *Crocuta crocuta* | | | | | | | |
| *Hyaena brunnea* | | ?A20 | | | | +A20 | +A19,+F19 |
| **Viverridae** | | | | | | | |
| *Herpestes auropunctatus* | | +A21 | | | | | |
| *Helogale parvula* | | +A22 | +C22 | | | | |
| **Ursidae** | | | | | | | |
| *Ursus arctos* | +U32 | | | | +U32 | | |
| **Felidae** | | | | | | | |
| *Felis catus* | | | | +U26,+C26 | | | |
| **Mustelidae** | | | | | | | |
| *Meles meles* | | +A27 | | | | | +F28 |
| *Lutra lutra* | | | | +A29 | | | |
| *Mustela erminea* | | ?A30 | | | | | |
| *Mustela vison* | ?A31 | | | | | | +V30 |

Despite a wealth of speculation there are few definitive examples of the function of given social odours within carnivore societies. In compiling this table I have had to use the rather relaxed criteria of implied rather than proven functions. For example, proof that members of a species can distinguish individuals on the basis of their scent does not in itself answer the question of what is the functional role of that scent (or the part played in that role, if any, of individual recognition). Having demonstrated various discriminations, functional questions become, in part, transposed to broader issues, e.g. what is the adaptive significance within the species' society of recognizing strangers, individuals, status, receptivity under circumstances in which social odours occur? Although many patterns of scent marking could imply a function in connection with spatial (generally territorial) organization, here the sole criterion is a difference in the pattern of marking between any of (a) the interior, (b) the perimeter, and (c) beyond the border of territories. As implied proof of discriminating receptivity or hormonal state I have required evidence of a differential response (normally duration of investigation) to two odours, or of variation in their chemical composition; I have disallowed, as sole evidence, a change in marking frequency associated with mating season. In the broad category of familiar *versus* alien discriminations I have included evidence of distinctions between the odours of self *vs* other and between familiar *vs* unfamiliar. Each entry on the table has three components: (i) proof (+) or circumstantial (?) support of an effect, or its disproof (−); (ii) the social odour: A = anal glands; C = cheek glands; F = faeces; S = vaginal secretions; V = ventral gland; U = urine; (iii) citation number. Conclusive demonstrations of the functions of carnivore social odours are sadly few. However, it is auspicious that all but two of the 34 papers cited on this table have appeared since Eisenberg and Kleiman's major review in 1972.

1. Doty and Dunbar (1974)
2. Dunbar (1977)
3. Preti et al. (1976)
4. Beach and Gilmore (1949)
5. Goodwin et al. (1979)
6. Uexkull and Sarris (1931)
7. Lehner (1978)
8. Harrington (1979)
9. Barrette and Messier (1980)
10. Peters and Mech (1975)
11. Jordan et al. (1967)
12. Brown and Johnston (1982)
13. Harrington (1981)
14. Macdonald (1979b)
15. Moehlman (pers. comm.)
16. Macdonald (1979a)
17. Blizzard and Perry (1979)
18. Henry (1977)
19. Kruuk (1972)
20. Mills et al. (1980)
21. Gorman (1976)
22. Rasa (1973)
23. Bailey et al. (1979)
24. Henry (1979)
25. Jorgenson et al. (1978)
26. Verberne and de Boer (1976)
27. Östborn (1976)
28. Kruuk (1978)
29. Gorman et al. (1978)
30. Erlinge et al. (1982)
31. Brinck et al. (1978)
32. Tschanz et al. (1970)
33. Yamamoto and Hidaka (in press)
34. Bekoff and Wells (1980)
35. Hart (1974)

or collectively on the same odorous object (Rieger 1979), could also result in increased similarities in the odours of members of a group. A dominant individual marking members of its group may be behaving in a way directly analogous to one marking its territory, namely placing its symbol on a resource. To the extent that a given odour does identify the individuals of a group one might predict that an outsider trying to ingratiate itself or to avoid detection should anoint itself with that odour. Similarly, it may be advantageous for subordinate members of a group to solicit allomarking from a dominant to emphasize their membership.

Clues to an animal's sex can be found in its odour. Firstly, many species of carnivore show sexual dimorphism in the extent to which glands are developed, or the frequency with which odours are deposited. It is possible that the volume and pattern of marking alone can indicate sex. Posture may also contain this information: female bush dogs urinate from a handstand posture whereas males do not, hence males' marks are lower than females' (Kleiman 1974). Chemists have searched for male or female factors in social odours: Whitten *et al.* (1980) believed (wrongly as it happened) that 2-methylquinoline was only found in the urine of male red foxes. Nevertheless, irrespective of reproductive condition, behavioural tests on cats (Verberne and de Boer 1976) and dogs (e.g. Dunbar 1977) show that the sexes can be distinguished on the basis of odours from each of several sources. However, most results concerning discrimination between the sexes involve the detection of receptivity as opposed to gender *per se* (see below). If an ability to recognize individuals on the basis of odour is eventually demonstrated as being widespread amongst carnivores, then presumably that will carry with it the conclusion that many other individual traits are simultaneously identifiable, as the characteristics associated with each individual.

*Status*        Information on dominance status, in common with that on sex, is contained within the frequency and pattern of marking in many species (Table 15.6). For example, stoats mark more with both anal and ventral glands if they are socially dominant (although the differential is greater for ventral glands) (Erlinge *et al.* 1982). Similarly, socially dominant female red foxes mark with urine more frequently than do their subordinates (Macdonald 1979*a*), and the same probably applies to most canids. Among wolves, it has been shown not only that dominant pack members mark more than subordinate ones, but also that non-territory holders mark less than resident wolves and apparently strive to hide their odours (Peters and Mech 1975; Rothman and Mech 1979). Although there is information on differences in the frequency and pattern of marking of dominant animals, and although known dominant individuals may be recognizable, there are no demonstrations of odorous information on dominance *per se*. That is to say, dominant members of a society have not been shown, for example, to produce more or less of a particular chemical. However, Rasa (1973) has shown that dwarf mongooses can make the discrimination necessary to interpret the information on status contained within the pattern of cheek-gland marking. Dominant dwarf mongooses rub more vigorously on marking sites with their cheek glands and hence presumably deposit more secretion. Rasa showed that other mongooses could distinguish between marks with varying amounts of cheek secretions. Indeed, Rasa's demonstration that the cheek gland secretions contain (in their pattern of deposition) information on status, whereas the anal glands of the same species contain information on identity, illustrates how separate odours can be specialized for distinct purposes. Among dwarf mongooses it is apparently necessary to transmit information on both identity and status. Within a group, knowledge of identity might seem to make signalling of status redundant, assuming as is likely, that the animals can associate the two. However, a stranger can discern an individual's status without knowing its identity. In this case separate information is communicated by different glands. In other cases, doubtless, many different types of information can be gleaned from the odours of one source.

Scent-marking is often associated with visual displays in aggressive contexts. Bekoff (1979*b*) emphasizes this for dogs cocking their legs, and dominant individuals of most canids have been seen to urinate ostentatiously in view of a vanquished rival. There is no evidence that when dissociated from the visual signal, the odours of these marks contain additional information on status. However, a dominant individual marking in the presence of its subordinate may be reinforcing the links between its odour, identity, and status: when the subordinate encounters the same odour in the absence of the other animal it may respond on the basis of its memory of the marker's status.

*Reproduction*   To the extent that the most often repeated observation throughout this review has been an increase in marking frequency during courtship, information on reproductive condition is widespread in both the pattern and composition of scent marks. The experiments of Doty and Dunbar (1974) and Dunbar (1977) reveal that male domestic dogs sniff for consistently different lengths of time at odours from different sources (e.g. urine *vs* anal sac *vs* vaginal secretion) and from different sexes and from bitches in different hormonal states. In that different sniffing durations indicate discrimination, dogs can clearly recognize oestrous bitches on the basis of smell. Of course, equal sniffing durations would not necessarily mean the dogs could not distinguish between samples. In fact, Goodwin *et al.* (1979) believe they have isolated a single factor in the bitch's vaginal secretion that signals oestrus. Similarly, Verberne and de Boer (1976) show that male domestic cats can differentiate both the urine and the cheek rubbings of females in oestrus from those that are not.

Although there are no clear-cut demonstrations of the role of given odours in reproductive behaviour, the implication is overwhelming in that heightened frequency of marking is so universally associated with courtship amongst the carnivores (Table 15.6). Gorman *et al.* (1978) and Gorman (1980), for example,

Table 15.6. Social context of scent-marking among carnivores

| Species | Marking frequency = status dependent | Victor marks after aggressive interaction | Marking during border clashes | Marking as a group activity | Allo-marking | Over-marking | Marking frequency high in mating season | Odour of conspecific stranger stimulates marking |
|---|---|---|---|---|---|---|---|---|
| **Canidae** | | | | | | | | |
| *Canis familiaris* | U7,8,9,11,38 | U1,2,3,48 | U3 | | | U4,5 | U1,5,6 | U4,3 |
| *Canis lupus* | U15,17 | U7,9,10,11,38 | U9, 8 | U38 | | U12,38 | U12,38 | U9 |
| *Canis latrans* | U22 | U13,15,16 | U13 | U13 | | U14,16,17 | U17 | U13,17 |
| *Canis aureus* | U22 | U19,21 | FU21 | | U18,19 | U18,20 | | |
| *Canis mesomelas* | U23 | | | | | | | |
| *Cuon alpinus* | U23 | | | FU23 | | U23 | | |
| *Lycaon pictus* | U30–33 | U30–33 | U33 | U30–33 | | U30–33 | U30–33 | U33 |
| *Speothos venaticus* | | U11 | | | | | U11 | |
| *Otocyon megalotis* | | | | | U24 | U24 | | |
| *Alopex lagopus* | U25 | | | | | | | |
| *Vulpes vulpes* | U26 | U26,27 | GU26 | | U26 | U26 U28 | U26 | U26 |
| *Cerdocyon thous* | | | | | | | | U26 |
| **Hyaenidae** | | | | | | | | |
| *Crocuta crocuta* | | | FG29,53 | FG29 | | | | |
| *Hyaena brunnea* | | | | | | | | G34 |
| **Viverridae** | | | | | | | | |
| *Nandinia binotata* | G59 | | | | | | | |
| *Arctictis binturong* | | | | | | | GU51 G35 | |
| *Herpestes auropunctatus* | | | | | | | | |
| *Helogale parvula* | G36 | | G36 | G36 | G36 G50 | G36 | | G36 |
| *Cynictis penicillata* | G50 | | | | | | | G50 |
| *Suricata suricatta* | G37 | | | FUG37 | G37 G50 | | | |
| *Mungos mungo* | | | | | | | | |
| *Mungotictis decemlineata* | | G55 | | | | | G55 | |
| **Felidae** | | | | | | | | |
| *Panthera leo* | U33 | | U39 U33 | | | U39 | U39,40 | |
| *Aonyx jubatus* | GU41,42, F49 | | | | | U33 | U33 | |
| *Felis catus* | | | | | G42,43 | GU42 | | U33 |

## Mustelidae

| | 1 | 2 | 3 | 4 | 5 | 6 | 7 |
|---|---|---|---|---|---|---|---|
| *Mustela erminea* | G44 | | G44 | | G44 | | G44 |
| *Lutra lutra* | | | | G45 | G45 | | G44 |
| *Pteranura brasiliensis* | | | FU54 | | | | |
| *Meles meles* | | | | G46,47 | | | |

## Procyonidae

| | 1 | 2 | 3 | 4 | 5 | 6 | 7 |
|---|---|---|---|---|---|---|---|
| *Ailurus fulgens* | | | | | G56 | | |
| *Ailurus melanoleea* | | | | | G57 | | |
| *Procyon lotor* | G58 | | | | | | |

The use of social odours among carnivores is associated with various social circumstances, some of which are summarized on this table. High status individuals tend to mark at a higher rate than do their subordinates (column 1); within social groups the victor of an agonistic interaction often marks whereas the loser does not (column 2); during territorial conflicts between neighbours either or both parties may engage in intensive marking (column 3); some scent marking activities result in the excited participation of some or all group members simultaneously (column 4); within a social group, members may mark on each other's bodies (column 5); members of a mated pair, especially during the mating season, may mark in tandem on the same spot (i.e. overmarking) (column 5); the peak frequencies of scent marking by several species of carnivore coincide with the period of female receptivity (column 6); finally (column 7), the scent of an intruding stranger stimulates marking on or near the site of the intruder's mark. In addition, careful sniffing is a universal feature of carnivore greeting and many other social interactions. (F = faeces; U = urine; G = glandular secretions.)

1. Scott and Fuller (1975)
2. Beck (1973)
3. Ghosh (1981)
4. Uexküll and Sarris (1931)
5. Beach and Gilmore (1949)
6. Dunbar (1977)
7. Schenkel (1967)
8. Woolpy (1968)
9. Peters and Mech (1975)
10. Mech (1970)
11. Kleiman (1966)
12. Rothman and Mech (1979)
13. Camenzind (1978)
14. Barrette and Messier (1980)
15. Lehner (1978)
16. Bowen and McTaggart Cowen (1980)
17. Wells and Bekoff (1981)
18. Golani and Keller (1975)
19. Golani and Mendelssohn (1971)
20. van Lawick and van Lawick Goodall (1970)
21. Macdonald (1979b)
22. Moehlman (1983 and pers. comm.)
23. Jonsingh (1980, 1982)
24. Lamprecht (1979)
25. Hersteinsson and Macdonald (1982)
26. Macdonald (1977b, 1980)
27. Tembrock (1957)
28. Montgomery and Lubin (1978)
29. Kruuk (1972)
30. van Lawick and van Lawick Goodall (1970)
31. Frame and Frame (1976)
32. Malcolm (1979)
33. Frame and Frame (1981)
34. Mills et al. (1980)
35. Gorman (1976, 1980)
36. Rasa (1973)
37. Personal observation
38. Zimen (1981, 1982)
39. Schaller (1972)
40. Bertram (1978)
41. Leyhausen (1979)
42. Verberne and de Boer (1976)
43. Macdonald et al. (In preparation.)
44. Erlinge et al. (1982)
45. Gorman (1980)
46. Neal (1977)
47. Kruuk (1978)
48. Doty and Dunbar (1974)
49. Corbett (1979)
50. Kingdon (pers. comm.)
51. Wemmer and Murtaugh (1981)
52. Bearder and Randall (1978)
53. Duplaix (1980)
54. Albignac (1976)
55. Roberts (1981)
56. Kleiman (1983)
57. Ough (1982)
58. Charles-Dominique (1978)

showed how synchronized bursts of anal-sac marking by members of a pair of otters coincided with periods of mating. Similarly, the supracaudal gland of male red foxes is more active in the mating season and, amongst the Canidae, intensive token urinating and over-marking are probably universal within courting pairs. Heightened marking frequency by females around oestrus is common among carnivoran families using various odour sources e.g. anal gland marking of female small Indian mongooses as the breeding season approaches (Gorman 1980, Fig. 15). However, these coincidences only provide clues to the general topic of olfactory conversation, rather than demonstrating particular functions. Indeed, there is no test amongst the Carnivora of Darwin's (1887) idea that the quality or quantity of a male's odour influences his mating success.

*Spatial information*    Traditionally the spatial pattern of scent marks has been interpreted in terms of territoriality; the scent of a resident animal was thought to repel intruders (Hediger (1944, 1949, 1955)). This idea has proved hard to test. The common inference that simply because marking occurs within territories it functions in their defence is absurd—animals may mark in their territory for no better reason than that they spend all their time within its borders. Similarly one animal may avoid another's territory for reasons having nothing to do with the scent marks therein. However, there is evidence that resident animals on temporary excursions from their territories may stop urine marking altogether, e.g. coyotes, red foxes, and silver-backed jackals (Bowen and McTaggart Cowan 1980; Macdonald 1979a; Moehlman, personal communication). In these cases urine marking is genuinely confined to the territory, but still not necessarily functioning in its maintenance. There is mounting evidence that within the territories of several species of carnivores marking behaviour differs qualitatively and/or quantitatively between places where intruders are infrequent and those where they pose a greater risk. This distinction is often, but not invariably, between marking at a territory perimeter and at its interior: spotted hyaenas, golden jackals, and Eurasian badgers are amongst the species which may ring their territories with latrine sites which are rarely found away from the perimeter (Kruuk 1972, 1978; Macdonald 1979b). These latrines may signal the location of a border to an intruder, and their greater density and strategic locations may increase their chances of being encountered.

Mills *et al.* (1980) demonstrated that brown hyaenas paste with their anal pouches at a higher rate per kilometre travelled when near the borders of their territories. However, these hyaenas spent less time foraging at the periphery of their territories, so that in total fewer marks were left at territorial borders. Consequently, although the resident pasted a higher rate towards the border, an intruder would encounter marks at a lower rate in those areas. In order to assess whether the density of odour is greater at the perimeter or the interior of a range one needs to know the rate of marking, the longevity of each mark, and the duration and frequency of visits to each area. Thus Peters and Mech (1975) show that wolves mark more often per kilometre travelled in the vicinity of territorial borders but this would not result in a greater cumulative density of marks unless the wolves also travelled there at least as frequently as elsewhere. Even within the border zone, some carnivores behave so as to increase the frequency with which their marks may be encountered in those areas where the risk of intrusion is greatest. Barrette and Messier (1980) found that coyotes marked at the highest rate where trespassers were most common, and Kruuk (1978) described how badger latrines were less clearly focused along sections of border which amount to no-man's-land, where foraging is poor and hence incursions presumably infrequent. Together, these observations indicate that under territorial circumstances some species of carnivore behave so as to increase the likelihood of a mark being encountered in zones where the risk of intrusion may be highest. These zones need not be on the perimeter—the greatest accumulation of marks may be in the places visited most frequently (e.g. red fox, Macdonald 1979a) and heavily visited places may be rich in the resources most likely to attract trespassers.

For scent marking to be involved in territorial maintenance does not necessarily require the rate or the density of marks to be greatest in perimeter zones. The dispersion of territorial marks would rather be patterned to increase the chances of intercepting an intruder, which suggests that this pattern will in turn reflect that of resources within the territory. The spatial pattern of marking does vary widely between the species of carnivore. For example, the proportion of faeces found in each of four annuli of equal width dividing the home ranges of four medium sized species varied greatly (Macdonald 1980, Fig. 5). For the four populations in question the trend was from border to central middens in the order golden jackal, badger, fox, and otter. However, when the data were corrected to the density of faeces in each annulus the patterns changed somewhat: for example, although a greater proportion of red fox scats were found in the outer annuli, a greater density were found in the inner annuli.

Although it does not weaken the circumstantial link between patterns of scent marking and territoriality, none of the foregoing constitutes evidence of scent deterring intruders, far less stopping them in their tracks. In field experiments female red foxes demonstrated a repellant effect of urine marks, by turning back into their 'territory' when an alien urine mark was encountered at the perimeter, but sometimes continuing on an excursion when no strange mark was encountered (Macdonald 1979a). In contrast, vixens encountering alien urine from donors of either sex within their 'territory' invariably over-marked it. The spatial pattern of vixen token urinations changed in the autumn, with previous borders to the marking range breaking down. There is some evidence from wild foxes that the late autumn, when the young disperse, is a period of territorial reshuffling. The observations of Peters and Mech (1975) and Jordan, Shelton, and Allen (1967) indicate that wolf urine marks may also have an aversive effect on intruders, and may intimidate them. There is no species of carnivore for which there is evidence that scent marks constitute an impenetrable barrier to trespassers.

There are two longstanding explanations for the distribution of scent marks. Hediger (1944, 1949, 1955), citing von Uexküll and Kriszat (1934), proposed that scent marks serve an explicitly territorial function in that they discourage intruders by representing the owner in its absence, whereas Lyall-Watson (1964) suggested that they increased the marker's familiarity with its own range, helping it to orientate. Lyall-Watson's explanation implies territorial function only in so far as increased familiarity results in greater confidence and Mykytowycz, Hesterman, Gamball, and Dudzinski (1976) formalized this idea by demonstrating that a male rabbit's dominance was affected by whether or not he was in an atmosphere tainted with his own scent (see Eibl-Eblesfeldt's (1970, p. 311) reference to badgers and Kleiman 1966). Both these hypotheses, and their derivatives, are very difficult to refute, nor are they mutually exclusive. Amongst carnivores there is only scant evidence of an aversive effect of scent marks (viz foxes and wolves above. However, the deterrent principle but not the detail of Hediger's (1955) *pars pro toto* idea could be upheld if, as argued in Macdonald (1977), smelling a scent mark eroded an intruder's confidence and made it more likely to flee if intercepted. Any aversive effect of a scent mark may be expressed differently depending on whether the receiver is a transient or a neighbour and whether or not it can afford to avoid the area. Johnson's (1973) conclusion that scent marks may be more an attractant than a deterrent does not in itself contradict an aversive function, since the message, or likely sites, must be sought out in order for it to be read. Some age–sex classes among most species of carnivore are itinerant, and in areas where territories are contiguous these individuals have little choice but to be perpetual trespassers and presumably to flee when challenged by a territory holder irrespective of whether they have recently encountered his marks. However, they might plot their course through each territory by avoiding areas of dense marks (cf Rothman and Mech 1979). Even this option may be partly closed if, for example, scent marks are concentrated in the only good foraging areas (e.g. brown hyaenas, Mills *et al.* 1980). In this case, even if territories were not contiguous, a transient might travel in no-man's-land but have no choice but to trespass when foraging. It is tempting to conclude that an increase in marking (either density or rate) along sections of territorial border more likely to be contested by neighbours indicates that the signals are intended for would-be intruders. However, on the Lyall-Watson hypothesis these could be areas where the individual had to be especially careful to orient correctly. Indeed, at a jointly marked border between territorial neighbours there seems to be no satisfactory logic for disentangling the predictions of these two hypotheses. In general the suggestion that scent marks help the occupier to orientate is made less compelling by the impression given in the field by most carnivores of knowing exactly where their marking sites are.

If they rarely meet, each resident may communicate its continued occupancy to neighbours by scent marking at their shared border. To the extent that failure to replenish these marks prompts an intrusion they serve an explicitly territorial function. There is evidence that territorial coyotes, foxes, and tigers (Bowen 1978; Macdonald 1977; Sargeant 1972; Sunquist 1979) quickly respond to changes in the ranging behaviour of neighbours and expand into vacated areas. So, scent marking may have a territorial function in confirming the *status quo* between neighbours. There seem to be limited circumstances under which it is likely to repel a transient intruder, although it might affect intruders' movements (even to the disadvantage of the resident). It seems probable that territorial advertisement is only one of several functions of scent marking and where itinerant non-residents are concerned it is likely to operate more commonly by influencing the outcome of an encounter perhaps by diminishing the confidence of the receiver than by turning him in his tracks. Consequently communication directed at intruders probably underlies only part of the spatial pattern of marking sites.

Gosling (1982) has recently reviewed the observations which gave rise to eight different hypotheses which seem to explain why mammals scent mark within their territories. He suggests that all the evidence actually conforms to the predictions of one hypothesis concerned with competitor assessment; 'territory residents provide marks so that intruders can assess their status in any subsequent encounter'. Thus when an interloper meets another animal whose odour 'matches' with that of the local scent marks the intruder knows that it has met a territory holder and retreats. The territory holder thus benefits from marking in his territory by diminishing the risk of an encounter with an outsider escalating dangerously. If the territory holder did not declare his tenancy and by marking inform all would-be rivals that the territory was occupied, then every time two strangers met they would have to contend ownership directly, both risking serious injury. Put briefly, all but the strongest intruders know that it is hopeless to challenge an incumbent. In general, unless there is an advantage to meetings with non-territory-holders, an intruder in a scent marked area might do best to avoid any encounter, without delaying to ascertain the identity of individuals it meets. In contrast, a newcomer to an unmarked area should be conspicuous in the attempt to establish itself as the resident. Gosling's meticulously argued case has the advantage of providing a functional explanation for scent marking directly linked to territorial defence and yet not requiring scent marks themselves to deter intruders. Furthermore, there is no reason to suppose that some, if not most, social odours deployed in territories do not serve other functions in addition to facilitating competitive assessment.

There is mounting evidence for inter- and intraspecific variation in the spatial patterns of scent-marking by carnivores (Macdonald 1980). At least part of this variation may not relate so much to the concent of the message being sent, but to the requirements of ensuring that it is received. For example, marks are often found aloft conspicuous objects, near trail junctions or habitat borders. These sites may be chosen as those where other individuals may pause to read the message, and the availability of such sites may thus influence the pattern of marking. For example, Duplaix (1980) describes a different pattern of marking by giant river otters in two riverine habitats, and these differences may stem from either or both of requirements for different messages in different habitats and the varying problems of sending the same message in the two areas.

*Temporal information*    Dwarf mongooses are able to discriminate between odours (anal pouch secretions) of different ages (Rasa 1973), and if this ability is widespread then there is a possibility of conveying temporal information. It seems highly likely that this is a common ability, perhaps not only on the basis of changing concentrations of volatiles, but also chemical changes in the aging mark. Wolves, for example, respond differently to urine marks of different ages (Peters and Mech 1975), as must all other species that periodically 'top-up' their marks if they are to avoid a vortex of over-marking. The transmission of temporal information was central to Leyhausen's interpretation of territorial marking by cats (Leyhausen and Wolff 1959; de Boer 1977) and for which Eaton (1970) found some support from observations on cheetah. Eaton found that cheetah were unlikely to follow trails recently marked by other individuals, whereas they did travel those marked several days previously.

*Foraging*    Henry (1977) first demonstrated that the red fox's habit of urinating on excavated caches reduced by about one-fifth subsequent investigation time at such sites, and could thus be interpreted as functioning to minimize wasted foraging effort. The same effect may be widespread amongst Canidae since Harrington (1981, 1982*a*) has demonstrated it in wolves and coyotes (evidence cited in Harrington 1979). A different relationship between urine-marking and caching is often assumed and occasionally written, namely that a urine mark helps the animal to relocate its cache. This idea was proposed for red foxes (Korytin and Solomin 1969) and can be abandoned since Macdonald (1976) showed that red foxes' caches are never marked before their rediscovery and Harrington (1981, 1982*a*) has observed the same for wolves and coyotes. In any case, it is contradictory to suppose that individuals would simultaneously hide something and signal its whereabouts.

Book-keeping can only be one of several functions of token urination by red foxes, as evidenced by the fact that there are circumstances where only a minority of marks are on cache sites, or indeed on any site tainted with the odour of prey (Macdonald 1979*a*, 1980; Henry 1979), and the same applies to wolves (Harrington 1981). Furthermore, there are unresolved puzzles, for example, in general, neither subordinate red foxes nor wolves token mark at all, irrespective of cache sites and so they can at best only benefit from the book-keeping of others. Furthermore, where there is actually some food in a token-marked cache the mark does not prevent its discovery (e.g. Harrington 1981, p. 283). Indeed,

although marked cache sites are investigated for less time, they are still very sub-
stantially investigated (12.7 s in the case of wolves; Harrington 1981).

As Macdonald (1980) and Harrington (1981) argue, this is probably because
such sites also function as 'normal' marking sites since emptied caches constitute
odoriferously conspicuous sites. The book-keeping hypothesis has great appeal
since it is unlikely that a canid would investigate a place smelling of food without taking that food, if at all possible, and this fact is doubtless as evident to
other foxes as it is to biologists.

Red foxes and pariah dogs (Macdonald 1979*a*; Ghosh 1981) token urinate on
large food items. Ghosh noted that pariah dogs will urinate on a carcass when
they have eaten their fill. He suspects that the token mark signifies ownership.
However, there is no experimental proof of this; on the contrary, coyotes will
steal and eat food urine marked by another individual in this way (Harrington
1982*a*) as will red foxes (personal observation).

*Inter-specific communication*   It is common to find the marks of several
species of carnivores at a given site. Although these species may be competitors,
the coincidence of their marking sites may say more about the medium than the
message: certain sites are good places from which to disseminate olfactory
signals, irrespective of the species or the signal content. An inescapable cost of
scent-marking and any other form of communication, is that the wrong person
may get the message. Hence predators may detect the odour of their prey, and
*vice versa*. Stoddart (1979) has shown how voles avoid traps tainted with the
odour of weasel and Mech (1977) and Rogers *et al.* (1979) showed that deer
yards often lie in the hinterland between wolf territories. Indeed, if the little
studied foot glands of canids are as important as casual observation suggests,
then they must provide invaluable subversive information to the prey community.

The habit of several species, including aardwolves and most small felids, of
burying their faeces may be a means of minimizing the risk of detection of their
frequent haunts. In contrast, one interspecific function of scent-marking that
does not require experimental validation is the repellent affect of the skunk's
anal-sac secretion.

*Variation*

Even the present fragmentary picture of olfactory communication among carnivores suffices to emphasize the variation within and between species in the sources
of social odours and their deployment. Phylogenetic accident doubtless underlies
the presence of an anal antechamber among mongooses and its absence amongst
dogs. Indeed, the absence of a perineal gland among mongooses may emphasize
their distinction from the civets and genets and in so doing may erode the validity
of the family Viverridae. However, aside from other generalizations, the detail of
interspecific variation remains mystifying: why, for example, amongst the
viverrid subfamily Hemigalinae, should the fanaloka not have a perineal gland

Table 15.7. Relationships between social organizaton, spatial patterns of scent-marking and the dispersion of resources

| Species | Social system | Marking pattern | Resource pattern |
|---|---|---|---|
| **Canidae** | | | |
| Canis familiaris | Loose-knit; territorial [1] | UF thoughout range; U at higher rate in areas visited less often. UF accumulate at feeding sites [1] | Offal available at some predictable sites but otherwise unpredictably scattered throughout range [1] |
| Canis lupus | Groups (approx. 6); territorial (approx. 200 km²) [2] | UF throughout range on trails; U at higher rate at borders [3]; F accumulate at feeding sites and rendezvous sites [24] | Deer and smaller prey available unpredictably through range [2] |
| Canis latrans | Groups (2–5); territories (7–20 km²) [4–7] | UF throughout range, but accumulate at carcass and rest sites. U at higher rate where neighbours intrude [5,36]; occasional small latrines [22,23] throughout; large latrine (500 faeces) at possible den [22] | Prey unpredictably avaible throughout range, but deer or elk carcasses last several days [4–7] |
| Canis aureus | (i) Loose-knit groups (10–20); territorial (10 ha) [8] | F throughout range, but latrines (3–105 faeces) mainly at border; latrines bigger and more numerous where clash with neighbours; F accumulate at feeding site [8] | Few, highly predictable feeding sites [8] |
| | (ii) Spatial group (2–5); territorial (200 ha) [9] | UF throughout range; no latrines [9] | Unpredictable prey scattered widely throughout territory [9] |
| Cuon alpinus | Cohesive group (8); territorial (40 km²) [30] | FU throughout range but F in latrines along trails, especially junctions [30] | Ungulates, widely dispersed in forest and tall grass [30] |
| Speothos venaticus | Cohesive groups [31] | F in latrines | ? |
| Nyctereutes procyonoides | Solitary foraging in overlapping ranges of uncertain affinity ? spatial groups [32,33] | F in latrines within range, near trails and temporary feeding places [32,33] | Omnivore, probably patchy food including fruit [32,33] |
| Chrysocyon brachyurus | Pairs, territorial (27 km²) [34] | FU throughout range, mainly on trails. Small latrine near rest sites [34] | Small vertebrate prey and fruit, widely dispersed [34] |

| | | | |
|---|---|---|---|
| *Vulpes cinereoargenteus* | Solitary foraging, pairs or ? spatial groups [35,36] | FU generally throughout range but F in latrine (220 faeces) below fruit tree [35] | Omnivore, probably on patchy food [35,36] |
| *Vulpes vulpes* | Spatial groups (2–6); territorial (45 ha) [10] | UF throughout range and accumulate in most frequently used areas [11]; F accumulate on food | Prey spatio-temporally patchy in availability throughout territory [10] |
| **Hyaenidae** | | | |
| *Crocuta crocuta* | | | |
| (i) | Loose-knit stable clans (30–80) hunting as unstable co-operative units (2–20); territorial (10–50 km²) [12] | FG around perimeter; latrines of conspicuous white faeces [12] | Large ungulates at high density evenly through territory [12] |
| (ii) | Unstable associations (1–20); non-territorial (>1000 km²) [12] | F in latrines at trail junctions [12] | Migratory prey at low density [12] |
| (iii) | Stable groups (2–10); territorial [13] | Some F in latrines, most latrines at trail junctions (144 faeces), others at dens or drinking sites; smaller latrines at temporary feeding sites, larger (295 faeces) at permanent feeding sites [13] | Large ungulates [14] at moderate density throughout range [13] |
| *Hyaena hyaena* | | | |
| (i) | Solitary pairs; ? territorial (44–77 km²) [15] | F scattered at random [15] | Small, widely dispersed prey [15] |
| (ii) | Small family group (4) [16] | F single, except around feeding site where latrines (2–13 faeces) and G accumulate [16] | Single highly predictable feeding site amidst otherwise widely scattered unpredictable prey |
| *Hyaena brunnea* | Spatial groups (2–6); territorial (250–540 km²) [17] | F in small latrines throughout range 9[18]; G throughout range, higher rate at perimeter, higher density in interior [19] | Low availability carrion and fruit widely dispersed through range [17,18] |
| *Proteles cristatus* | Pairs; territorial (150 ha) [20] | F in latrines within range. G throughout range, higher rate at border [20] | Termites, highly spatio-temporally patchy [20,21] |
| **Viverridae** | | | |
| *Viverra civetta* | Spatial groups; assumed territorial [25,26] | All F in latrines along trails; G throughout at civetries | Omnivorous |
| *Nandinia binotata* | Spatial groups (♀ and subad. daughters); territorial (45 ha) [27] | FG especially on and near fruiting trees [27] | Omnivore-frugivore [27] |

Table 15.7 (cont.)

| Species | Social system | Marking pattern | Resource pattern |
|---|---|---|---|
| *Viverra tangalunga* | Solitary foraging, spatial system unknown (range 100 ha) [28] | F in latrines often filling crevices within range; G in foraging areas [28] | Omnivore [28] |
| *Paradoxurus hermaphroditus* | Solitary foraging, spatial system unknown [29] | F on branch in latrines within range [29,38] | Omnivore [29] |
| *Helogale parvula* | Cohesive groups (8) [37] | FG near dens, F in latrines near dens, within range [37] | Small prey and insects widely spread in range [37] |
| *Crossarchus alexandri* | | F in latrines ? within range [39] | |
| *Suricata suricatta* | Cohesive groups (3–20); ? territorial [40,24] | G throughout range; F in latrines at den sites [40,24] | Insects and small vertebrates [42], widely dispersed and sometimes patchy availability; dens throughout range vital for escape from predators [24] |
| *Cynogale penicillata* | Spatial groups (8); territorial (5–6 ha) [41] | G (face) at dens; G (back) at border; F in latrines on trails at borders | Insects and small vertebrates [42] |
| **Felidae** | | | |
| *Lynx rufus* | Solitary, ♂ territories overlap several ♀♀ [43] | U throughout on trails, F in latrines within territory | Small mammalian prey, widely dispersed |
| *Felis sylvestris* | Solitary, ♂ territories overlap several ♀♀ [44] | FU throughout range; F greater density in foraging areas [44] | Small mammalian prey, widely dispersed but patchy [44] |
| **Ursidae** | | | |
| *Ursus arctos* | Solitary forager; ♂ territories overlap several ♀♀ | FU throughout range [55]. Some F in latrines [54] | Omnivore |
| **Mustelidae** | | | |
| *Mustela nivalis* | Solitary forager; territorial, ♂ overlaps several ♀♀ [50] | F in latrines within territory [45] | Small mammals |
| *Mustela frenata* | Solitary forager | F in latrines within territory [45] | Small mammals |
| *Poecilogale albinucha* | Solitary [51] | F in latrines [51] | |
| *Pteroneura brasiliensis* (i) | Loose-knit groups (2–6); ? territorial [46] | FGU in communal latrines along rivers within territories [46] visited in rotation [56] | Forest streams and rivers [46] |
| (ii) | Loose-knit groups [46] | F on stones in lakes within sight of each other | Open lakes [46] |

| | | | |
|---|---|---|---|
| *Aonyx capensis* | ? Small spatial groups; ? territorial [47] | F in communal latrines [46,47] | Rivers and coastline [47] |
| *Lutra lutra* | (i) Pairs; territorial [49] | F in latrines close to den [49] | Coastal waters; forages alone off-shire, den site on shore [49] |
| | (ii) Pairs; territorial [50] | F accumulates at junctions between rivers and lakes [50] | Inland lakes and rivers [50] |
| *Spilogale interrupta* | ? | Some F in latrines near den [48] | Omnivore |
| *Meles meles* | Spatial groups (4–12); territorial (22–200 ha) [52] | G throughout range; F in latrines near den and around border, especially where risk of intrusion [52] | Omnivore, specializing on earth-worms, a spatio-temporally patchy prey [53] |

The spatial organization of carnivore communities is affected by the distribution of key resources, be they some or all of prey, den sites, drinking sites or mates (e.g. Kruuk 1975; Macdonald 1983). Either because potential receivers of a message are likely to travel where resources are concentrated, and/or because the message is related to the resources themselves (e.g. serving some territorial function) one might expect scent marks to be deployed in a way which varied with resource dispersion. This table summarizes information on the social organization of species whose scent marks are known to show some spatial pattern; other species where no pattern is known or which so far appear to mark throughout their home ranges are not included on the table. All species using faecal latrines are listed. Although the data are few, marks do seem to be concentrated in the vicinity of resources or in a pattern which might maximize the probability of intercepting a conspecific. The most compelling examples are those where different populations of the same species deploy scent marks in different patterns. (F = faeces; G = glandular secretions; U = urine.)

1. Ghosh (1981)
2. Mech (1973)
3. Peters and Mech (1975)
4. Bekoff and Wells (1980)
5. Wells and Bekoff (1981)
6. Bowen and McTaggart Cowan (1980)
7. Bowen (1978, 1982)
8. Macdonald (1979b)
9. Moehlman (1983, personal communication)
10. Macdonald (1981)
11. Macdonald (1979a, 1980)
12. Kruuk (1972)
13. Bearder and Randall (1978)
14. Bearder (1977)
15. Kruuk (1976)
16. Macdonald (1978)
17. Mills (1982)
18. Mills (1978)
19. Mills et al. (1980)
20. Kruuk and Sands (1972)
21. Bothma and Nel (1980)
22. Camenzind (1978)
23. Ozoga and Hergar (1966)
24. Personal observation
25. Bearder and Randall (1978)
26. Randall (1977)
27. Charles-Dominique (1978)
28. Macdonald and Wise (1981)
29. Bartels (1964)
30. Jonsingh (1980, 1982)
31. Kleiman (1972)
32. Ikeda (1982)
33. Ikeda et al. (1979)
34. Deitz (1981)
35. Trapp (1978)
36. Trapp and Hallberg (1975)
37. Rasa (1973)
38. Wemmer (1977b)
39. Kingdon (1978)
40. Ewer (1963)
41. Earle (1981)
42. Lynch (1980)
43. Bailey (1974)
44. Corbett (1979)
45. Quick (1951)
46. Duplaix (1980)
47. Arden-Clark (personal communication)
48. Crabb (1948)
49. Kruuk and Hewson (1978)
50. Erlinge (1968)
51. Alexander and Ewer (1959)
52. Kruuk (1978)
53. Kruuk and Parish (1982)
54. Mealey (1980)
55. Tschanz et al. (1970)
56. Laidler and Laidler (1983)

while the banded palm civet does, and why should the lips and ridges of these glands vary in the way they do amongst the genets? Doubtless, time will provide explanation for the adaptive significance of at least some of these differences, and for others, such as why the supracaudal gland of the Arctic fox is bigger than that of the red fox, and those of both species are shorter than that of the grey fox. The findings of Sands *et al.* (1977) on variation in the foot glands of dogs, wolves, and coyotes depending on their geographical range, and hence the necessities of walking on ice, show the diversity of factors that must be taken into account. Brinck *et al* (1983) show that phylogeny explains some of the chemical variation in mustelid secretions.

As with other aspects of their behavioural ecology (Kruuk 1975; Macdonald 1983), there is considerable intraspecific variation in patterns of scent-marking by carnivores (reviewed by Macdonald 1980). For example, in different ecological circumstances Kruuk (1972) found that spotted hyaenas either used middens at territory borders or at trail junctions, and in yet a different habitat, Bearder and Randall (1978) describe further variation. The various studies of feral domestic cats show varying tendencies to bury their faeces or to leave them either aloft conspicuous objects or in middens, and comparison of the work of Kruuk and Hewson (1978), Gorman *et al* (1978) and Erlinge (1968) reveal how in different environments otters deploy their spraints differently. Duplaiz's (1980) findings even suggest that as individual otters move from one habitat to the next they may alter their pattern of marking. Golden jackals in one population in Israel ringed their territory with faecal latrines, sometimes totalling dozens of droppings in a neat pile, yet not far away other populations never used such sites at all. I have speculated elsewhere that border latrines are the prerogative of animals living in large groups, simply because they are the only ones able to generate enough faeces (Macdonald 1980), but by and large data are too scant to substantiate this and other generalizations on social odours amongst the carnivores. Table 15.7 summarizes cases where scent marks are deployed in patterns which appear to vary with the distribution of important resources and perhaps with the economics of their defence.

### Acknowledgements

I am grateful to Drs. M. Beckoff, J. Birks, G. Kerby, H. Kruuk, G. Moran, A. Rasa, and C. Wemmer for checking sections of this chapter and to Geoffrey Carr, Heribert Hofer, Martyn Gorman and Stefan Natynczuk for helpful comments on the entire manuscript.

# References

Albignac, R. (1969). Notes ethologiques sur quelques carnivores Malagaches: le *Galidia elegans. Terre Vie* **23**, 202–15.
— (1970). Notes ethologiques sur quelques carnivores Malagaches: le *Fossa fossa. Terre Vie* **24**, 383–93.
— (1976). Econologie de *Mungotictis decemlineata* dans les forêts decid'ies de l'ouest de Madagascar. *Terre Vie* **30**, 347–76.
Albone, E. S. (1975). Dihydroactinidiolide in the supracaudal gland secretion of the red fox. *Nature, Lond.* **256**, 575.
— (1977). The chemical ecology of mammals—a new focus for chemical research. *Chem. Br.* **13**, 92–9.
— Eglington, G., Walker, J. M., and Ware, G. C. (1974). The anal sac secretion of the red fox (*Vulpes vulpes*); its chemistry and microbiology. A comparison with the anal sac secretion of the lion (*Panthera leo*). *Life Sci.* **14**, 387–400.
— and Flood, P. F. (1976). The supracaudal scent gland of the red fox, *Vulpes vulpes. J. chem. Ecol.* **2**, 167–75.
— and Fox, M. W. (1971). Anal gland secretion of the red fox. *Nature, Lond.* **233**, 569–70.
— Gosden, P. E., Ware, G. C., Macdonald, D. W., and Hough, N. G. (1978). Bacterial action and chemical signalling in the red fox, *Vulpes vulpes*, and other mammals. *Flavor chemistry of animal foods* (ed. R. W. Bullard). *Proc. Am. chem. Soc.* **67**, 78–91.
— and Grönneberg, T. O. (1977). Lipids of the anal sac secretions of the red fox, *Vulpes vulpes*, and of the lion, *Panthera leo. J. Lipid Res.* **18**, 474–9..
— and Perry, G. C. (1976). Anal sac secretion of the red fox, *Vulpes vulpes*: volatile fatty acids and diamines: implications for a fermentation hypothesis of chemical recognition. *J. chem. Ecol.* **2**, 101–11.
— Robins, S. P., and Patel, D. (1976). 5-Aminovaleric acid, a major free amino acid component of the anal sac secretion of the red fox, *Vulpes vulpes. Comp. Biochem. Physiol.* **55B**, 483–6.
Alexander, A. J. and Ewer, R. F. (1959). Observations on the biology and behaviour of the smaller African polecat (*Poecilogale albinucha*). *Afr. wildl. J.* **13**, 313–20.
Altman, D. von (1972). Verhaltensstudien an Mahenwolfen, *Chrysocyon brachyurus. Sool. Gart., Lpz.* N.F. **41**, 278–6.
Anderson, K. K. and Bernstein, D. T. (1975). Some chemical constituents of the scent of the striped skunk (*Mephitis mephitis*). *J. chem. Ecol.* **1**, 493–501.
Anisko, J. J. (1976). Communication by chemical signals in Canidae. In *Mammalian olfaction, reproductive processes and behavior* (ed. R. L. Doty). Academic Press, New York.
Anon. (1982). Quiz re. 'tail-gland hyperplasia'. *On Practice (Vet. Rec. Suppl.)* **4**, 36.
Apps, P. J. (1981). Behavioural ecology of the feral house cat (*Felis catus*, Linnaeus) on Dassen Island. M.Sc. thesis, University of Pretoria.
Ashdown, R. R. (1968). Symposium on canine recto-anal disorders. I. Clinical anatomy. *J. small anim. Pract.* **9**, 315–22.
Bailey, S., Bungan, P. J., and Page, J. M. J. (1979). Variation in the levels of some components of the volatile fraction of urine from captive red foxes (*Vulpes vulpes*) and its relationships to the state of the animal. In *Chemical signals* (ed. D. Müller-Schwarze and R. M. Silverstein) pp. 391–403. Plenum Press, London.

Bailey, T. N. (1974). Social organisation in a bobcat population. *J. Wildl. Mgmt* **38**, 435–446.

Barrette, C. and Messier, F. (1980). Scent marking in free-ranging coyotes, *Canis latrans. Anim. Behav.* **28**, 814–19.

Bartels, E. (1964). On *Paradoxurus hermaphroditus* (Horsfield 1824). *Beufortia* **19**, 193–201.

Bates, R. B. and C. W. Sigel (1963). Terpenoids, cis-trans and trans-cis nepatalactones. *Experientia* **19**, 564–5.

Beach, F. A. and Merari, A. (1970). Coital behaviour in dogs. V. Effects of oestrogen and progesterone on mating and other forms of social behaviour in the bitch. *J.C.C.P. Monogr.* **70**, 1–22.

— and Gilmore, R. N. (1949). Response of male dogs to urine from females in heat. *J. Mammal.* **30**, 391–2.

Bearder, S. K. (1977). Feeding habits of spotted hyaenas in a woodland habitat. *E. Afr. J. Wildl.* **15**, 263–80.

— and Randall, R. M. (1978). The use of fecal marking sites by spotted hyáenas and civets. *Carnivore* **1**, 32–8.

Beck, A. L. (1973). *The ecology of stray dogs: a study of free-ranging urban animals*, York Press, Baltimore.

Bekoff, M. (1979a). Ground scratching by male domestic dogs: a composite signal. *J. Mammal.* **60**, 847–8.

— (1979b). Scent marking by free ranging domestic dogs: olfactory and visual components. *Biol. Behav.* **4**, 123–39.

— (1980). Accuracy of scent mark identification for free-ranging dogs. *J. Mammal.* **61**, 150.

— and Diamond, J. (1976). Precopulatory and copulatory behavior in coyotes. *J. Mammal.* **57**, 372–5.

— and Wells, M. C. (1980). The social ecology of coyotes. *Scient. Am.* (April), 130–8.

Bertram, B. C. R. (1978). *A pride of lions*. Dent, London.

Bigalke, R. (1953). A note on the behaviour of a group of five spotted hyaenas. *Fauna Flora* **3**, 1–4.

Bilz, R. (1940). *Pars pro toto. Ein Beitrag zur Pathologie Meschlicher Affekte und Organfunktionen*. Thieme, Leipzig.

Birks, J. D. S. (1981). Home range and territorial behaviour of the feral mink (*Mustela vison* Schreber) in Devon. Ph.D. thesis, University of Exeter.

Blackman, M. W. (1911). The anal glands of *Mephitis mephitis. Anat. Rec.* **5**, 491–515.

Blizzard, R. A. and Perry, G. C. (1979). Response of captive male red foxes, *Vulpes vulpes*, to some conspecific odours. *J. chem. Ecol.* **5**, 869–80.

Boer, J. N. de (1977). The age of olfactory cues functioning in chemocommunication among male domestic cats. *Behav. Processes* **2**, 209–25.

Bothma, J. du P. and Nel, J. A. J. (1980). Winter food and foraging of the aardwolf, *Proteles cristatus*, in the Namib-Naukluft park. *Madoqua* **12**, 141–9.

Bowen, W. D. (1978). Social organisation of the coyote in relation to prey size. Ph.D. thesis, University of British Columbia.

— and McTaggart Cowan, I. (1980). Scent marking in coyotes. *Can. J. Zool.* **58**, 473–80.

— (1982). Home range and spatial organisation of coyotes in Jasper National Park, Alberta. *J. Wildl. Mgmt* **46**, 201–216.

Brahmachary, R. L. (1979). The scent marking of tigers. *Tiger Paper* 6, 19–20. FAO Regional Office for Asia and Far East.

Brinck, C., Erlinge, S., and Sandell, M. (1983). Anal sac secretion in mustelids: a comparison. *J. chem. Ecol.* 9, 727–45.

— Gerell, R., and Odham, G. (1978). Anal pouch secretion in mink, *Mustela vison. Oikos* 30, 68–75.

Brown, D. S. and Johnston, R. E. (1982). Individual discrimination on the basis of urine in dogs and wolves. Paper presented at the 3rd Int. Symp. on Chemical Signals in Vertebrates, Syracuse, New York.

Budgett, H. M. (1933). *Hunting by scent.* Eyre and Spottiswoode, London.

Bullard, R. W. (1982). Wild canid associations with fermentation products. *Ind. & eng. chem., Prod Res. & Dev.* 21, 646.

Cabrera, A. and Yepes, J. (1940). Mamiferos sur-americanos (vida, costumbres y descripcion). *Historia Natural Ediar.* Compañia Argentina de Editores, Buenos Aires.

Camenzind, F. J. (1978). Behavioral ecology of coyotes on the National Elk Refuge, Jackson Wyoming. In *Coyotes* (ed. M. Bekoff) pp. 267–93. Academic Press, New York.

Chantry, D. F. (1982). Social context of dog–dog communication. Paper to Society for Companion Animal Studies, Paris, 13–14 March.

Chapman, F. B. (1946). An interesting feeding habit of skunks. *J. Mammal.* 27, 379.

Charles-Dominique, P. (1978). Ecology and social behaviour of the African palm civet, *Nandinia binotata*, in Gabon: with a comparison of sympatric prosimians. *Terre Vie* 32, 477–528.

Chattin, L. (1874). *Ann. Sci. Nat.* (5) **XIX**, 106–9.

Corbet, G. B. and Hill, J. E. (1980). *A world list of mammalian species*, Brit. Mus. Nat. Hist. and Comstock Publ. Assocs. Cornell University Press.

Corbett, L. C. (1979). Ph.D. thesis, University of Aberdeen, Scotland.

Crabb, W. D. (1948). Ecology of the spotted skunk. *Ecol. Mong.* 18, 203–32.

Crump, D. G. (1978). 2-Propylthiotane, the major malodorous substance from the anal gland of the stoat. *Tetrah. Lett.* 5233–4.

— (1980a). Thiotanes and dithiolanes from the anal gland of the stoat (*Mustela erminea*). *J. chem. Ecol.* 6, 341–7.

— (1980b). Anal sac secretion of the ferret (*Mustela putorius, forma furo*). *J. chem. Ecol.* 6, 837–44.

Darwin, C. (1887). *The descent of man and selection in relation to sex.* Murray, London.

— (1900). *Journal of researches into the natural history and geology of the countries visited on the voyage of HMS "Beagle" round the world.* Ward Lock, London.

Deitz, J. M. (1981). Ecology and social organisation of the maned wolf (*Chrysocyon brachyurus*). Ph.D. thesis, Michigan State University.

Donovan, C. A. (1967). Some clinical observations on sexual attraction and deterrence in dogs. *Vet. Med.* 62, 1047–51.

— (1969). Canine anal glands and chemical signals (pheromones). *J. Am. vet. med. Ass.* 155, 1995–6.

Dorp, D. A. van, Klok, R., and Nugteren, D. H. (1973). New macrocyclic compounds from the secretions of the civet cat and the musk rat. *Recueil* 92, 915–28.

Doty, R. L. and Dunbar, I. (1974). Attraction of beagles to conspecific urine, vaginal and anal sac secretion odours. *Physiol. Behav.* 12, 825–33.

Dücker, G. (1965). Das Verhalten der Viverriden. *Handb. Zool., Berl.* 8, 1–48.

— (1971). Gefangenschaftesbeobachtungen an Pardelrollern, *Nandinia binotata,* (Reinwardt). *Z. Tierpsychol.* 28, 77–89.

Dunbar, I. F. (1977). Olfactory preferences in dogs: the response of male and female to conspecific odours. *Behav. Biol.* 20, 471–81.

Duplaix, N. (1980). Observations on the ecology and behaviour of the giant river otter, *Pteronura brasiliensis,* in Suriname. *Rev. Ecol. (Terre Vie)* 34, 496–620.

Earlé, R. A. (1981). Aspects of the social and feeding behaviour of the yellow mongoose, *Cynictis penicillata* (G. Cuvier). *Mammalia* 45, 143–52.

Eaton, R. L. (1970). Group interactions, spacing and territoriality in cheetahs. *Z. Tierpsychol.* 27, 481–91.

— (1973). *The world's cats.* World Wildlife Safari, I. Winston, Oregon.

Egoscue, H. J. (1962). Ecology and life history of the kit fox in Toole Country, Utah. *Ecology* 43, 481–97.

Eibl-Eibesfeldt, I. (1970). *Ethology: the biology of behaviour.* Holt, Rinehart, and Winston, N.Y.

Eisenberg, J. F. and Kleiman, D. G. (1972). Olfactory communication in mammals. *Ann. Rev. Ecol. Syst.* 3, 1–32.

— and Lockhart, M. (1972). An ecological reconnaissance of Wilpattu National Park. *Smithsonian Contrib. Zool.* 101, 1–118.

Erlinge, S. (1968). Territoriality of the otter, *Lutra lutra,* L. *Oikos* 19, 81–98.

— Sandell, M., and Brinck, C. (1982). Scent marking and its territorial significance in stoats, *Mustela erminea. Anim. Behav.* 30, 811–18.

Estes, R. D. and Goddard, J. (1967). Prey selection and hunting behaviour of the African wild dogs. *J. Wildl. Manag.* 31, 52–69.

Evans, J. (1839). *J. Asiatic Soc. Bengal* VIII, 408.

Ewer, R. F. (1963). The behaviour of the meerkat, *Suricatta suricata,* Schreber. *Z. Tierpsychol.* 20, 570–607.

— (1973). *The carnivores.* Weidenfield & Nicolson, London.

— and Wemmer, C. (1974). The behaviour in captivity of the African civet, (*Civettictis civetta* Schreber). *Z. Tierpsychol.* 34, 359–94.

Fagre, D. B., Butler, B. A., Howard, W. E., and Teranishi, R. (1981). Behavioural responses of coyotes to selected odors and tastes. In *Worldwide Furbearer Conference Proceedings* (ed. J. A. Chapman and D. Pursley) pp. 966–83. R. R. Donnelley and Sons, Falls Church, Virginia.

Fiedler, E. (1957). Beobachtungen zum Markierungsverhalten einiger Säugetiere. *Z. Säugetierk.* 22, 57–76.

Flower, W. H. (1869). On the anatomy of the proteles, *Proteles cristatus* (Sparrman). *Proc. zool. Soc. Lond.* 4–37.

Fox, M. W. (1971). *Behaviour of wolves, dogs, and related canids.* Cape, London.

Frame, G. W. and Frame, L. H. (1981). *Swift and enduring: cheetahs and wild dogs of the Serengeti.* Dutton, New York.

Frame, L. H. and Frame, G. W. (1976). Female African wild dogs emigrate. *Nature, Lond.* 263, 227–9.

Fritzell, G. K. (1978). Aspects of raccoon (*Procyon lotor*) social organization. *Can. J. Zool.* 56, 261–71.

Gairdner, J. (1915). *J. nat. Hist. Soc. Siam.* 4, 253.

Gerell, R. (1970). Home range and movements of the mink, *Mustela vison*. *Oikos* 21, 160-73.

Ghosh, B. (1981). Some aspects of the behavioural ecology of stray dogs (*Canis familiaris*) in urban and rural environments. Ph.D. thesis. Burdwan University, India.

Gier, H. T. (1975). Ecology and social behavior of the coyote. In *The wild canids* (ed. M. W. Fox) pp. 247-62. Van Nostrand Reinhold, New York.

Goethe, F. (1938). Beobachtungen über das Absetzen von Witterungsmarken beim baummarder. *Deutsche Jäger* 13, 211-13.

Golani, I. and Keller, A. (1975). A longitudinal field study of the behavior of a pair of golden jackals. In *The wild canids* (ed. M. W. Fox) pp. 303-5. Van Nostrand Reinhold, New York.

— and Mendelssohn, H. (1971). Sequences of precopulatory behaviour of the jackal (*Canis aureus* L.). *Behaviour* 38, 169-92.

Goodwin, M., Gooding, K. M., and Regnier, F. (1979). Sex pheromone in the dog. *Science, NY* 203, 559-61.

Gorman, M. L. (1976). A mechanism for individual recognition by odour in *Herpestes auropunctatus* (Carnivora: Viverridae). *Anim. Behav.* 24, 141-5.

— (1980). Sweaty mongooses and other smelly carnivores. *Symp. zool. Soc. Lond.* 45, 87-105.

— Jenkins, D., and Harper, R. J. (1978). The anal sacs of the otter (*Lutra lutra*). *J. Zool., Lond.* 186, 463-74.

— Nedwell, D. B., and Smith, R. M. (1974). An analysis of the contents of the anal scent pockets of *Herpestes auropunctatus* (Carnivora: Viverridae). *J. Zool., Lond.* 172, 389-99.

Gosden, P. and Ware, G. C. (1977). The aerobic flora of the anal sac of the red fox. *J. appl. Bact.* 41, 271-5.

Gosling, L. M. (1981). Demarcation in a gerenuk territory: an economic approach. *Z. Tierpsychol.* 56, 305-22.

— (1982). A reassessment of the function of scent marking in territories. *Z. Tierpsychol.* 60, 89-118.

Grau, (1935), *Tierärtzl. Rundschau.* 41, 351-4. [Cited in Ashdown 1968.]

Greer, M. B. and Calhoun, M. L. (1966). *Am. J. Vet. Res.* 27, 773-81.

Grinnell, J., Dizon, J., and Linsdale, J. M. (1937). *Furbearing mammals of California—their natural history, systematic status, and relations to man*, Vol. 2, pp. 377-777. University of California Press, Berkeley.

Haglund, B. (1966). Winter habits of the lynx (*Lynx lynx* L.) and wolverine (*Gulo gulo* L.) as revealed by tracking in the snow. *Viltrevy* 4 (3), 87-299.

Hall, E. R. (1926). The abdominal skin of *Martes*. *J. Mammal.* 33, 419-28.

Hamilton, P. H. (1976). The movements of leopards in Tsavo National Park, Kenya, as determined by radio-tracking. M.Sc. thesis. Nairobi University, Kenya.

Harrington, F. H. (1979). Urine-marking and its relation to mating in coyotes. Paper presented to Anim. Behav. Soc. Wood's Hole Mass.

— (1981). Urine-marking and caching behaviour in the wolf. *Behaviour* 76, 280-8.

— (1982a). Urine marking at food and caches in captive coyotes. *Can. J. Zool.* 60, 776-82.

— (1982b). Pseudo-urination by coyotes. *J. Mamm.* 63, 501-3.

Hart, B. L. (1974). Environmental and hormonal influences on urine marking behaviour in the adult male dog. *Behav. Biol.* 11, 167-76.

— and Haugen, C. M. (1971). Scent marking and sexual behaviour maintained in anosmic male dogs. *Comm. Behav. Bull.* **6**, 131–5.

Hediger, H. (1944). Die Bedeutung von Miktion und Defakation bei Wildtieren. *Z. Psychol.* **3**, 170–82.

— (1949). Säugetier-Territorien und ihre Markierung. *Bijdr. Dierkd.* **28**, 172–84.

— (1955). *The psychology and behaviour of animals in zoos and circuses.* Butterworths Scientific Publications, London.

Hefetz, A., Ben-Jaacov, R., Yom-Tov, Y., and Lloyd, H. A. The anal gland secretion of the African mongoose, *Herpestes ichneumon*: chemistry and function. *J. Chem. Ecol.* (In press.)

Henry, J. D. (1977). The use of urine marking in the scavenging behaviour of the red fox (*Vulpes vulpes*). *Behaviour* **61**, 82–105.

— (1979). The urine marking behaviour and movement patterns of red foxes (*Vulpes vulpes*) during a breeding and post-breeding period. In *Chemical signals* (ed. D. Müller-Schwarze and R. M. Silverstein) pp. 11–27. Plenum Press, London.

Hersteinsson, P. and Macdonald, D. W. (1982). Some comparisons between red and Arctic foxes, *Vulpes vulpes* and *Alopex lagopus*, as revealed by radio tracking. *Symp. zool. Soc., Lond.* **49**, 259–89.

Hildebrand, M. (1952). The integument of the Canidae. *J. Mammal.* **33**, 419–28.

Hornocker, M. G. (1969) Winter territoriality in mountain lions. *J. wildl. Mgmt.* **33**, 457–64.

Ikeda, H. (1982). Socio-ecological study on the raccoon dog, *Nyctereutes procyonoides viverrinus*, with reference to the habitat utilization pattern. Ph.D. thesis. Kyushu University, Japan.

— Eguchi, K., and Ono, Y. (1979). Home range utilization of a raccoon dog, *Nyctereutes procyonoides viverrinus*, Temminck, in a small islet in Western Kyushu. *Jap. J. Ecol.* **29**, 35–48.

Jacobi, A. (1938). Der Seeotter. *Monogr. Wildsäuget.* **6**, 1–93.

Jenkins, D., Makepeace, D. and Gorman, M. L. (1981). Two dimensional electrophoresis of otter (*Lutra lutra*) anal gland secretions. *J. Zool., Lond.* **195**, 558–9.

Johnson, R. P. (1973). Scent marking in mammals. *Anim. Behav.* **21**, 521–35.

Jonsingh, A. T. J. (1980). Ecology and behaviour of the dhole or Indian wild dog *Cuon alpinus* pallas 1811, with specific reference to predator–prey relations in Bandipur. Ph.D. thesis. Indian Institute of Science, Bangalore, India.

— (1982). Reproductive and social behaviour of the dhole *Cuon alpinus* (Canidae). *J. Zool., Lond.* **198**, 443–63.

Jordan, P., Shelton, P., and Allen, D. L. (1967). Numbers, turnover, and social structure of the Isle Royale wolf population. *Am. Zool.* **7**, 233–52.

Jorgenson, J. W., Novotny, M., Carmack, M., Copeland, G. B., Wilson, S. R., Katona, S., and Whitten, W. K. (1978). Chemical scent constituents in the urine of the red fox (*Vulpes vulpes*) during the winter season. *Science, NY* **199**, 796–8.

Kaufmann, J. (1962). Ecology and social behavior of the coati, *Nassua narica*, on Barro Colorado Island, Panama. *Univ. Calif. Publ. Zool.* **60**, 95–222.

Kingdon, J. (1978). *East African mammals*, Vol. IIIa. Academic Press, London.

Kleiman, D. G. (1966). Scent marking in the Canidae. *Symp. zool. Soc. Lond.* **18**, 167–77.

— (1972). Social behaviour of the maned wolf, *Chrysocyon brachyurus* and the bush dog, *Speothos venaticus*: a study in contrasts. *J. Mammal.* **53**, 791–806.

— (1974). Scent marking in the binturong *(Arctictis binturong)*. *J. Mammal.* **55**, 224–7.

— (1983). Ethology and reproduction of captive giant pandas, *Ailuropoda melanolenca*. *Z. Tierpsychol.* **62**, 1–46.

Koenig, von L. (1970). Zur Fortpflanzung und Jugendenwicklung des Wustenfichses *(Fennecus zerda*, Zimm. 1780). *Z. Tierpsychol.* **27**, 205–46.

Korytin, S. A. and Solomin, S. N. (1969). Materialy po etiologii psovykh. *Trudy vses. nauchno-issled. Inst. Zhevotonogo Syr'ya Psu Hniny* **22**, 235–70. (*Biol. Abstracts* **51**, 23886.)

Kruuk, H. (1972). *The spotted hyaena, a study of predation and social behaviour.* University of Chicago Press.

— (1975). Functional aspects of social hunting in càrnivores. In *Function and evolution in behaviour* (ed. G. Baerends, C. Beer, and A. Manning) pp. 119–41. Oxford University Press.

— (1976). Feeding and social behaviour of the striped hyaena *(Hyaena vulgaris* Desmarest). *E. Afr. wildl. J.* **14**, 91–111.

— (1978). Spatial organisation and territorial behaviour of the European badger, *Meles meles*. *J. Zool., Lond.* **184**, 1–19.

— and Hewson, R. (1978). Spacing and foraging of otters *(Lutra lutra)* in a marine habitat. *J. Zool., Lond.* **185**, 205–12.

— and Parish, T. (1982). Factors affecting population density, group size and territory size of the European badger. *Meles meles. J. Zool., Lond.* **196**, 31–9.

— and Sands, W. A. (1972). The aardwolf *(Proteles cristatus*, Sparrman 1783) as a predator of termites. *E. Afr. Wildl. J.* **10**, 211–27.

Kuhlhorn, F. (1940). Beitrag zur Systematic der Sudamerikanischen Caniden. *Arch. für Naturgesch.*, Neu. Ser. **9**, 194–223.

Kuhnel, W. (1971). Die Glandulae rectales (Proctodaealdrüsen) beim Kaninchen. *Z. Zellforsch. mikrosk. Anat.* **118**, 127–41.

Laidler, K., and Laidler, L. (1983). *The river wolf.* Allen and Unwin, London.

Lamprecht, J. (1979). Field observations on the behaviour and social system of the bat-eared fox, *Otocyon megalotis*, Desmarest. *Z. Tierpsychol.* **49**, 260–84.

van Lawick, H. (1974). *Solo: the story of an African wild dog.* Houghton Mifflin Co., Boston.

— and van Lawick Goodall, J. (1970). *The innocent killers.* Collins, London.

Langguth, A. (1969). Die südamerikanishchen Canidae unter besonderes Benichsteichteigung des Mähenwolfes, *Chrysocyon brachyurus*, Illiger. *Z. wiss. Zool.* **179**, 1–188.

Le Boeuf, B. J. (1967). Interindividual associations in dogs. *Behaviour* **29**, 268–95.

Lehner, P. N. (1978). Coyote communication. In *Coyotes: biology, behavior and management* (ed. M. Bekoff) pp. 128–62. Academic Press, New York.

Lewin, R. A. and Robinson, M. (1979). The greening of polar bears in zoos. *Nature, Lond.* **278**, 445–7.

Leyhausen, P. (1965). The communal organisation of solitary mammals. *Symp. zool Soc. Lond.* **14**, 249–63.

— (1971). Dominance and territoriality as complements in mammalian social structure. In *Behavior and environment* (ed. H. Esser) pp. 22–33. Plenum Press, New York.

— (1979). *Cat behaviour: the predatory and social behaviour of domestic and wild cats.* Garland Press, London.

— and Wolff, R. (1959). Das Revier einer Hauskatz. *Z. Tierpsychol.* **16**, 666–70.

Liberg, O. (1980). Spacing patterns in a population of rural free-roaming domestic cats. *Oikos* **35**, 336–49.

— (1981). Predation and social behaviour in a population of domestic cats: an evolutionary perspective. Ph.D. thesis. University of Lund, Sweden.

Lindemann, W. (1955). Über die Jugendentwicklung beim Luchs (*Lynx lynx* Kerr) und bei der Wildkatze (*Felis s. silvestris* Schreb.). *Behaviour* **8**, 1–45.

Linhart, S. B. and Knowlton, F. F. (1975). Determining the relative abundance of coyotes by scent station lines. *Wildl. Soc. Bull.* **3**, 119–24.

Lloyd, H. A. (1978). Aspects of the ecology of black and grizzly bears in coastal British Columbia. M.Sc. thesis, University of British Columbia, Vancouver.

Lockie, J. D. (1966). Territory in small carnivores. *Symp. zool. Soc. Lond.* **18**, 143–65.

Lorenz, K. (1954). *Man meets dog.* Methuen, London.

Lyall-Watson, M. (1964). The ethology of food-hoarding in mammals—with special reference to the green acouchi, *Myoprocta pratti*, Pocock. Ph.D. thesis, University of London.

Lynch, C. D. (1980). Ecology of the suricate, *Suricata suricatta*, and yellow mongoose, *Cynictis penicillata*, with special reference to their reproduction. *Memoirs van die Nationale Museum, Bloemfontein* R.S.A. **14**, 1–145.

Macdonald, D. W. (1976). Food caching by red foxes and some other carnivores. *Z. Tierpsychol.* **42**, 170–85.

— (1977). The behavioural ecology of the red fox *Vulpes vulpes*: a study of social organisation and resource exploitation. D.Phil. thesis, University of Oxford.

— (1978). Observations on the behaviour and ecology of the striped hyaena, *Hyaena hyaena*, in Israel. *Israel J. Zool.* **27**, 189–98.

— (1979*a*). Some observations and field experiments on the urine marking behaviour of the red fox, *Vulpes vulpes. Z. Tierpsychol.* **51**, 1–22.

— (1979*b*). The flexible social system of the golden jackal, *Canis aureus. Behav. Ecol. Sociobiol.* **5**, 17–38.

— (1980). Patterns of scent marking with urine and faeces amongst carnivore communities. *Symp. zool. Soc. Lond.* **45**, 107–39.

— (1981). Resource dispersion and the social organisation of the red fox, *Vulpes vulpes.* In *Proceedings of the Worldwide Furbearer Conference*, Maryland (ed. J. Chapman and D. Pursely) Vol. II, pp. 918–49. R. R. Donnelly and Sons, Falls Church, Va.

— (1983). The ecology of carnivore social behaviour. *Nature, Lond.* **301**, 379–89.

— and Apps, P. J. (1978). The social behaviour of a group of semi-dependent farm cats, *Felis cattus Carnivere Genet. Newsl.* **3**, 256–68.

— and Moehlman, P. D. (1982). Cooperation, altruism, and restraint in the reproduction of carnivores. In *Perspectives in ethology* (ed. P. P. G. Bateson and P. Klopfer) pp. 433–66. Plenum Press, London.

— and Wise, M. J. (1981). Notes on the behaviour of the Malay civet, *Viverra tangalunga. Sarawak Mus. J.* **48**, 291–2.

McElvain, S. M., Bright, R. B., and Johnson, P. (1941). The constituents of the volatile oil of catnip. I. Nepetalic acid, nepetalactone and related compounds. *J. Am. chem. Soc.* **63**, 1558–63.

Malcolm, J. R. (1979). Social organization and communal rearing of pups in African wild dogs (*Lycaon pictus*). Ph.D. thesis, Harvard University.

Matthews, L. H. (1939). Reproduction in the spotted hyena, *Crocuta crocuta* (Erxl). *Phil Trans. Roy. Soc.* **B230**, 1-78.

Mealey, S. P. (1980). The natural food habits of grizzly bears in Yellowstone National Park, 1973-74. In *Bears—their biology and management. 4th Int. Conf. on Bear Research*, 1977 (ed. C. J. Marinka and K. L. McArthur) pp. 281-92. Bear Biology Association.

Mech, L. D. (1970). *The wolf*. Natural History Press, New York.

— (1973). Wolf numbers in the Superior National Forest. *USDA Serv. Res. Pap.* **97**, 1-10.

— (1977). Wolf-pack buffer zones as prey reservoirs. *Science, N.Y.* **198**, 320-1.

Miller, M. E., Christensen, G. C., and Evans, H. E. (1964). *Anatomy of the dog*. Saunders, Philadelphia.

Mills, M. G. L. (1978). The comparative socio-ecology of the Hyaenidae. *Carnivore* **1**, 1-7.

— (1982). The mating system of the brown hyaena, *Hyaena brunnea*, in the Southern Kalahari. *Behav. Ecol. Sociobiol.* **10**, 131-6.

— Gorman, M. L., and Mills, M. E. J. (1980). The scent marking behaviour of the brown hyaena, *Hyaena brunnea*. *S. Afr. J. Zool.* **15**, 240-8.

Mivart, St. G. (1882*a*). On the classification and distribution of the Aeluroidea. *Proc. zool. Soc., Lond.* **1882**, 135-208.

— (1882*b*). Notes on some points in the anatomy of the Aeluroidea. *Proc. zool. Soc. Lond.* **1882**, 459-520.

Moehlman, P. D. (1983). Socioecology of silverbacked and golden jackals, *Canis mesomelas* and *C. aureus*. In *Recent advances in the study of mammalian behavior* (ed. J. F. Eisenberg and D. G. Kleiman). Am. Soc. Mammal. Special Publication, No. 7.

Montagna, W. and Parks, H. F. (1948). A histochemical study of the glands of the anal sac of the dog. *Anat. Rec.* **100**, 297-318.

Montgomery, G. G. and Lubin, Y. D. (1978). Social structure and food habits of crab-eating fox (*Cerdocyon thous*) in Venezualan Uanis. *Acta cient. Venez.* **29**, 382-3.

Morris, R. and Morris, D. (1966). *Men and pandas*. Hutchinson, London.

Müller-Schwarze, D., Volkman, N., and Zemanek, K. (1977). Osmetrichia: specialized scent releasing hairs in black-tailed deer. *J. Ultrast. Res.* **59**, 223-30.

Murie, A. (1936). Following fox trails. *Univ. Mich. Misc. Publ.* **32**, 7-45.

Murie, J. (1871). On the female generative organs, viscera and fleshy parts of *Hyaena brunnea*. *Trans. zool. Soc., Lond.* **7**, 503-12.

Murphy, E. L., Flath, R. A., Black, D. R., Mon, T. R., Teranishi, R., Timm, R. M., and Howard, W. E. (1978). Isolation, identification, and biological activity assay of chemical fractions from estrus urine attractive to the coyote. *Am. chem. Soc. Symp. Ser.* **67**, 66-67.

Mykytowycz, R., Hesterman, E. R., Gambale, S., and Dudzinski, M. L. (1976). A comparison of the effectiveness of the odours of rabbits, *Oryctolagus cuniculus*, in enhancing territorial confidence. *J. chem. Ecol.* **2**, 13-24.

Neal, E. (1948). *The badger*. Collins, London.

— (1977). *Badgers*. Blandford Press, Poole.

Ortmann, R. (1960). Die Analregion der Saugetiere. *Hand. Zool.* **VIII** 3(7), 1-68.

Östborn, H. (1976). Doftmarkering hos graveling. *Zool. Revy*. **38**, 103–12.

Ough, W. D. (1982). Scent marking of captive raccoons. *J. Mammal.* **63**, 318–19.

Owens, D. A. and Owens, M. J. (1979). Notes on social organization and behaviour in brown hyaenas (*Hyaena brunnea*). *J. Mammal.* **60**, 405–8.

Ozoga, J. J. and Harger, E. M. (1966). Winter activities and feeding habits of northern Michigan coyotes. *J. Wildl. Mgmt.* **30**, 809–18.

Palen, G. F. and Goddard, G. V. (1966). Catnip and oestrus behaviour in the cat. *Animal Behav.* **14**, 372–7.

Panaman, R. (1981). Behaviour and ecology of free-ranging female farm cats (*Felis catus* L.). *Z. Tierpsychol.* **56**, 59–73.

Peters, R. P. and Mech, L. D. (1975). Scent-marking in wolves. *Am. Scient.* **63**, 628–37.

Petskoi, P. G. and Kolporskii, V. M. (1970) (trans.) Neck glandular structure in animals of the family Mustelidae. *Zool. Zhur.* **49**, 1208–9.

Pocock, R. I. (1908). Warning colouration in the, musteline Carnivora. *Proc. zool. Soc. Lond.* 1908, 944–59.

— (1911). Some probable and possible instances of warning characteristics amongst insectivorous and carnivorous mammals. *Ann. Mag. Nat. Hist.* **8**, 750–7.

— (1914a). On the feet of domestic dogs. *Proc. zool. Soc. Lond.* 1914, 478–84.

— (1914b). On the feet and external features of the Canidae and Ursidae. *Proc. zool. Soc. Lond.* 914–41.

— (1915a). On the feet and glands and other external characters of the paradoxurine genera *Paradoxurus, Arctictis, Arctogalidea,* and *Nandinia. Proc. zool. Soc. Lond.* 1915, 387–412.

— (1915b). On some of the external characters of *Cynogale bennetti* Gray. *Ann. Mag. nat. Hist.* **15**, 351–60.

— (1915c). On the feet and hands and other external characters of the Viverrinae, with the description of a new genus. *Proc. zool. Soc., Lond.* 1915, 131–49.

— (1915d). On some of the external characters of the genus *Linsang*, with notes upon the genera *Poiana* and *Eupleres. Ann. Mag. nat. Hist.* **16**, 341–51.

— (1915e). On some external characters of *Galidia, Galidictis* and related genera. *Ann. Mag. nat. Hist.* **16**, 351–6.

— (1916a). On some of the external characters of *Cryptoprocta. Ann. Mag. nat. Hist.* **17**, 413–25.

— (1916b). On some of the external structural characteristics of the striped hyaena (*Hyaena hyaena*) and related genera and species. *Ann. Mag. nat. Hist.* **17**, 330–43.

— (1916c). On the external characters of the mongooses (Mungotidae). *Ann. Mag. nat. Hist.* 349–74.

— (1918). Further notes on some external characters of the bears (Ursidae). *Ann. Mag. nat. Hist.* **1**, 375–84.

— (1920a). On the external and cranial characters of the European badger (*Meles*) and of the American badger (*Taxidea*). *Proc. zool. Soc. Lond.* 1920, 423–36.

— (1920b). On the external characters of the ratel (*Mellivora*) and the wolverine (*Gulo*). *Proc. zool. Soc. Lond.* 1920, 179–87.

— (1921a). On the external characters and classification of the Mustelidae. *Proc. zool. Soc. Lond.* 1921, 803–7.

— (1921*b*). On the external characters of some species of Lutrinae (otters). *Proc. zool. Soc. Lond.* 1921, 535–46.

— (1921*c*). The external characters and classification of the Procyonidae. *Proc. zool. Soc. Lond.* 389–422.

— (1925). The external characters of an American badger (*Taxidea taxus*) and an American mink (*Mustela vison*), recently exhibited in the Society's gardens. *Proc. zool. Soc. Lond.* 1925, 17–27.

— (1927*a*). The external characters of a bush dog (*Speothos venaticus*) and of a maned wolf (*Chrysocyon brachyurus*), exhibited in the Society's gardens. *Proc. zool. Soc. Lond.* 1927, 307–21.

— (1927*b*) The external characters of the South African striped weasel, (*Poecilogale albinucha*). *Proc. zool. Soc. Lond.* 1927, 125–33.

— (1933). The rarer genera of oriental Viverridae. *Proc. zool. Soc. Lond.* 966–1035.

— (1939). *The fauna of British India including Ceylon and Burma: Mammalia.* Taylor and Francis, London.

Poglayen-Neuwall, I. (1962). Beitrage zu einen Ethogram des Wickelbaren (*Potos flavus* Schreber). *Z. Säugetierk.* 27, 1–44.

Poole, T. B. (1967). Aspects of aggressive behaviour in polecats. *Z. Tierpsychol,* 24, 351–69.

Powell, R. A. (1981). *Martes pennanti. Mammal. Species* 156, 1–6.

— (1982). *The fisher. Life history, ecology, and behavior.* University of Minnesota Press.

Preti, G., Meutterties, E. L., Furman, J., Kenelly, J. J., and Johns, B. (1976). Volatile constituents of dog (*Canis familiaris*) and coyote (*Canis latrans*) anal sacs. *J. chem. Ecol.* 2, 177–87.

Pulliainen, E. (1982). Scent marking in the pine marten (*Martes martes*) in Finnish Forest Lapland in winter. *Z. Säugetierk.* 47, 91–99.

Pulliainen, E. and Ovaskainen, P. (1975). Territory marking by a wolverine (*Gulo gulo*) in northeastern Lapland. *Ann. Zool. Fennici* 12, 268–70.

Quick, H. F. (1951). Notes on the ecology of weasels in Gunnison County, Colorado. *J. Mammal.* 32, 281–90.

Ralls, K. (1971). Mammalian scent marking. *Science, NY* 171, 443–9.

Randall, R. M. (1977). Aspects of the ecology of the civet, *Civettictus civetta* (Schreber, 1878). M.Sc. thesis, University of Pretoria.

Rasa, O. A. E. (1973). Marking behaviour and its significance in the African dwarf mongoose, *Helogale undulata rufula. Z. Tierpsychol.* 32, 449–88.

Reid, J. B. (1982). Aspects of dog eliminative behaviour in street and park. Paper to Society for Companion Animal Studies, Paris, 13–14 March 1982.

Rensch, B. and Dücker, G. (1959). Die Spiele von Mungo und Ichneumon. *Behaviour* 14, 185–213.

Rieger, I. von (1977). Markierungsverhalten von Streifen-hyänen, *Hyaena hyaena*, im Zoologischen Garten Zurich. *Z. Säugetierk.* 42, 307–17.

— (1979). Scent rubbing in carnivores. *Carnivore* 2, 17–25.

— and Walzkoenig, D. (1979). Markieren Karzen beim Wagenreiben? *Z. Säugetierk.* 44, 319–20.

Roberts, M. S. (1981). Reproductive biology of the giant panda, *Ailurus fulgens.* M.S. thesis, University of Michigan.

Rogers, L. L., Mech, L. D., Dawson, D. K., Peek, J. M. and Korb, M. (1979). Deer distribution in relation to wolf pack territory edges. *J. wildl. Mgmt.* 44, 253–8.

Roth, H. E. (1980). Defecation rates of captive brown bears. In *Bears—their biology and management. 4th Int. Conf. on Bear Research*, 1977 (ed. C. J. Martinka and K. L. McArthur) pp. 249-53. Bear Billogy Association.

Rothman, J. and Mech, L. D. (1979). Scent marking in lone wolves and newly formed pairs. *Anim. Behav.* **27**, 750-60.

Sands, M. W., Coppinger, R. P., and Phillips, R. J. (1977). Comparisons of thermal sweating and histology of sweat glands of selected canids. *J. Mammal.* **58**, 74-8.

Sargeant, A. B. (1972). Red fox spatial characteristics in relation to waterfowl predation. *J. wildl. Mgmt.* **36**, 225-36.

Saunders, J. K. (1963). Movements and activities of the lynx in Newfoundland. *J. wildl. Mgmt.* **27**, 390-400.

Schaffer, J. (1940). *Die Hautdrüsenorgane der Säugetiere*. Urban & Schwarzenberg, berlin.

Schaller, G. B. (1967). *The deer and the tiger*. University of Chicago Press.

—— (1972). *The Serengeti lion: a study of predator-prey relations*. University of Chicago Press.

Schenkel, R. (1947). Ausdrucks-studien und Wölfen. *Behaviour* **1**, 81-129.

—— (1967). Submission: its features and functions in the wolf and dog. *Am. Zool.* **7**, 319-29.

Scidknecht, H., Witz, I., Enzmann, F., Grund, N., and Ziegler, M. (1976). Mustelan, the malodorous substance from the gland of the mink (*Mustela vison*) and the polecat (*Mustela putorius*). *Angew. Chem. Inst. Ed. Engl.* **15**, 242-3.

Scott, J. P. and Fuller, J. L. (1965). *Genetics and social behavior of the dog*. University of Chicago Press.

Seidensticker, J. C., Hornocker, M. G., Wiles, W. V., and Messick, J. P. (1973). Mountain lion social organization in the Idaho primitive area. *J. wildl. Mgmt.* **35**, 1-60.

Seitz, A. (1955). Untersuchungen uber angeborene Verhaltensweisen bei Caniden. III. Beobachtungen an Marder hunden (*Nyctereutes procyonoides* Gray). *Z. Tierpsychol.* **12**, 463-89.

Seton, E. T. (1909), *Life histories of northern animals* **11**, 677-1267. Charles Scribner and Son, N.Y.

Sokolov, V. E., Albone, E. S., Flood, P. F., Heap, P. F., Kagen, M. Z., Vasilieva, V. S., Roznov, V. V., and Zinkevich, E. P. (1980). Secretion and secretory tissues of the anal sac of the mink (*Mustela vison*): chemical and histological studies. *J. chem. Ecol.* **6**, 805-25.

—— and Khorlina, I. M. (1976). Mammalian pheromones: study of the volatile acid composition of vaginal secretions in the mink, *Mustela vison*, Briss. *Doklady Akad. Nauk USSR* **228**, 225-7.

—— —— Glovnya, R. V., and Zhuravleva, I. L. (1974). Change in the composition of the amines in the volatile substances of the vaginal secretions of the American mink (*Mustela vison*) depending on the sexual cycle. *Doklady Akad. Nauk USSR* **216**, 220-2.

Southwood, T. R. E. (1966). *Ecological methods, with particular reference to the study of insect populations*. Methuen, London.

Spannhof, I. (1969). The histophysiology and function of the anal sac of the red fox (*Vulpes vulpes* L.). *Forma Funct.* **1**, 26-45.

Sprague, R. H. and Anisko, J. J. (1973). Elimination patterns in the laboratory beagle. *Behaviour* **47**, 257–67.

Stains, H. J. (1956). The raccoon in Kansas. *Misc. Publ. Mus. nat. Hist. Univ. Kansas* **10**, 1–76.

Stoddart, D. M. (1979). Some responses of a free living community of rodents to the odors of predators. In *Chemical signals* (ed. D. Müller-Schwarze and R. M. Silverstein) pp. 1–10. Plenum, London.

Story, H. E. (1945). The external genitalia and perfume gland in *Arctictis binturong. J. Mammal.* **26**, 64–6.

Stuart, C. (1977). Analysis of *Felis lybica* and *Genetta genetta* scats from the central Namib desert, South West Africa. *Zool. Africana* **12**, 239–41.

Stubbe, M. (1970). Evolution of anal marking organs in mustelids. *Biol. Zentralbl.* **89**, 213–23.

Sunquist, M. (1979). The movements and activities of tigers (*Panthera tigris tigris*) in Royal Chitawan National Park, Nepal. Ph.D. thesis, University of Minnesota.

Tembrock, G. (1957). Zur Ethologie des Rotfuchses (*Vulpes vulpes* L) uter besonderer Berüksichtigung der Fortplanzung. *Zool. Garten, Lpz.* **23**, 289–532.

Teranishi, R., Murphy, E. L., Stern, D. J., Howard, W. E., and Fagre, D. B. (1981). Chemicals useful as attractants and repellants for coyotes. In *Worldwide furbearer conference proceedings* (ed. J. A. Chapman and D. Pursley) pp. 1839–51. R. R. Donnelley and Sons, Falls Church, Virginia.

Thompson, D. Q. (1952). Travel, range and food habits of timber wolves in Wisconsin. *J. Mammal.* **33**, 429–42.

Thompson, E. T. (1923). The mane on the tail of the gray fox. *J. Mammal.* **33**, 429–42.

Tinbergen, N. (1965). Von den Vorratskammern des Rotfuchses (*Vulpes vulpes* L.). *Z. Tierpsychol.* **22**, 119–49.

Todd, (1962). The inheritance of the catnip response in domestic cats. *J. Hered.* **53**, 54–6.

Toldt, K. (1907). Studien über das Haarkleid von, *Vulpes vulpes* L. Wein. *Ann. Naturhist. Hofmuseum* **22**, 197–269.

Trapp, G. R. (1978). Comparative behavioural ecology of the ringtail and the grey fox in Southwestern Utah. *Carnivore* **1**, 3–31.

— and Hallberg, D. L. (1975). Ecology of the grey fox (*Urocyon cinereoargenteus*); a review. In *The wild canids* (ed. M. W. Fox) pp. 164–78. Van Nostrand Reinhold, New York.

Tschanz, B., Mayer-Hotzapfel, M., and Bachmann, S. (1970). Das Informationssystem bein Braunbären. *Z. Tierpsychol.* **27**, 47–72.

Uexküll, J. V. von and Kirszat, G. (1934). *Streifzuge durch die Umwelten von Tieren und Menchen.* Berlin.

— and Sarris, E. G. (1931). Das Duftfeld des Hundes. *Z. Hundeforschung* **1**, 55–68.

Ustinov, S. K. (1971). The brown bear in the Maritime Ridge by Lake Baikal: numbers, food, habits, behaviour. *Abstr. Int. Theriol. Congress.* **2**, 260–1.

Verberne, G. and de Boer, J. N. (1976). Chemocommunication among domestic cats. *Z. Tierpsychol.* **42**, 86–109.

— and Leyhausen, P. (1976). Marking behaviour of some Viverridae and Felidae: time-interval analysis of the marking pattern. *Behaviour* **58**, 192–253.

Vosseler, J. (1929a). Vom Binturong (*Arctictis binturong* Ruffl.). *Zool. Gart. Lpz.* 1, 296-302.
— (1929b). Beitrag zur Kenntnis der Fossa (*Cryptoprocta ferox*) und ihrer Fortpflanzung. *Zool. Gart. Lpz.* 2, 1-9.
Ware, G. C. and Gosden, P. E. (1980). Anaerobic microflora of the anal sac of the red fox (*Vulpes vulpes*). *J. chem. Ecol.* 6, 97-102.
Weber, M. (1927). *Die Säugetiere.* Gustav Fischer, Jena.
Wells, M. C. and Bekoff, M. (1981). An observational study of scent marking in coyotes, *Canis latrans. Anim. Behav.* 29, 332-50.
Wemmer, C. M. (1977a). Scent marking and anointing: behavioural parallelisms in mammals. *Am. Zool.* 11, 623.
— (1977b). Comparative ethology of the large spotted genet (*Genetta tigrina*) and some related Viverrids. *Smithson. Contr. Zool.* No. 239, 1-93.
—and Murtaugh, J. (1981). Copulatory behaviour and reproduction in the binturong, *Arctictis binturong. J. Mammal.* 62, 342-52.
— and Scow, K. (1977). Communication in the Felidae with emphasis on scent marking and contact patterns. In *How animals communicate* (ed. T. Sebeok) pp. 749-66. Indiana University Press, Bloomington.
— and Watling, R. (in press). Ecology and status of the Sulawesi palm civet (*Macrogalidia musschenbroekii* Schlegel). *Biol. Conserv.*
— West, J., Watling, R., Collins, L. and Lang, K. (1983). The external characters of the Sulawesi palm civet, *Macrogalidia musschenbroekii* Schlegel, 1879. *J. Mammal.* 64, 133-6.
Wheeler, J. W., Endt, D. W. von, and Wemmer, C. (1975). 5-Thiomethylpentan-2,3-dione, a unique product from the striped hyaena. *J. Am. chem. Soc.* 97, 441-2.
Whitten, W. K., Wilson, M. C., Wilson, S. R., Jorgenson, J. W., Novotny, M., and Carmack, M. (1980). Induction of marking behaviour in wild red foxes (*Vulpes vulpes* L.) by synthetic urinary constituents. *J. chem. Ecol.* 6, 49-55.
Whittle, N. (1981). Reaction of tigers to the scent of conspecifics. *J. Zool., Lond.* 194, 263-5.
Wilson, M. C., Whitten, W. K., Wilson, S. R., Jorgenson, J. W., Novotny, M., and Carmack, M. (1979). Marking behaviour in wild red foxes in response to synthetic volatile urinary compounds. In *Chemical signals* (ed. D. Müller-Schwarze and R. M. Silverstein) pp. 29-38. Plenum Press, London.
Wilson, S. R., Carmack, M., Novotny, M., Jorgenson, J. W., and Whitten, W. K. (1978). Delta-3-isopentenyl methyl sulfide: a new terpenoid in the scent mark of the red fox (*Vulpes vulpes*). *J. org. Chem.* 43, 4675-6.
Woolpy, J. (1968). The social organization of wolves. *Nat. Hist.* 72, 46-55.
Yamamoto, I. and Hikada, T. (In press.) Utilization of 'latrines' in the raccoon dog, *Nyctereutes procyonoides. Acta Zool. Fennica.*
Young, S. P. and Goldman, E. A. (1944). *The wolves of North America.* American Wildlife Institute.
Zannier, F. (1965). Verhaltensuntesuchungen an der Zwergmanguste, *Helogale undulata rufula,* im Zoologischen Garten, Frankfurt am Main. *Z. Tierpsychol.* 22, 672-95.
Zimen, E. (1981). *The wolf: his place in the natural world.* Souvenir Press, London.
— (1982). A wolf pack sociogram. In *Wolves of the world* (ed. F. H. Harrington and P. C. Paquet) pp. 282-322. Noyes Publ. Park Ridge, New Jersey.

# 16 The marine mammals: orders Cetacea, Pinnipedia, and Sirenia

RICHARD E. BROWN

The marine mammals include the Cetacea (Odontoceti, or toothed whales, and Mysticeti, or baleen whales); the Pinnipedia (seals, sea lions, and walruses), and the Sirenia (dugong and manatee).

## Cetacea

There are 38 genera of cetaceans (90 species) in eight families, all of which are completely aquatic, with their front limbs modified to form flippers and their hind limbs vestigial or absent. The skin of cetaceans is naked and, except for the mammary glands of the female, there are no epidermal glands (Walker 1975). Eisenberg (1981) has summarized what is presently known about the socio-ecology of the cetaceans and Lowell and Flanigan (1980) have described many of the findings on chemoreception.

### Olfactory sensitivity

While Slijper (1962, p. 242) stated that the sense of olfaction is practically non-existent in the Cetacea, it now appears that there is a continuum of degeneration of the olfactory receptors, olfactory nerves, and cerebral olfactory areas.

The olfactory epithelium is reduced in all Cetacea (Yablokov, Bel'Kovich, and Borison 1972) and adult bottle-nosed dolphin (*Tursiops truncatus*) have no olfactory epithelium (Manderson 1968). The fin whale (*Balaenoptera physalus*) has rudimentary nasopalatine canals, but no chemoreceptors (Quay and Mitchell 1971).

The beluga whale (*Delphinapterus leucas*) shows a complete absence of olfactory structures, but may sense chemical stimuli through chemosensory receptors at the base of the tongue (Kleinenberg, Yablokov, Bel'Kovich, and Tarasevich 1969, pp. 180-2). Likewise, the sperm whale (*Physeter catodon*) has no olfactory nerves or bulbs, but may possess chemoreceptors at the base of the tongue (Berzin 1972, pp. 145-51). In some Cetacea, chemosensory information may also be transmitted through the trigeminal nerve (Jacobs, Morgane, and McFarland 1971).

Olfactory nerves and bulbs have not been found in any odontocete species (Breathnach 1960, p. 192). Rudimentary olfactory peduncles (nerves) occur in bottle-nosed whales (*Hyperoodon rostratus*), white whales (*Delphinapterus leucas*), and sperm whales (*Physeter catodon*) (Breathnach 1960; Lowell and Flanigan 1980).

Foetal spotted dolphins (*Stenella caeruleoalbus*) show some development of olfactory bulbs and tracts, but these atrophy and are absent in adults (Kamiya and Pirlot 1974). Likewise, the foetal olfactory peduncles of bottle-nosed dolphins (*Tursiops truncatus*), common dolphins (*Delphinus delphis*), and the harbour porpoise (*Phocoena phocoena*) atrophy before adulthood (Breathnach 1960).

In the Odontoceti the anterior olfactory nucleus and the nucleus of the lateral olfactory tract are absent (Breathnach 1960).

Mysticetes have both medial and lateral olfactory tracts (Breathnach 1960). The olfactory tracts and bulbs of these whales resemble those of terrestrial mammals, but are somewhat reduced in size (Jacobs *et al.* 1971).

*Taste*

Cetaceans do not have the typical mammalian papillae, or taste buds, on their tongues, but many species of odontocetes have a depression, or pit, at the base of the tongue, which is lined with sensory epithelium (Yablokov *et al.* 1972). Sokolov and Volkova (1973) report finding such a chemosensory pit in the tongues of the blue-white dolphin (*Stenella caeruleoalbus*), the common dolphin (*Delphinus delphis*), the bottle-nosed dolphin (*Tursiops truncatus*), and the harbour porpoise (*Phocoena phocoena*). This epithelium resembles the papillae on the tongue of land mammals, and nerves from this tongue cavity innervate the gustatory centres in the brain, the nucleus solitarius and the ventromedial complex of the thalamus (Kruger 1959). These tongue pits have not been found in the mysticete whales and it is thought that they lack a sense of taste (Fobes and Smock 1981).

There seem to be no studies on taste sensitivity or taste preference in Cetacea. Brown (1960) reported that a single captive pilot whale (*Globicephala scammoni*) 'appeared to relish' Pacific baraccuda (*Sphyraena argentea*) and Kellogg (1961) used food preferences in his experiments on echolocation with bottle-nosed dolphins, suggesting taste sensitivity in this species, but careful psychophysical studies seem to be lacking.

*Skin glands*

Cetacea lack cutaneous glands (Ling 1974), but many species of odontocetes have anal or perianal glands. The grey whale (*Eschrichtius robustus*) has a large postanal sac, or 'stink sac', which produces a slight mid-ventral swelling in the tail (Durham and Beierle 1976; Beierle, Degnen, Beierle, Patten, and Durham 1976). Anal glands have been reported in the beluga whale (*Delphinapterus leucus*) by Kleinenberg *et al.* (1969), paired anal pores in the Atlantic bottle-nosed dolphin (*Tursiops truncatus*) by Caldwell and Caldwell (1977) and perianal glands in the common porpoise (*Phocaena*) by Schaffer (1940). Schaffer also reports 'Talgdrüsen' in bottle-nosed whales (*Hyperoodon*) and eye-lid glands in the common dolphin (*Delphinus*).

Many odontocetes have excretory ducts in the lobes of the preputial gland which release secretions into the urinary canal (Kleinenberg *et al.* 1969). Odontocetes release small quantities of urine at frequent intervals so that 'one may thus assume that the passing of a school of whales leaves a band of dissolved substances in the water' (Kleinenberg *et al.* 1969, p. 182).

*Olfaction and social behaviour*

The role of chemosensory communication in Cetacea is unclear. Sokolov and Kuznetsov (1971) trained Black Sea dolphins (*Tursiops truncatus*) to discriminate between plain sea water and sea water mixed with indole, camphor, or trimethylamine, showing that these cetaceans could make chemosensory discriminations. Kuznetsov (1974) performed similar experiments with the common dolphin (*Delphinus delphis*) and the harbour porpoise (*Phocoena phocoena*).

Some authors suggest that the contents of the anal and perianal glands of the Cetacea may be released into the water and used as chemical signals which other whales may detect through their taste receptors (Kuznetsov 1974; Caldwell and Caldwell 1977; Herman and Tavolga 1980). Kuznetsov (1978, see Bullock and Gurevich 1979) reported that Black Sea dolphins responded to solutions of dolphin prostate glands and perineal glands and suggested that these glands might be used for communication.

While these studies suggest that some Cetacea may have the ability to use chemical signals, there is no strong evidence for a social function of these chemicals. A number of social functions for chemosensory signals in Cetacea have been suggested, but these are all based on anecdotal observations. These functions include the recognition of conspecifics (Evans and Bastian 1969), orientation in the environment (Sokolov and Kuznetsov 1971), maintaining the integrity of the group and 'tract laying' in migration (Kleinenberg *et al.* 1969). Defecation, especially in mysticetes, may be used to indicate the presence of conspecifics.

Some species may use urine to detect the sex, reproductive status or readiness to mate in conspecifics (Norris and Dahl 1980). Bottle-nosed dolphins, for example, nuzzle the genitalia of their partner during courtship (Tavolga and Essapian 1957; Puente and Dewsbury 1976), but no studies have been done to investigate whether oestrous female cetacea produce urinary or vaginal odours which are attractive to males. Finally, because some species, such as the beluga, flee from areas in which whales have been killed or frightened, Yablokov *et al.* (1972) have suggested that these whales may use some sort of alarm odour.

Because of the extremely vocal nature of cetaceans and the belief that they are all anosmic, chemical communication has not been seriously investigated in the Cetacea. One would expect that those whales with the more complete olfactory nervous systems, such as the baleen whales, would use olfactory signals more than those with reduced olfaction, such as the toothed whales. Unfortunately the baleen whales are too large to keep in captivity and experiment upon. Research on chemical communication in Cetacea, therefore, may best be done with those small odontocetes, such as bottle-nosed and white whales, which show some olfactory tracts and which can be kept in captivity.


## Pinnipedia

The pinnipedia include six genera (12 species) of sea lions and fur seals (Otariidae), the single species of walrus (*Odobenus rosmarus*), and 13 genera (18 species) of true seals (Phocidae). Pinnipeds are aquatic carnivores which have all four limbs modified into flippers. Although well adapted for swimming and clumsy on land, most species of pinnipeds spend large amounts of time on land (Walker 1975). The socio-ecology of the pinnipedia has been summarized by Eisenberg (1981).

### Olfactory sensitivity

Scheffer (1958) and King (1964) state that the pinnipeds have a poor sense of smell. 'Studies of the Northern fur seal on land suggest that the sense of smell is poorly developed as compared with that of say, the sea otter' (Scheffer 1958, p. 15). Kuzin and Sobolevsky (1976), on the other hand, have described the olfactory epithelium of the fur seal as similar to that of terrestrial mammals and state that the sense of smell is well developed in the fur seal and plays an important part in social behaviour. Harrison and Kooyman (1968) concluded that the olfactory bulbs and nerves of pinnipedia are reduced in size as compared to those terrestrial carnivores.

### Taste

The taste buds of the pinnipedia consist of the four typical mammalian papillae (Bradley 1971), but the number of taste buds is reduced in comparison to land mammals (King 1964; Kubota 1968).

### Skin glands

Sweat glands and sebaceous glands are well developed in the pinnipedia (Ling 1974). In the fur seals (Otariidae) the sweat glands are 'relatively enormous organs extending well below the base of the hair follicles' (Ling 1965, p. 561). These glands are probably responsible for the musky odour of fur seals (Ling 1974) and may function to regulate their skin temperature (Ling 1965). In the Phocidae (true seals), temperature regulation is achieved by the thick layer of blubber and sweat glands and more simple tubules which are 'rather insignificant' (Ling 1965). In the walrus the sweat glands are large in summer, but atrophy during the autumn (Ling 1965, 1974). Some of the largest sweat glands in the walrus are on the snout and these may produce an odorous secretion which can be used for individual identification (Ling 1974).

*Olfaction and social behaviour*

Both the Otariidae and the Phocidae appear to use olfactory cues in mating and mother–infant interactions. Female California sea lions (*Zalophus californianus*) suckle only their own pups and rebuff strange infants. The rebuff may consist of an open-mouthed threat or the female may seize the pup in her teeth and toss it into the air (Peterson and Bartholomew 1967, p. 41). Females must find their own pups among hundreds of animals. Distance recognition may occur by vocalization or by visual contact, but final identification appears to be done by smell. Females nose and sniff pups when they first meet and then nuzzle the pup before accepting or rejecting suckling attempts (Peterson and Bartholomew 1967, p. 43). It thus appears that female Californian sea lions identify their pups by odour.

Female grey seals (*Halichoerus grypus*), also suckle only their own pups. For about an hour after the birth of a pup, the cow runs her nose along its body. 'It is presumed that the cow is smelling the pup during these physical contacts, as dilation of her nostrils is frequently observed, and is thus learning the individual odour of the pup' (Burton, Anderson, and Summers 1975, p. 199). Whereas female *Halichoerus* respond to the vocalizations of alien pups, they do not allow them to suckle, and smell appears to be the most important sense used in the maintenance of the mother–pup bond (Burton *et al.* 1975). Where grey seals live at very high densities, or when a herd of seals is disturbed, and the mother cannot establish an olfactory identification of her pup, the mother–pup bond may not form and the pup may be abandoned to starve (Fogden 1971; Burton *et al.* 1975).

Olfactory recognition of pups also appears to be used by Stellar sea lions (Ono 1972); fur seals (Stirling 1970; Vladimirov 1975), and the Weddell seal (Kaufman, Siniff, and Reichle 1975). Weddell seals may use olfactory cues to recognize their own pups under water, as females are aggressive to strange pups even under water (Kaufman *et al.* 1975). Muzzle-to-muzzle contact, nuzzling, and sniffing occur in all species of pinnipeds and this suggests that olfaction may be used for individual recognition or the recognition of family members (Evans and Bastian 1969). The strong scent of male fur seals may function to attract females or to mark territories (Hamilton 1956). Male California sea lions sniff at the genitals of females in heat before copulation (Peterson and Bartholomew 1967, p. 32).

There is some suggestion that fur seals and harp seals may use olfactory signals in orientation and navigation to breeding grounds or in migration (Sergeant 1970; Kinne 1975; Kuzin and Sobolevsky 1976). Finally, grey seals react with alarm to the odours of humans (Burton *et al.* 1975); thus, olfactory cues may play a role in predator detection by seals as well as in social behaviour.

The few observations on olfactory communication in the pinnipedia suggest that olfactory signals are far more important for both the Otariidae and the

Phocidae than was formerly believed. Fobes and Smock (1981) suggest that the amphibious marine mammals use olfactory signals more than those which are completely aquatic.

Unfortunately, most of the reports of olfactory signals in the social behaviour of the pinnipedia are based on anecdotal or unsystematic observations. More research on the olfactory signals of the pinnipedia may show the importance of the sense of smell for social communications in these animals.

**Sirenia**

There are two living genera of Sirena: dugongs with one living species (*Dugong dugon*) and manatees (*Trichechus*) with three living species. The Sirenia are aquatic herbivores which have evolved from the same ancestors as the elephants. They have no hind limbs and their fore limbs are modified as flippers. Dugongs are found in the Old World from Africa to India and Australia. Manatees occur from Florida to northern South American and in western Africa (Walker 1975; Kingdon 1971). Sirenia occur most often singly, in pairs or in small 'family' groups but may be found in groups as large as 20 animals (Walker 1975; Kingdon 1971). The young are cared for by the mother for up to one year and may be carried on her back. What little is known about the social behaviour of the Sirenia has been summarized by Eisenberg (1981).

*Olfactory sensitivity and taste*

The olfactory lobes and nerves of the manatee are reduced in comparison to terrestrial mammals (Lowell and Flanigan 1980) and Hartman (1969) has suggested that the sense of smell is reduced or absent in the manatee. Kingdon (1971), however, states that the dugong has a good sense of smell. The tongues of both the manatee and dugong contain papillae which may function as taste buds (Bradley 1971).

*Skin glands, olfaction, and social behaviour*

The Sirenia have only rudimentary sebaceous glands (Ling 1974). The role of olfaction in sirenian social behaviour seems to be unknown. Nose-to-nose and nose-to-body contact in manatees may involve smell (Moore 1956), and the selection of preferred foods may be done by smell or taste (Lowell and Flanigan 1980) but there are no observations indicating the use of olfactory signals in courtship, mother–infant interactions, or other social behaviours.

**Summary**

As recently as 1962 it was thought that the marine mammals had, at best, a rudimentary sense of smell. But recent research has suggested that certain whales and most pinnipeds not only have a usable sense of smell but that they may also use olfactory signals in courtship and mother–infant recognition.

Further research is required at all levels of olfactory communication in marine mammals. Which glands secrete odorous compounds, how sensitive are animals to these compounds and in what social situations are they used? Can marine mammals identify individuals, family members (relatives) and conspecifics by their odour? Are there fear odours? Are odorous secretions used for territory marking or in agonistic encounters by pinnipeds on terrestrial mating grounds? These and many other questions no longer seem to be meaningless when applied to marine mammals and there is an open field for research on olfaction and social behaviour with these animals.

## References

Beierle, J. W., Degnen, C., Beierle, J. J., Patten, D. R., and Durham, F. E. (1976). An analysis of the fluid contents in the postanal sac of the gray whale (*Eschrichtius robustus*). *Bull. Sth. Calif. Acad. Sci.* **75**, 5–10.

Berzin, A. A. (1972). *The sperm whale.* Israel Program for Scientific Translations, Jerusalem.

Bradley, R. M. (1971). Tongue topography. In *Handbook of sensory physiology*, Vol. IV *Chemical senses*, Part 2 *Taste* (ed. L. M. Beidler) pp. 1–30. Springer, Berlin.

Breathnach, A. S. (1960). The cetacean central nervous system. *Biol. Rev.* **35**, 187–230.

Brown, D. H. (1960). Behaviour of a captive Pacific pilot whale. *J. Mammal.* **41**, 342–9.

Bullock, T. H. and Gurevich, V. S. (1979). Soviet literature on the nervous system and psychobiology of cetacea. *Int. Rev. Neurobiol.* **21**, 47–127.

Burton, R. W., Anderson, S. S., and Summers, C. F. (1975). Perinatal activities in the Gray seal, *Halichoerus grypus*. *J. Zool., Lond.* **177**, 197–201.

Caldwell, D. K. and Caldwell, M. S. (1977). Cetaceans, In *How animals communicate* (ed. T. A. Sebeok) pp. 794–808. Indiana University Press, Bloomington.

Durham, F. E. and Beierle, J. W. (1976). Investigations on the postanal sac of the gray whale, *Eschrichtius robustus*. *Bull. Sth. Calif. Acad. Sci.* **75**, 1–5.

Eisenberg, J. F. (1981). *The mammalian radiations.* The University of Chicago Press.

Evans, W. E. and Bastian, J. (1969). Marine mammal communication: social and ecological factors. In *The biology of marine mammals* (ed. H. T. Anderson) pp. 424–75. Academic Press, New York.

Fobes, J. L. and Smock, C. C. (1981). Sensory capacities of marine mammals. *Psychol. Bull.* **89**, 288–307.

Fogden, S. C. L. (1971). Mother–young behaviour at Gray seal breeding beaches. *J. Zool., Lond.* **164**, 61–92.

Hamilton, J. E. (1956). Scent of otariids. *Nature, Lond.* **177**, 900.

Harrison, R. J. and Kooyman, G. L. (1968). General physiology of pinnipedia. In *The behavior and physiology of pinnipeds* (ed. R. J. Harrison). Appleton-Century-Crofts, New York.

Hartman, D. S. (1969). Florida's manatees: mermaids in peril. *Nat. geogr. Mag.* **136**, 343–53.

Herman, L. M. and Tavolga, W. N. (1980). The communication system of cetaceans. In *Cetacean behaviour: mechanisms and functions* (ed. L. M. Herman). Wiley, New York.

Jacobs, M. S., Morgane, P. J., and McFarland, W. L. (1971). The anatomy of the brain of the bottlenose dolphin (*Tursiops truncatus*). Rhinic lobe (rhinencephalon). I. The paleocortex. *J. comp. Neurol.* **141**, 205–72.

Kamiya, T. and Pirlot, P. (1974). Brain morphogenesis in *Stenella caeruleoalba*. *Scient. Rep. Whales Res. Inst.* **26**, 245–53.

Kaufman, G. W., Siniff, D. B., and Reichle, R. (1975). Colony behaviour of Weddell seals, *Leptyonychotes weddelli*, at Hutton Cliffs, Antarctica. *Rapp. P.-v. Réun. Cons. perm. int. Explor. Mer* **169**, 228–46.

Kellogg, W. N. (1961). *Porpoises and sonar.* University of Chicago Press.

King, J. E. (1964). *Seals of the world.* British Museum (Natural History), London.

Kingdon, J. (1971). *East African mammals*, Vol. 1, Academic Press, London.

Kinne, O. (1975). Orienting in space: animals: mammals. In *Marine ecology*, Vol. 2 *Physiological mechanisms*, Part 2 (ed. O. Kinne) pp. 702–852. Wiley, London.

Kleinenberg, S. E., Yablokov, A. V., Bel'Kovich, B. M., and Tarasevich, M. N. (1969). *Beluga (Delphinapterus leucas). Investigation of the species.* Israel Program for Scientific Translations, Jerusalem.

Kruger, L. (1959). The thalamus of the dolphin (*Tursiops truncatus*) and comparison with other mammals. *J. comp. Neurol.* **111**, 133–94.

Kubota, K. (1968). Comparative anatomical and neurohistological observations on the tongue of the northern fur seal (*Callorhinus ursinus*). *Anat. Rec.* **161**, 257–65.

Kuzin, A. Y. and Sobolevsky, Y. I. (1976). Morphological and functional characteristics of the fur seal's respiratory system. *Proc. 6th All-Union Conf. Study of Marine Mammals*, Kiev. Joint Publishing Service, Arlington, Virginia.

Kuznetsov, V. B. (1974). A method of studying chemoreception in the Black Sea bottlenose dolphin (*Tursiops truncatus*). In *Morphology, physiology and acoustics of marine mammals*. Nauka Publishing House, Moscow.

— (1978). Ability to communicate chemically and to transform information about chemical stimuli in the Black Sea *Tursiops*. *VII – aya Vses. Knof. Morsk. Mlekopitayushchim.* Simpherspol 178–80. [In Russian.]

Ling, J. K. (1965). Functional significance of sweat glands and sebaceous glands in seals. *Nature, Lond.* **208**, 560–2.

— (1974). The integument of marine mammals. In *Functional anatomy of marine mammals*, Vol. 2 (ed. R. J. Harrison) p. 336. Academic Press, New York.

Lowell, W. R. and Flanigan, W. F. Jr, (1980). Marine mammal chemoreception. *Mammal Rev.* **10**, 53–9.

Moore, J. C. (1956). Observations of manatees in aggregations. *Am. Mus. Novit.* **1811**, 1–24.

Norris, K. S. and Dahl, T. P. (1980). The structure and functions of cetacean schools. In *Cetacean behavior: mechanisms and functions* (ed. L. M. Herman) Wiley, New York.

Ono, K. A. (1972). Mother–pup interaction in the Stellar sea lion (*Eumetopias jubatus*). *Proc. 9th Ann. Conf. Biological Sonar and Diving Mammals.* Stanford Research Institute, Menlo Park, California.

Peterson, R. S. and Bartholomew, G. A. (1967). The natural history and behavior of the Californian sea lion. *Am. Soc. Mammal.* Special Publication No. 1.

Puente, A. E. and Dewsbury, D. A. (1976). Courtship and copulatory behavior of bottlenosed dolphins (*Tursiops truncatus*). *Cetology* **21**, 1-9.

Quay, W. N. and Mitchell, E. D. (1971). Structure and sensory apparatus of oral remnants of nasopalatine canals in the fin whale (*Balaenoptera physalus L.*). *J. Morph.* **134**, 271-80.

Schaffer, J. (1940). *Die Hautdrüsenorgane der Säugetiere*. Urban and Schwarzenberg, Berlin.

Scheffer, V. B. (1958). *Seals, sea lions and walruses*. Stanford University Press.

Sergeant, D. E. (1970). Migration and orientation in harp seals. In *Proc. Ann. Conf. Biological Sonar and Diving Mammals*, pp. 123-31. Stanford Research Institute, Menlo Park, California.

Slijper, E. J. (1962). *Whales*. Hutchinson, London.

Sokolov, V. E. and Kuznetsov, V. B. (1971). Chemoreception in the Black Sea dolphin. *Doklady Akad. Nauk SSR* **201**, 998-1000. [Transl. by Consultants Bureau, New York, 768-70.]

— and Volkova, O. V. (1973). Structure of the dolphin's tongue. In *Morphology and ecology of marine mammals* (ed. K. K. Chapskii and V. E. Sokolov). Israel Program for Scientific Translations, Jerusalem.

Stirling, I. (1970). Observations on the behavior of the New Zealand fur seal (*Arctocephalus forsteri*). *J. Mammal.* **51**, 766-78.

Tavolga, M. C. and Essapian, F. S. (1957). The behavior of the bottlenose dolphin (*Tursiops truncatus*): mating, pregnancy, parturition and mother-infant behavior. *Zoologica* **42**, 11-32.

Vladimirov, V. A. (1975). Female fur seal recognition of her pup. In *Marine mammals*, Part I. Publishing House Naukova dumka.

Walker, E. P. (1975). *Mammals of the world*, 3rd edn. The Johns Hopkins Press, Baltimore.

Yablokov, A. V., Bel'Kovich, V. M., and Borison, V. I. (1972). *Whales and dolphins*. Israel Program for Scientific Translations, Jerusalem.

# 17 Armadillos, sloths, anteaters, and pangolins: orders Edentata and Pholidota

RICHARD E. BROWN

## Edentata

The order Edentata includes three families: the armadillos (*Dasypodidae*), the sloths (*Bradypodidae*), and the anteaters (*Myrmecophagidae*), all of which are found exclusively in the Americas. Although differing greatly in external characteristics, the edentata are the only mammals to possess xenarthrales, extra articulations between the lumbar vertebrae. Only the anteaters are toothless. Edentata have primitive brains with large olfactory bulbs and few neocortical convolutions. Olfactory communication is probably important in the social behaviour of Edentata but little is known about this (Walker 1975; Grzimek 1975; Eisenberg 1981).

### Myrmecophagidae

There are three genera of anteaters found from Mexico to South America, each with only one species: the giant anteater (*Myrmecophaga tridactyla*), the Tamandua or collared anteater (*Tamandua tetradactyla*), and the silky or two-toed anteater (*Cyclopes didactylus*). All anteaters lack teeth and feed on ants, termites, and other insects which are licked up by the saliva-coated tongue. The giant anteater is diurnal and terrestrial, while the other two species are nocturnal and arboreal. They are all solitary except during the breeding season and for the occurrence of mother-young pairs. The young are carried on the mother's back or cling to her tail for several months and may forage with the mother until two years of age (Walker 1975; Eisenberg 1981).

Tamandua have apocrine sebaceous glands in the eyelids and the anogenital area (Machida, Giacometti, and Perkins 1966) and an anal gland (Pocock 1924*b*; Schaffer 1940). The anal gland may be the source of the characteristic unpleasant odour emitted by the Tamandua when it is excited (Walker 1975; Grzimek 1975). Giant anteaters (*Myrmecophaga*) also have an anal gland and nasal glands, and males may have preputial glands (Pocock 1924*b*; Schaffer 1940). The pygmy anteater (*Cyclopes didactylus*) has a cheek, or facial, gland (Schaffer 1940). Giant anteaters also respond to the odour of their own saliva (McAdam and Way 1967) and may use this for communication.

The giant anteater rapidly learns to discriminate between artificial odours (peppermint, eucalyptus, camphor, and formic acid) in a T-maze for food reward (McAdam and Way 1967), but no information is available on the ability of anteaters to discriminate among conspecifics by their odours.

## Bradypodidae

There are two genera of Bradypodidae found in Central and South America: the three-toed sloths (*Bradypus*), which have about six species, and the two-toed sloths (*Choloepus*), which have two species. All sloths are strictly arboreal nocturnal animals which hang beneath branches, moving slowly arm-over-arm to feed on leaves, blossoms, and fruits. Sloths are not able to stand or walk on the ground. Although the olfactory sense of sloths does not appear to be as acute as that of other Edentata (Grzimek 1975), it has been suggested that sloths depend on smell and touch more than on vision to locate their food (Walker 1975).

Both *Bradypus* and *Choloepus* have large anal glands (Schaffer 1940). Sloths also have a dorsal gland which may be associated with a coloured patch of fur, possibly the result of the secretion of the gland staining the fur (Schaffer 1940). Male sloths mark branches with secretions from their anal glands and their nostrils but the function of these marks is unknown. Eisenberg (1981) suggests that the nasal secretions may eminate from the Harderian glands of the orbit and that these may be homologous with the eye secretions of the tenrecoid insectivores (see Chapter 4).

Young sloths cling to the hair on their mother's breast immediately after birth and remain with her until nine months of age, but nothing seems to be known about the role of olfaction in mother–young relationships or in other social interactions between sloths.

## Dasypodidae

The outstanding characteristic of the Dasypodidae (armadillos) is their covering of armour-like skin on the back, sides, head, and limbs. There are nine genera (20 species) of armadillos, all of which are primarily nocturnal burrowing animals which feed on insects, plants, small vertebrates, and carrion. Armadillos are mainly solitary but sometimes are seen in pairs and in small bands. Armadillos are 'olfactory specialists', using scent to locate food and, presumably, for social communication (Walker 1975; Grzimek 1975; Eisenberg 1981).

Some species, such as the six-banded armadillo (*Euphractus sexcinctus*) and the hairy armadillo (*Euphractus villosus*), have dorsal glands which secrete through the carapace by means of two to four holes (Schaffer 1940; Grzimek 1975). The hairy armadillo (*Euphractus villosus*) has three glandular pits in its pectoral armour (see Fig. 17.1). The dorsal glands, which secrete through these pits, appear at birth and are equally developed in both sexes (Pocock 1913). The smell is 'nauseous' and Pocock suggests that it may function to protect the armadillos from their enemies or to enable them to identify and track group members.

All armadillo species have anal glands (Grzimek 1975). The nine-banded armadillo (*Dasypus*) has paired anal glands with 'folded walls invested in muscular tissue and containing a yellow nauseous secretion' (Pocock 1924b, p. 1021).

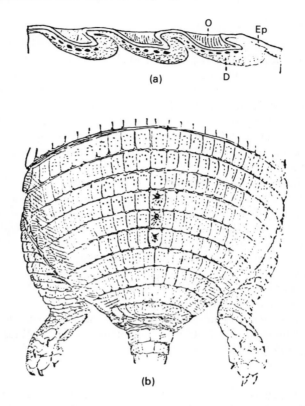

Fig. 17.1. (a) Longitudinal section of the three glands in a newborn hairy armadillo (*Euphractus villosus*). Abbreviations: Ep = epidermis; D = glandular
thickened dermis; O = orifice of pit. (b) The three glandular pits on the pelvic
carapace of a newborn hairy armadillo. (Reprinted from Pocock (1913) with
permission from the Zoological Society of London.)

*Dasypus* also have ear glands, glands of the eye lid, and glands on the sole of the
foot (Schaffer 1940).

Nothing seems to be known about the role of olfaction in the social behaviour
of the Dasypodidae.

## Pholidota

The order Pholidota consists of one genus (*Manis*) with seven species; four in
Africa and three in Asia. These scaly anteaters have no teeth and were at one
time classified as Edentata (Walker 1975; Kingdon 1971; Grzimek 1975). The
most outstanding physical characteristic of the pangolins is the scales which
cover the body, except for the underside, so that the animal resembles a pinecone.

Some species are completely terrestrial (e.g. *M. gigantea* and *M. temminicki*), while others are primarily arboreal (e.g. *M. tetradactyla* and *M. javanica*), and still others are semi–arboreal (e.g. *M. tricuspis*). Except for *M. tetradactyla*, all pangolins are nocturnal, sleeping in dens, often excavated from termite nests, or in the forked branches of trees during the day and foraging at night. Pangolins are solitary except during the breeding season when they may nest in pairs. They have large home ranges (Pages 1975) with well travelled paths from their burrows to their feeding sites. They appear to locate their food (ants and termites) by scent (Walker 1975; Kingdon 1971) and mark their trails and other areas of the home range with urine, faeces, or glandular secretions (Kingdon 1971; Eisenberg 1981).

The defensive behaviour of pangolins consists of escaping to their burrow or rolling in a ball, protecting their soft underside with their tails. Pangolins roll up so tightly that it is difficult to unroll them. If a predator persists in attacking, it is sprayed with urine and anal gland secretion. These jets of ill-smelling liquid are an effective deterrent against dogs and possibly other animals (Pages 1968; Walker 1975; Kingdon 1971).

The social life of pangolins appears to be dominated by the sense of smell (Kingdon 1971; Delany and Happold 1979) but details of olfactory communication are sketchy. All species of pangolins possess anal glands. These are present in both sexes and in sub-adults as well as adults (Pocock 1924*a*; Schaffer 1940; Pages 1968). Males, however, have larger, more complex anal glands than females. These glands secrete into anal sacs in which the secretion can be stored before it is ejected for defence (Pages 1968). In addition to its defensive function, the anal sac secretion may be used for sex or individual recognition, as pangolins sniff each other's anal regions when they meet (Kingdon 1971; Pages 1972*a*). The anal glands of the giant pangolin (*M. gigantea*) are filled with 'a white waxy substance with a powerful odour like stale chicken dung' (Kingdon 1971). Secretions from the anal sacs as well as the sebaceous and sudoriferous circumanal glands may play a role in aggressive and sexual behaviour (Schaffer 1940; Pages 1968, 1970, 1972*a*) (see Figs. 17.2 and 17.3).

Some pangolins, such as the Chinese pangolin (*Manis pentadactyla*), have glands on the soles of the feet and a gland on the ventral surface of the neck (Pocock 1926; Schaffer 1940) and pangolins have huge salivary glands which are used to coat the tongue with mucus for catching insects. It is possible that saliva acts as a chemical signal, but this has not·been investigated.

Pangolins scent mark by rubbing the anogenital area on trees and other objects, by squirting urine, and by spraying the contents of the anal sac (Pages 1968) (see Fig. 17.4). These odours play an important role in the social behaviour of pangolins, being used for sex, rut, dominance, and possibly individual identification. Female odours may change over the oestrous cycle (Pages 1968).

Pangolins have one young per year and show considerable maternal behaviour. Arboreal species carry the young on the mother's tail soon after it is born, but

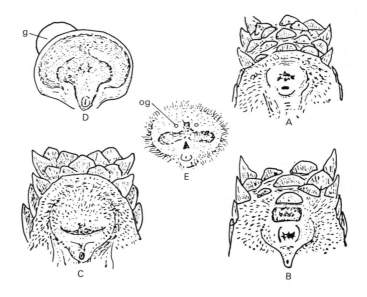

Fig. 17.2. Anal and genital areas of Pholidota. A. Anal-genital area of a female *Manis javanica*. B. Anal-genital area of a young male *Manis pentadactyla* showing the oblong glandular depression behind the anal eminence and the disarrangement of the scales at the base of the tail. C. Anal-genital area of an adult male *Manis tricuspis* showing the wide orifice of the anal sac. D. Anal sac of a male *Manis tricuspis* dilated to show one of the anal glands (g). E. Anal sac of a female *Manis tricuspis* expanded to show the orifices of the anal glands (og). (Reprinted from Pocock (1924a) with permission from the Zoological Society of London.)

in terrestrial species the young remains in the burrow for two to four weeks before being carried on the mother's tail (Walker 1975; Kingdon 1971; Grzimek 1975). The mother curls around its young when sleeping and for protection, and odours may be used for mother–infant recognition (Pages 1972b).

## References

Delany, M. J. and Happold, D. C. D. (1979). *Ecology of African mammals.* Longman, New York.

Eisenberg, J. F. (1981). *The mammalian radiations.* University of Chicago Press.

Grzimek, B. (1975). *Grzimek's animal life encyclopedia*, Vol. 11. Van Nostrand Reinhold, New York.

Kingdon, J. (1971). *East African mammals*, Vol. 1. Academic Press, London.

McAdam, D. W. and Way, J. S. (1967). Olfactory discrimination in the giant anteater. *Nature, Lond.* **214**, 316–17.

Machida, H., Giacometti, L., and Perkins, E. (1966). The skin of the lesser anteater (*Tamandua tetradactyla*). *J. Mammal.* **47**, 280–6.

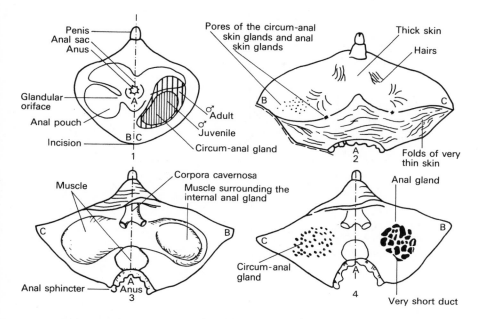

Fig. 17.3. The anal-genital region of the pangolins *Manis tricuspis* and *Manis longicaudata* showing the locations of the anal gland, the anal pouch, the perianal gland and the circumanal glands. 1. The anogenital area with a simple diagram of the glands on the right and a more detailed diagram on the left. 2. An external view of the anogenital area after an incision from the anus along the line A–B–C. 3. An internal view of the same region as the previous sample. 4. An internal view of this region after dissection. (Reprinted from Pages (1968).)

Pages, E. (1968). Les glandes odorantes des pangolins arboricoles (*M. tricuspis et M. longicaudata*): morphologie, développement et rôles. *T. Gabonica* **4**, 353–400.

— (1970). Sur l'écologie et les adaptations de l'orycterope et des pangolins sympatriques du Gabon. *Biol. Gabonica* **6**, 27–92.

— (1972*a*). Comportement aggressif et sexual chez les pangolins arboricoles (*Manis tricuspis et M. longicaudata*). *Biol. Gabonica* **8**, 1–62.

— (1972*b*). Comportement maternal et développement du jeune chez un pangolin arboricole (*M. tricuspis*). *Biol. Gabonica* **8**, 63–120.

— (1975). Étude éco-éthologique de *Manis tricuspis* par radio-tracking. *Mammalia* **39**, 613–41.

Pocock, R. I. (1913). On the dorsal glands in armadillos. *Proc. zool. Soc. Lond.* 1913, 1099–103.

— (1924*a*). The external characters of the pangolins (Manidae). *Proc. zool. Soc. Lond.* 1924, 707–23.

— (1924*b*). The external characters of the South American edentates. *Proc. zool. Soc. Lond.* 1924, 983–1031.

— (1926). The external character of an adult female Chinese pangolian (*Manis*

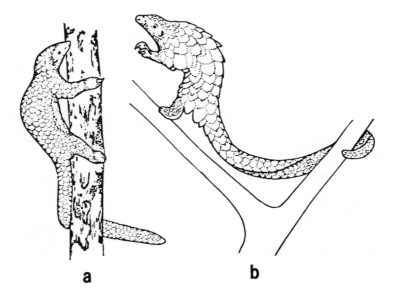

Fig. 17.4. (a) A young *Manis tricuspis* marking a branch with the secretions from the circumanal glands. (b) A female *Manis longicaudatus* marking a branch with a mixture of a drop of the secretion from the glands of the anal pouch and a small amount of urine. (Reprinted from Pages (1968).)

*pentadactyla*) exhibited in the Society's Garden. *Proc. zool. Soc. Lond.* 213–20.

Schaffer, J. (1940). *Die Hautdrüsenorgane der Säugetiere*. Urban Schwarzenberg, Berlin.

Walker, E. P. (1975). *Mammals of the world* (3rd edn.) Johns Hopkins Press, Baltimore.

# 18 The primates I: order Anthropoidea

GISELA EPPLE

## Introduction

Chemical signals are important means of social and sexual communication in most mammalian groups (see other chapters in this volume). Our most complete understanding of the role of chemical communication in socio-sexual behaviour and in the regulation of reproductive physiology is derived from a great number of studies on rodents, which have been reviewed recently by Aron (1979); Cheal (1975); Fass and Stevens (1977); Macrides, Bartke, and Svare (1977); and Thiessen (1976). In comparison to rodents, little is known about the role of chemical communication in primates. In contrast to many other mammalian species, primates, and particularly Old World monkeys and apes, possess rich, multimodular communication systems of combining chemical signals, touch, large vocabularies of variable calls, and complex repertoires of postures and facial expressions (Marler 1975). These enable the individual to express fine details of its motivational state and thus to function as a member of a complex society.

While communication by visual and vocal signals has been studied intensively during recent years, our knowledge of primate chemical communication lags far behind. This is not surprising, considering the limitations placed on the human observer by his or her own sensory capacities. Because of a relatively poorly developed sense of smell, the human observer is much more likely to record a visual or vocal pattern than an odour, and much of the chemical communication which may occur in a primate group escapes notice during direct observation. In spite of these limitations, evidence for the use of chemical signals in primate communication has accumulated during recent years. Much of this evidence is derived from the fact that many species possess specialized skin glands which are interpreted as scent glands, show scent-marking behaviours resulting in the deposition of the secretions from these glands or of excretions such as faeces or urine, and sniff as well as lick while exploring conspecifics and their scent marks.

Chemical signals in primates are perhaps predominantly odours. The frequent sniffing of these signals suggests that olfaction is the major sense involved in their perception. In addition to the main olfactory system, perception of chemical communicants via the vomeronasal organ has to be considered, since a functioning vomeronasal organ and distinctly formed accessory olfactory bulbs are present in many prosimians and South American monkeys (cf. Epple and Moulton 1978; Maier 1981). Some primates, when investigating urine or scent marks, show behaviours which may bring chemical stimuli into the mouth and the nasopalatine ducts, from where they could reach the vomeronasal organ. *Nycticebus coucang*, for example, sniffs the urine of conspecifics while showing rapid oscillating protrusions of the tongue (Seitz 1969). Several species of South American marmoset monkeys frequently sniff urine, scent marks, or the bodies of conspecifics while showing rhythmical protrusions of the tongue, whose tip may touch the stimulus. *Saimiri sciureus* males show an almost identical behaviour when sniffing the urine and the genitals of females (personal observations). In marmosets, the sniffing behaviour is occasionally followed by a facial expression very reminiscent of the 'Flehmen' face, which is supposedly involved in perception of stimuli via the vomeronasal organ (Knappe 1964). *Lemur catta* shows a similar facial expression during investigation of conspecific chemical stimuli (Jolly 1966). Some species not only sniff and lick the various chemical stimuli provided by conspecifics but also ingest excretions and secretions. This suggests that taste might play a role in their perception and/or that some of the stimuli affect the receiving organism via ingestion. Spider monkey males, for instance, frequently taste and drink the urine of females in all stages of the reproductive cycle (Klein 1971).

The vomeronasal and main olfactory system are both relatively well developed in most prosimians and in South American monkeys. In Old World monkeys and apes, the vomeronasal organ is believed to be absent, although it may sometimes be present during foetal life. The degree of development of the main olfactory structures tends to be inversely proportional to the degree of neocorticalization, of hand mobility and of frontality of orbits (cf. Epple and Moulton 1978; Keverne 1978; Maier 1981; Schilling 1979). The fact that intensive sniffing and licking of the bodies of conspecifics in social, and particularly in sexual situations, is so widespread amongst primates, and is conspicuous even in most catarrhini (Keverne 1978), suggests that chemical cues are used by most primate species. The importance of chemical communication relative to other sensory modalities, however, seems to vary widely among species. The extent of differentiation of the peripheral and central olfactory and accessory olfactory systems, as well as the presence of specialized scent glands and of marking behaviour in a given species may be a rough indicator of the relative importance of chemical signals in the communicatory system of the species. In the following, I shall briefly summarize some of the evidence for chemical communication and discuss a few examples in more detail. Chapter 19 uses the work of Epple and Smith on a South American monkey as a case in point and as an example of the problems and complexities encountered in the analysis of chemo-communicatory systems.

### Families and genera of living primates

Much of the material presented in this chapter deals with the presence of specialized scent glands and scent-marking behaviour in various primate species. Therefore the following taxonomical list, adapted from Hershkovitz (1977), is provided for the orientation of the reader. One of the criteria which separate prosimian and simian primates is the complexity of the nasal apparatus (Cave 1967).

Based on the anatomy of the rhinarium and of the nasal cavity, most of the prosimians are regarded as Strepsirhini (complex rhinarium of insectivore type), while the simian primates (New and Old World monkeys and apes) comprise the Haplorhini (simple rhinarium). Reductions in the nasal anatomy of the tarsiers as compared to the other prosimians have resulted in nasal structures similar to those of South American monkeys, the reason for placing this prosimian with the Haplorhini (cf. Cave 1967; Hershkovitz 1977).

## ORDER PRIMATES

I. **Suborder Strepsirhini**
    A. Superfamily Lemuroidea
       1.  Family Lemuridae
      (a)  Subfamily Cheirogaleinae (mouse lemurs, dwarf lemurs, fork-striped lemur)
          *Microcebus, Cheirogaleus*
          *Allocebus, Phaner*
      (b)  Subfamily Lemurinae (lemurs)
          *Lepilemur, Hapalemur, Lemur, Varecia*
       2.  Family Indridae (avahi, siafkas, indri)
          *Avahi, Propithecus, Indri*
    B. Superfamily Lorisoidea
       1.  Family Galagidae (bush-babies)
          *Galago, Euoticus*
       2.  Family Lorisidae (lorises, angwantibo, potto)
          *Loris, Nycticebus, Arctocebus, Perodicticus*
    C. Superfamily Daubentonioidea
       1.  Family Daubentoniidae (aye-aye)
          *Daubentonia*

II. **Suborder Haplorhini**
    I'. *Infraorder Platyrrhini*
    A. Superfamily Tarsioidea
       1.  Family Tarsiidae (tarsiers)
          *Tarsius*
    II'. *Infraorder Platyrrhini*
       1.  Family Callitrichidae (marmosets, tamarins)
          *Cebuella, Callithrix, Saguinus, Leontopithecus*
       2.  Family Callimiconidae (Goeldi's monkeys)
          *Callimico*
       3.  Family Cebidae
         (a)  Subfamily Saimiriinae (squirrel monkeys)
            *Saimiri*
         (b)  Subfamily Aotinae (night monkeys)

*Aotus*
   (c)  Subfamily Callicebinae (titi monkeys)
      *Callicebus*
   (d)  Subfamily Alouattinae (howler monkeys)
      *Alouatta*
   (e)  Subfamily Pitheciinae (sakis, uacaris)
      *Pithecia, Chiropotes, Cacajao*
   (f)  Subfamily Cebinae (capuchins)
      *Cebus*
   (g)  Subfamily Atelinae (woolly and spider monkeys)
      *Lagothrix, Ateles, Brachyteles*

III'. *Infraorder Catarrhini*

A. Superfamily Cercopithecoidea
   1.  Family Cercopithecidae
      (a)  Subfamily Cercopithecinae (talapoin, guenons, patas monkey, mangabeys, macques, baboons)
         *Miopithecus, Cercopithecus, Erythrocebus, Allenopithecus, Cercocebus, Macaca, Papio, Theropithecus*
      (b)  Subfamily Colobinae (langurs, guerezas)
         *Presbytis, Pythagrix, Rhinopithecus, Nasalis, Colobus*

B. Superfamily Hominoidea
   1.  Family Hylobatidae (gibbons, siamang)
      *Hylobates, Symphalangus*
   2.  Family Pongidae (orang-utan, chimpanzee, gorilla)
      *Pongo, Pan, Gorilla*
   3.  Family Hominidae (man)
      *Homo*

**Sources of chemical signals**

The presence of morphologically specialized skin glands, interpreted as scenting organs, and the observation of stereotyped behaviour patterns concerned with depositing skin secretions or excretions into the environment, suggested the use of chemical communicants by primates long before much about primate behaviour was known (e.g. Brinkmann 1909, 1923; Schaffer 1940; Sonntag 1924). Not surprisingly, these and other visual components of chemosignalling still make up the bulk of our knowledge about primate chemical communication. In comparison, very little is known about the communicatory contents of chemical signals, their exact functions in controlling behaviour and their chemical nature.

As in other mammalian species, the primate body has many sources from which chemical signals may be derived. Urine, faeces, genital discharge, saliva, and skin secretions provide chemical communicants by themselves, in combination with each other or with environmentally derived odoriferous material (cf. Epple 1974*a*, *b*, 1976; Goldfoot 1981; Keverne 1978; Michael and Bonsall 1977; Nolte 1958; Schilling 1979). In most species, the skin of the general body surface is well supplied with sebaceous (holocrine) glands and apocrine sweat glands, both associated with hair follicles (Montagna 1972). The volar surfaces of hands and feet and the naked ventral tail end of the prehensile-tailed platyr-rhines are rich in eccrine sweat glands, which are also present on the hairy body surface of some platyrrhines and of all Old World simians (Montagna 1972). While the primary function of these glands is not the production of communica-tory signals, their secretions or bacterial breakdown products may significantly contribute to the general body odour of an individual.

In some body areas, sebaceous and/or apocrine glands tend to be larger or accumulate in greater numbers than in others, without forming macroscopically distinct scent glands. In general, sebaceous glands are larger and more numerous in the face, external auditory canal, and anogenital areas (Montagna 1972). For example, microscopic concentrations of apocrine and sebaceous glands occur in the normal-appearing epidermis in the circumgenital and perianal areas of some prosimians, platyrrhines, and catarrhines (cf. Montagna 1972; Schilling 1979) while in other species, such as some lemurs and in marmosets and tamarins, these areas are characterized by specialized, morphologically conspicuous scent glands (cf. Perkins 1966, 1968, 1969*a,b,c*; Schilling 1979; see Table 18.1). Similarly, accumulations of large apocrine glands on the mid-chest are found in many species. In some Old World simians, such as some baboons (*Papio anubis, P. doguera*), the green monkey (*Cercopithecus aethiops*), Syke's monkey (*C. mitis*), and macaque (*Macaca speciosa*), apocrine and eccrine glands accumulate on the chest without forming morphologically distinct scent glands (Machida, Perkins, and Montagna 1964; Montagna and Yun 1962*a*; Montagna, Machida, and Perkins 1966*a,b*). In other species, however, specialized gular, neck, sternal, or epigastric scent glands are found. These occur in many prosimians, all New World monkeys, some Old World monkeys (i.e. the drill and mandrill) and even in two hominoids, the gibbon and the orang-utan (see Table 18.1).

Table 18.1 lists the occurrence of specialized scent glands in the primates. Simple accumulations of skin glands, such as those described above, are not included in the table. The glands are classified according to their location on the body. It should be kept in mind, however, that similarly named scent glands located in the same area may show distinct morphological species differ-ences, which cannot be expressed in a table. The 'sternal gland' of many calli-trichids, for example, consists of a glandular field associated with a small brush of stiff coarse hair (Epple and Lorenz 1967), that of *Ateles* is represented by a pair of small raised apocrine fields, also associated with specialized hair (Schwarz 1937; Wislocki and Schultz 1925), and that of the orang-utan consists of a small invaginated pit on the mid-chest (Schultz 1921). Glands located in the same body areas also show differences in their histological composition, as indicated in Table 18.1. Moreover, the statement that apocrine glands are present in the scent organs of various species does not express the fact that there are great species differences in the types of apocrine gland among primates (cf. Ellis and Montagna 1964; Montagna 1972).

Table 18.1. The occurrence of scent glands in various areas of the primate body. + = a gland is present (+ without further specifications = a gland is present but details are not available; ? = the presence of a gland is suspected but unconfirmed; ♂ = the gland is present in males; ♀ = the gland is present in females (♂ > ♀ the gland is larger in males than in females); a = the gland has apocrine components; s = the gland has sebaceous components; e = the gland has eccrine components; i = interstitial cells are present.

| Species | Circumgenital (± suprapubic and perineal) areas | Circumanal area | Ventral area | Other areas | References |
|---|---|---|---|---|---|
| **Lemuridae** | | | | | |
| *Microcebus murinus* (lesser mouse lemur) | +, a, s, ♂ | | | | 28 |
| *M. coquereli* (Coquerel's mouse lemur) | +, a, s, ♂ | | neckgland? | | 28 |
| *Cheirogaleus major* (dwarf lemur) | +, a, ♂, ♀ | | | | 28 |
| *Phaner furcifer* (fork-crowned lemur) | | | neckgland s, ♂ > ♀ a, ♂ = ♀ | | 26 |
| *Hapalemur griseus* (gentle lemur) | | | | shoulder: brachial gland +, s, ♂, wrist: antebrachial gland +, a, i, ♂ > ♀ | 15 |
| *H. simus* (gentle lemur) | | | neckgland +, ♂, ♀ | shoulder: brachial gland +, wrist: antebrachial gland +. | 28 |
| *Lemur catta* (ring-tail lemur) | +, a, s, ♂, ♀ | +, s, ♂, ♀ | | shoulder: brachial gland +, s, ♂. wrist: antebrachial gland, +, a, i, ♂ > ♀ | 15 |

| Species | Circumgenital (± suprapubic and perineal) areas | Circumanal area | Ventral area | Other areas | References |
|---|---|---|---|---|---|
| **Lemuridae** (*cont.*) | | | | | |
| *L. fulvus* (brown lemur) | +, a, s, ♂, ♀ | + | | forehead +; wrist: antebrachial gland + ♂ (much less specialized than in *L. catta*). | 8, 28 |
| *L. albifrons* (white-fronted lemur) | | + | | | 28 |
| *L. mongoz* (mongoose lemur) | +, a, s, ♂ > ♀ | +, s, ♂ | | | 15, 28 |
| *L. macaco* (black lemur) | +, a, s, ♂ > ♀ | +, a, s, ♂, ♀ | | wrist: antebrachial gland +, a, s, ♂, ♀ (much less specialized than in *L. catta*) | 15, 27, 28 |
| *Varecia varoegatus* (ruffed lemur) | | | neckgland +, a, s, ♂ | | 26 |
| **Indriidae** | | | | | |
| *Avahi laniger* (avahi) | | | neckgland +, s, ♂ > ♀; a, ♂ = ♀ | | 26 |
| *Propithecus verreauxi* (sifaka) | | | neckgland +, a, s, ♂ | | 26 |
| *P. diadema* (sifaka) | | | neckgland +, a, s, ♂ | | 26 |
| *Indri indri* (indri) | ? | | neckgland ? | | 28 |
| **Galagidae** | | | | | |
| *Galago crassicaudatus* (greater galago) | +, a, s, ♂ > ♀ | +, s, ♂, ♀ | sternal gland +, a, ♂ = ♀; s, ♂ > ♀ | | 3, 28 |
| *G. senegalensis* (bush-baby) | ?, ♂, ♀ | | | | 28 |
| *G. demidovii* (bush baby) | +, a, ♂ > ♀ | | | | 28 |
| *Euoticus elegantulus* (bush-baby) | +, a, ♂ > ♀ | | | | 28 |

Table 18.1 (*cont.*)

| Species | Circumgenital (± suprapubic and perineal) areas | Circumanal area | Ventral area | Other areas | References |
|---|---|---|---|---|---|
| **Lorisidae** | | | | | |
| *Loris tardigradus* (slender loris) | +, a, ♂ > ♀ | | | elbow gland +, a, ♂, ♀ | 4, 28 |
| *Nycticebus coucang* (slow loris) | ?, ♂, ♀ | | | elbow gland +, a, ♂, ♀ | 4, 28 |
| *Perodicticus potto* (potto) | inguinal gland +, a > s, ♂ > ♀; pair of preclitorial glands +, s, ♀ | | | | 17, 28 |
| *Arctocebus calabarensis* (angwantibo) | +, a, s, ♂ > ♀ | | | | 18, 28 |
| **Tarsiidae** | | | | | |
| *Tarsius syrichta* (Philippine tarsier) | | + | epigastric gland +, s, a, ♂, ♀ | | 1, 9 |
| **Callitrichidae** | | | | | |
| *Cebuella pygmaea* (pygmy marmoset) | +, a, s, ♂, ♀ | | sternal gland +, a, ♂, ♀ (?) | | 3, 6, 20, 31 |
| *Callithrix jacchus* (common marmoset) | +, a, s, ♂, ♀ | | sternal gland +, a, s, ♂ > ♀ | | 6, 32 |
| *C. argentata* (silver marmoset) | +, a, s, ♂, ♀ | | sternal gland +, a, ♂, ♀ | | 6, 21 |
| *Saguinus fuscicollis* (saddle back tamarin) | +, a, s, ♀ > ♂ | | sternal gland +, a, ♂, ♀ | | 6, 19, 34 |
| *S. oedipus* (cotton top pinché) | +, a, s, ♀ > ♂ | | sternal gland +, a, ♂, ♀ | | 6, 22 |
| *S. o. geoffroyi* (red naped tamarin) | +, ♀ > ♂(?) | | sternal gland? ♂, ♀ | | 6, 34 |
| *S. mystax* (black-chested moustached tamarin) | +, ♂, ♀ | | sternal gland +, ♂, ♀ | | 6, 34 |

| Species | Circumgenital (± suprapubic and perineal) areas | Circumanal area | Ventral area | Other areas | References |
|---|---|---|---|---|---|
| **Callitrichidae** (*cont.*) | | | | | |
| *S. labiatus* (red-chested moustached tamarin) | +, ♂, ♀ | | sternal gland +, ♂, ♀ | | 34 |
| *S. midas* ssp (golden-handed and black-handed tamarins) | +, ♂, ♀ | | sternal gland +, ♂, ♀ | | 2, 6, 34 |
| *S. leucopus* (silvery-brown, bare-face tamarin) | +, ♂, ♀ | | sternal gland +, ♂, ♀ | | 34 |
| *Leonto-pithecus rosalia* (lion tamarin) | +, ♂, ♀ | | sternal gland +, ♂, ♀ | | 5 |
| **Callimiconidae** | | | | | |
| *Callimico goeldii* (Goeldi's monkey) | +, s, a, ♂, ♀ | +, s, a, ♂, ♀ | sternal gland +, s, a, ♂, ♀ | | 6, 23 |
| **Cebidae** | | | | | |
| *Saimiri sciureus* (squirrel monkey) | + | + | sternal gland +, a, ♂, ♀ | | 11, 13 |
| *Aotus trivirgatus* (night monkey) | +, a | | sternal gland +, a, ♂ > ♀ | ventral tail root: caudal gland +, s, a, ♂ > ♀ | 6, 7 |
| *Callicebus moloch* (titi monkey) | ? | | sternal gland +, ♂, ♀ | | 14, 34 |
| *Alouatta seniculus* (red howler monkey) | | | gular gland +, ♂, ♀ | | 6 |
| *Pithecia pithecia* (pale-headed saki) | | | gular-sternal gland +, a, s, ♂, ♀, (?) | | 6, 11 |
| *P. monachus* (monk saki) | | | gular-sternal gland +, ♂, ♀ | | 6 |

Table 18.1 (*cont.*)

| Species | Circumgenital (± suprapubic and perineal) areas | Circumanal area | Ventral area | Other areas | References |
|---|---|---|---|---|---|
| **Cebidae (*cont.*)** | | | | | |
| *Cacajao rubicundus* (red uakari) | | | epigastric gland +, a, ♂, ♀(?) | | 6, 24 |
| *Cebus apella* (black-capped capuchin) | | | epigastric gland +, ♂, ♀(?) | | 6 |
| *C. capucinus* (white-throated capuchin) | | | sternal-epigastric gland +, ♂, ♀ | | 6, 11 |
| *Lagothrix lagotricha* (woolly monkey) | | | sternal gland +, ♂, ♀ | | 6 |
| *Ateles geoffroyi* (black-headed spider monkey) | | | sternal gland +, a, s, ♂, ♀ | | 30, 33 |
| *Brachyteles arachnoides* (woolly spider monkey) | | | gular gland +, ♂, ♀(?) | | 6 |
| **Cercopithecidae** | | | | | |
| *Papio leucophaeus* (drill) | | | sternal gland +, a, s, e (?), ♂ > ♀ | | 10, 12 |
| *P. sphinx* (mandrill) | | | sternal gland +, a, s, e(?), ♂ > ♀ | | 10, 12 |
| *Theropithecus gelada* (gelada) | | | sternal gland ?, a, e | | 12 |
| **Hominoidae** | | | | | |
| *Hylobates moloch* (silvery gibbon) | | | sternal gland + | | 25 |
| *Pan paniscus* (chimpanzee) | | | | axillary gland +, a, s, e, ♂, ♀ | 16 |
| *Gorilla gorilla* (gorilla) | | | | axillary gland, a, s, e, ♂, ♀ | 16 |
| *Pongo pygmaeus* (orang-utan) | | | sternal gland ('pit') a, s, ♂, ♀ | | 29 |

Another set of variables which cannot properly be expressed in tabular form is the effect of age, sex, hormonal condition, and rank on the degree of differentiation of the scent glands. Scent glands often are not present in infants but develop shortly before or around the time of puberty (e.g. Schwarz 1937). However, little systematic work has been done on the effects of age on primate scent glands and a strict time correlation between puberty and glandular development has not been demonstrated (e.g. Sutcliffe and Poole 1978, for *Callithrix jacchus*). In some species the scent glands are sexually dimorphic. For example, the brachial–antebrachial complex of *Lemur catta* is fully developed only in the male (cf. Montagna 1962), while in the tamarins *Saguinus oedipus* and *S. fuscicollis*, the circumgenital–suprapubic glands of females are larger than those of males (Epple, Alveario, and Katz 1982, Wislocki 1930). In some species social status may also influence the gland size. Epple and Lorenz (1967) and Sutcliffe and Poole (1978), for example, have found that in *Callithrix jacchus* the sternal gland is best developed in adult, dominant males.

### Deposition, distribution, and function of chemical signals

A number of behaviours, not primarily performed as communicatory acts, may result in the discharge and/or deposition of chemical signals. The presence of faeces and urine in the environment probably provides a number of chemical cues. Urination and defecation are autonomic responses which often occur as reactions to fear-eliciting stimuli (Eisenberg and Kleiman 1972). Oppenheimer (1977) has reported that several species of cebids show this response in the wild and has suggested that the odour of the excrement could signal distress. For the Nilgiri langur (*Presbytis johnii*) Poirier (1970) has suggested that the odour of the faeces produced in copious amounts during territorial disputes between two groups may have a communicatory function. Moreover, the scent of their faecal deposits may also serve to mark the sleeping sites (Poirier 1970).

References
1. Arao and Perkins (1969).
2. Christen (1974).
3. Dixson (1976).
4. Ellis and Montagna (1964).
5. Epple (1972).
6. Epple and Lorenz (1967).
7. Hanson and Montagna (1962).
8. Harrington (1975).
9. Hill (1951).
10. Hill (1954).
11. Hill (1960).
12. Hill (1970).
13. Machida, Perkins and Hu (1967).
14. Mason (1966).
15. Montagna (1962).
16. Montagna (1972).
17. Montagna and Yun (1962*b*).
18. Montagna *et al.* (1966*a,b*).
19. Perkins (1966).
20. Perkins (1968).
21. Perkins (1969*a*).
22. Perkins (1969*b*).
23. Perkins (1969*c*).
24. Perkins, Arao, and Uno (1968).
25. Pocock (1925).
26. Rumpler and Andriamiandra (1971).
27. Rumpler and Oddou (1970).
28. Schilling (1979).
29. Schultz (1921).
30. Schwarz (1937).
31. Starck (1969).
32. Sutcliffe and Poole (1978).
33. Wislocki and Schultz (1925).
34. Personal observations.

The secretions from the various skin glands, regardless of whether these are specialized scent glands or not, are distributed over much of the body surface as an animal scratches and grooms himself. In this way, the secretions or their bacterial breakdown products, perhaps in combination with faecal and urinary residues, saliva, or genital discharge, contribute to the body odour emanating from an individual. Contact between the body and the physical environment may result in the deposition of some of this material. Contact between conspecifics during huddling, grooming and other social activities probably transfers some of the material from individual to individual. These behaviours, in addition to defecation and urination, result in a passive discharge or deposition of material which may carry communicatory information.

In addition to these passive modes of transfer, specialized, stereotyped scent-marking behaviours have evolved in many primates. These result in the application of material to items in the environment, the animal's own body and those of conspecifics. Table 18.2 lists the various types of scent-marking behaviours. The fine details of the motor patterns, however, cannot be expressed in tabular form. Schilling (1979) has recently described the scent-marking behaviour of prosimians in detail and the reader is referred to his review for further information on this primate group and to the references in the table for details on the other groups. Many marking patterns consist of rubbing movements during which faeces, urine, genital discharge, saliva, or glandular secretions, sometimes in combination with each other, are deposited. Some species have evolved highly complex marking patterns, involving the mixing of materials and their application to various parts of their own body, the environment, and sometimes the bodies of social partners. Since such complex behaviours cannot be described in a table, we shall discuss a few examples in detail. The behavioural functions of chemical signals have recently been reviewed by Epple (1972, 1974a,b, 1976), Oppenheimer (1977), and by Schilling (1979) and are further illustrated in Chapter 19 on *Saguinus fuscicollis*. As discussed in detail in the reviews cited above, chemical signals communicate information on the identity of the species and individual as well as its gender and its reproductive and emotional condition. They influence a great number of sociosexual behaviours, such as parent–infant interactions, intragroup and intergroup aggression, spacing, territoriality, courtship, pair bonding, and copulation. Instead of repeating this previously reviewed material, a discussion of these functions will be limited to a few examples.

*Example*—Lemur catta

The first example is the marking behaviour of the Madagascan ring-tailed lemur, which stands out for the complexity of its glandular apparatus as well as of its marking behaviour. *Lemur catta* has apocrine and sebaceous scent glands on the skin of the external genitalia and around the anus of both males and females (Montagna 1962). Glands which are perhaps secreting in both an apocrine and eccrine fashion are present on the volar surfaces of hands and feet (Montagna 1962). In addition, *Lemur catta* possesses a sebaceous brachial-gland organ near the shoulder which is only present in males, and an antebrachial scent organ at the wrist (Evans and Goy 1968; Montagna 1962). The antebrachial organ, better developed in males than females, consists of a glandular area associated with a strong horny spur, pierced by the openings of the apocrine ducts. The latter is absent in females. The glandular area is comprised of unusual apocrine glands and interstitial cells, which show histochemical and electronmicrosopical similarities to the steroidogenic interstitial cells of the testes. These cells may comprise an endocrine organ of unknown function (cf. Kneeland-Sisson and Fahrenbach 1967). The cytology of the interstitial cells varies in relation to the reproductive season. This variability is greater in females than in males, perhaps reflecting ovarian cyclicity (Kneeland-Sisson and Fahrenbach 1967). The antebrachial organ of males is testosterone dependent and its size undergoes seasonal variation (cf. Schilling 1979).

The complexity in the scent-marking behaviour of *Lemur catta* rivals that of its scent glands. Jolly (1966) and Evans and Goy (1968) provide detailed descriptions. Females rub their genitalia against the substrate either in a squatting position, by backing up and marking with the hind-legs extended and the feet planted on the ground, or by assuming a 'handstand' posture. During female marking the labia are deflected, their inner surfaces pressed against the substrate; in addition to the secretions of the genital skin glands, vaginal mucus and urine may be applied (Evans and Goy 1968). Males 'genital mark' less frequently than females, assuming the same postures. When males genital mark in a squatting position, a sub-penile concentration of sebaceous glands in the scrotal skin is specifically placed in contact with the marked surface, and glandular secretions are applied by a sex-specific side-to-side rubbing motion. Both sexes genital mark objects as well as social partners (Evans and Goy 1968). Males 'palmar mark' by pulling small branches through their hands, perhaps depositing secretions from the palmar sweat glands on to the twigs. The antebrachial organ, though present in both sexes, is only used for scent marking by males. Various forms of marking with this organ have been described by Evans and Goy (1968). 'Spur marking' by vigorously pulling the spur of the antebrachial organ across the object is a frequent and important pattern. The spur actually produces horizontal scars or scrapes on branches which may serve as visual signals as well. In another pattern the spurs are snapped across each other resulting in audible spur clicking.

'Shoulder rubbing' presumably results in the uptake of the sebaceous material accumulated in the ducts of the brachial glands on to the spur of the antebrachial organ, and mixing of secretions as the male touches his wrist to the brachial gland. The material accumulated on the antebrachial organ is transferred to the tail during 'tail marking', in which the tail is pulled forward between the hind legs and repeatedly pulled across the wrist, thus being anointed with the secretions

Table 18.2. Scent-marking behaviour in primates. In some cases (i.e. muzzle rubbing in some prosimians or rubbing of the ventral body surface) scent-marking function is inferred, although no specialized scent glands have been identified. + = material is deposited by rubbing the appropriate area against the substratum, by streaking urine or faeces or by rhythmic micturition; w = urine washing (see text); ? = rubbing or deposit of material believed to occur or to function as scent-marking behaviour but function unconfirmed; * = pattern is also used to scent mark the bodies of conspecifics; ▲ = most marmosets and tamarins mix a small amount of urine into the circumgenital scent mark, but urine marking alone is usually not shown; ● = in some Callitrichidae anogenital marking also involves a gland pad in the suprapubic area and can be extended into rubbing of the ventral surface (see Chapter 19).

| Species | Faeces | Urine | Skin glands | Other materials | References |
|---|---|---|---|---|---|
| **Lemuridae** | | | | | |
| *Microcebus murinus* (lesser mouse lemur) | + | +, w | anogenital +, muzzle +, ventral body surface + | saliva ? | 34 |
| *M. coquereli* (Coquerel's mouse lemur) | + | + | anogenital +, muzzle +, neck + | saliva + | 34 |
| *Cheirogaleus major* (dwarf lemur) | + (frequently!) | + | anogenital +, volar ?, muzzle + | | 34 |
| *C. medius* (dwarf lemur) | + | + | anogenital + | | 34 |
| *Phaner furcifer* (fork-crowned lemur) | ? | | neck +*, muzzle + | saliva ? | 34 |

| | | | | | |
|---|---|---|---|---|---|
| *Lepilemur* sp. (sportive lemur) | ? | + | anogenital ?, muzzle + | | 34 |
| *Hapalemur griseus* (gentle lemur) | ? | + | anogenital +, antebrachial +, brachial–antebrachial complex +, tail anointing +, palmar +, muzzle + | saliva ? | 34 |
| *H. simus* (gentle lemur) | | | neck +, antebrachial + | | 34 |
| *Lemur catta* (ring-tail lemur) | ? | ? | anogenital +*, antebrachial +, brachial–antebrachial complex +, tail anointing +, scalp +, palmar + | | 9, 34 |
| *L. fulvus* (brown lemur) | ? | ? | anogenital +*, scalp +, antebrachial ?, palmar +, palmar marking of tail + | | 13, 34 |
| *L. mongoz* (mongoose lemur) | | | anogenital +*, palmar ? | | 34 |
| *L. macaco* (black lemur) | + | | anogenital +*, muzzle +, palmar +, neck ?, palmar marking of tail + | | 32, 34 |
| *Varecia variegatus* (ruffed lemur) | | | anogenital ?, neck +, muzzle + | | 34 |
| **Indriidae** | | | | | |
| *Propithecus verreauxi* (sifaka) | + | + | anogenital +, neck + | | 34 |
| *Indri indri* (indri) | + | + | anogenital +, neck ?, ventral body surface + | | 34 |

Table 18.2 (*cont.*)

| Species | Faeces | Urine | Skin glands | Other materials | References |
|---|---|---|---|---|---|
| **Galagidae** | | | | | |
| *Galago crassi caudatus* (greater galago) | | +, w | anogenital +, sternal +, volar +, muzzle + | saliva ? | 1, 4, 34 |
| *G. senegalensis* (bush baby) | ? | +*, w | anogenital +, sternal +, volar +, muzzle + | saliva ? | 1, 34 |
| *G. demidovii* (bush baby) | + | +, w | anogenital +, volar +, muzzle + | saliva ? | 34 |
| *G. alleni* (bush baby) | | +, w | anogenital +, volar +, muzzle +, ventral body surface + | saliva ? | 34 |
| *Euoticus elegantulus* (bush baby) | | + | anogenital +, muzzle + | | 34 |
| **Lorisidae** | | | | | |
| *Loris tardigradus* (slender loris) | | +, w | volar ? | | 34 |
| *Nycticebus coucang* (slow loris) | + | + | anogenital +, neck +, volar +, muzzle + | | 34 |
| *Arctocebus calabarensis* (angwantibo) | + | + | anogenital +*, ventral body surface + | | 34 |
| *Perodicticus potto* (potto) | | + | anogenital +, volar ?, muzzle + | | 34 |

| | | | | |
|---|---|---|---|---|
| **Daubentoniidae** | | | | |
| *Daubentonia madagascariensis* (aye-aye) | + | anogenital +, volar ? | | 34 |
| **Tarsiidae** | | | | |
| *Tarsius bancanus* (Horsefield's tarsier) | + | anogenital ? | saliva + | 34 |
| *T. syrichta* (Phillipine tarsier) | +* | | | 34 |
| **Callitrichidae** | | | | |
| *Cebuella pygmaea* (pygmy marmoset) | +▲ | anogenital +*, sternal +, | saliva ? | 3 |
| *Callithrix jacchus* (common marmoset) | +▲ | anogenital +*, sternal +, muzzle + | saliva ? | 7, 8, 36, 39 |
| *C. argentata* (silver marmoset) | ?▲ | anogenital +*, sternal + | saliva ? | 7, 8, 39 |
| *Saguinus fuscicollis* (saddle-back tamarin) | +▲ | anogenital +•*, sternal + | | 7 |
| *S. oedipus* (cotton-top pinché) | +▲ | anogenital +•*, sternal + | | 7, 8, 10 |
| *S. o. geoffroyi* (red-naped tamarin) | +•; ♀ wipes moistened tail across suprapubic gland | anogenital +•*, sternal + | | 7, 8, 26 |

Table 18.2 (*cont.*)

| Species | Faeces | Urine | Skin glands | Other materials | References |
|---|---|---|---|---|---|
| **Callitrichidae** (*cont.*) | | | | | |
| *S. mystax* (black-chested moustached tamarin) | | ?▲ | anogenital +, sternal ? | | 2 |
| *S. labiatus* (red-chested moustached tamarin) | | +▲ | anogenital +, sternal + | | 39 |
| *S. midas* ssp (golden and black-handed tamarins) | | +▲ | anogenital +●, sternal ? | | 3, 39 |
| *S. leucopus* (silvery-brown bare-face tamarin) | | +▲, w | anogenital +●, sternal + | | 39 |
| *Leontopithecus rosalia* (lion tamarin) | | +▲ | anogenital +●, sternal + | | 7, 8, 21 |
| **Callimiconidae** | | | | | |
| *Callimico goeldii* (Goeldi's monkey) | | +▲, wipes urine-moistened tail across circumgenital glands and ventral surface | anogenital +●, sternal + | | 8, 20, 38 |

**Cebidae**

| Species | | | | References |
|---|---|---|---|---|
| *Saimiri sciureus* (squirrel monkey) | +, w, w-kick | anogenital +, sternal + | sneezing deposits nasal secretions in hands and on substratum | 8, 16, 17, 29, 35, 39 |
| *Aotus trivirgatus* (night monkey) | w | anogenital +, caudal +, | | 15, 24, 29 |
| *Callicebus moloch* (titi monkey) | | anogenital ?, sternal +, rubs hands across sternal gland | anoints chest with saliva after chewing certain leaves or inhaling tobacco smoke | 22, 25, 39 |
| *Alouatta palliata* (mantled howler monkey) | w, variable patterns of rubbing into fur with hands | | | 12, 23, 29 |
| *A. seniculus* (red howler monkey) | w, same as *A. palliata* | anogenital ?, back ?, neck and face ? | | 17, 27, 29 |
| *Cebus apella* (black-capped capuchin) | w, urinates on substrate and rubs back or neck in it | sternal-epigastric + | anoints own body with odoriferous foreign substance | 5, 28, 29 |
| *C. albifrons* (white-fronted capuchin) | w | sternal + | anoints own body like *C. apella* | 8, 29 |
| *C. capucinus* (white-throated capuchin) | w | sternal + | anoints own body like *C. apella* | 8, 29 |
| *Lagothrix lagotricha* (woolly monkey) | enlarged clitoris may function as in *Ateles* | sternal + | anoints chest with saliva, rubs wet chest on substrate | 8, 29, 33, 37 |

Table 18.2 (*cont.*)

| Species | Faeces | Urine | Skin glands | Other materials | References |
|---|---|---|---|---|---|
| **Cebidae** (*cont.*) | | | | | |
| *Ateles geoffroyi* (black-headed spider monkey) | | urinates in socio-sexual situations; enlarged clitoris serves as urine depository organ; rubs chest in urine puddle | anogenital +, sternal +, rubs hands and wrists across sternal gland | anoints chest with saliva | 6, 8, 18, 19, 29 |
| **Cercopithecidae** | | | | | |
| *Cercopithecus aethiops* (green monkey) | faecal deposits may mark sleeping sites | | chest ?, cheek ? | | 11, 31 |
| *Papio leucophaeus* (drill) | | | | rubs mango tree branches over mouth and chest, saturating sternal gland area with saliva | 14 |
| *Presbytis johnii* (Nilgiri langur) | faecal deposits may mark sleeping sites | | | | 30 |

of both brachial and antebrachial glands. 'Tail waving' often follows tail marking. The lemur arches its strikingly black and white tail over its back, and waves it in the vertical plane. In the wild, 'stink fights' between males have been observed by Jolly (1966). During these agonistic confrontations, series of palmar marking, tail marking, and tail waving may be performed for up to an hour.

The behavioural significance of scent marking in *Lemur catta* has been studied by several authors and is discussed in great detail by Schilling (1979). It probably results in the deposition of a number of messages. Scent from the brachial-antebrachial complex identifies individual males (Mertl 1975). This information deposited into the environment during marking and discharged into the air during the tail-waving display which males show during dominance interactions, may play a role in the establishment of intragroup dominance relationships as well as in territorial interaction (Mertl 1977; Schilling 1979). Another message, whose existence is inferred from the high frequency of genital marking by females during oestrus and the high sniffing activity of males at that time, concerns the reproductive condition of the female. The nature and the role of sexual odours, however, is not known. They may play a role not only in the correlation of sexual behaviour during the short breeding season of this species but also in synchronizing the reproductive physiology of both sexes (Evans and Goy 1968; Jolly 1967; Schilling 1979).

*Example*—Saimiri sciureus

The second example to be discussed is the marking behaviour of the squirrel monkey, *Saimiri sciureus*. This species shares a highly stereotyped 'urine washing' pattern with some prosimians (i.e. the mouse lemur, bush-babies of the genus *Galago*, and the slender loris; see Table 18.2) and several other South American

---

References
1. Bearder and Doyle (1974).
2. Box and Morris (1980).
3. Christen (1974).
4. Clark (1978).
5. Dobroruka (1972).
6. Eisenberg and Kuehn (1966).
7. Epple (1975).
8. Epple and Lorenz (1967).
9. Evans and Goy (1968).
10. French and Snowdon (1981).
11. Gartlan and Brain (1968).
12. Glander (1980).
13. Harrington (1975).
14. Hill (1944).
15. Hill (1960).
16. Hopf *et al*. (1974).
17. Kirschofer (1963).
18. Klein (1971).
19. Klein and Klein (1971).
20. Lorenz (1972).
21. Mack and Kleiman (1978).
22. Mason (1966).
23. Milton (1975).
24. Moynihan (1964).
25. Moynihan (1966).
26. Moynihan (1970).
27. Neville (1972).
28. Nolte (1958).
29. Oppenheimer (1977).
30. Poirier (1970).
31. Poirier (1972).
32. Rumpler and Oddou (1970).
33. Schifter (1968).
34. Schilling (1979).
35. Schwartz and Rosenblum (1980).
36. Sutcliffe and Poole (1978).
37. Ullrich (1954).
38. Wojcik and Heltne (1978).
39. Personal observation.

monkeys (i.e. the night monkey, the howler monkey, and the capuchin; see Table 18.2). The basic motor pattern is similar in all species. The animal urinates into the palm of one hand, and follows this by wiping the urine-moistened hand against the sole of the ipsilateral foot. This procedure is usually repeated with the hand and foot of the other side (for references see Table 18.2). In the South American monkeys, urine washing is often followed by complex and variable patterns of distributing urine to other areas of the body (Andrew and Klopmann 1974; Milton 1975; Oppenheimer 1977). In the squirrel monkey, this includes rubbing the urine-moistened hands together, rubbing the tailtip with the hands, or rolling on the back in drops of urine. Urine is also applied to the chest by 3–4 rapid kicks of the urine-moist foot over the sternal area ('urine-wash-kick'). This behaviour, which is only shown by the 'Roman type' of *Saimiri*, may follow urine washing and is occasionally followed by rubbing the area of the chest where an accumulation of glands is found (Hill 1960) over the substratum (Baldwin 1968; Hopf, Hartman-Wiesner, Kühlmorgen, and Mayer 1974; Kirschofer 1963; personal observations). Anogenital rubbing (= rump rubbing) and rubbing the back against surfaces not premoistened with urine may also represent scent-marking behaviours (Hennessy, Coe, Mendoza, Lowe, and Levine 1978). Anogenital rubbing may result in anal cleaning as well as in deposition of secretions from the specialized circumgenital and circumanal glands (Candland, Blumer, and Mumford 1980; Hill 1960; Hopf *et al.* 1974). In addition, sneezing behaviour, resulting in the deposition of nasal secretions, has also been interpreted as scent marking (Schwartz and Rosenblum 1980). The frequencies with which some of the marking patterns are performed as well as the frequencies of sniffing marked items and conspecifics are influenced by the breeding season, by location within the environment, and by association with social partners (Candland *et al.* 1980; Hennessy *et al.* 1978). In addition to these factors, social rank seems to influence the frequency of urine washing (Castell and Heinrich 1971; Talmage-Riggs and Anschel 1973) and environmental factors such as humidity and temperature also play a role (Castell and Maurus 1967).

Scent-marking behaviour may serve a number of functions. In the dense arboreal environment in which this small species forages, visual communication is limited and vocal communication may attract predators and strange conspecifics. For this reason, chemical communication may be one of the major means of maintaining group cohesion (Candland *et al.* 1980; Hennessy *et al.* 1978). Urinary cues may identify individuals in the sometimes very large troops of squirrel monkeys. A similar function has been suggested for the urine-washing behaviour of wild howler monkeys (Milton 1975). Furthermore, reproductive conditions may also be communicated by urinary cues. Baldwin (1970); Hennessy *et al.* (1978), and Latta, Hopf, and Ploog (1967) suggested that urinary odours communicate female receptivity. Males are attracted to female urine all year round but are particularly interested in it during the breeding season. Baldwin (1970) therefore suggests that, in addition to communicating reproductive condition, female urine also serves to synchronize and correlate the physiological events of the breeding season.

Urinary cues also seem to be involved in dominance interactions. Dominant males and females direct a highly ritualized 'genital display' at submissive recipients (Hopf *et al.* 1974). During the display, which is probably mostly a visual signal, a few drops of urine are discharged and sometimes directed at the face of the recipient. Urinary cues and/or the secretions of skin glands in the circumgenital area may identify the individual, inform about its motivation, and in this way function in dominance encounters (Epple 1974*b*). Baldwin (1968) also considers the possibility that the urine of the dominant male may exert an inhibiting influence on the gonadal hormone levels of the subdominant male.

Another area in which odours seem to be of importance is mother–infant attachment. In experiments in which infants were presented with a choice between their own mother and another mother, both anaesthetized and with their heads hooded, they recognized and preferred their own mothers. After both mothers had been washed with soap, however, the infants seemed no longer able to recognize their own mother (Kaplan 1978). In this experiment, the sources of the mothers' body odours were not identified, but urine may well have been one of them.

*Example*—Papio leucophaeus

The third example is taken from the Cercopithecoidea, a group in which scent glands and scent-marking behaviours are rare. Hill (1954, 1970), however, has reported the presence of mixed sebaceous, apocrine, and eccrine sternal glands associated with a specialized hair brush in male and female drills and mandrills, after the scent-marking behaviour of a male drill had alerted him to its presence (Hill 1944). The marking behaviour of this mature male was only observed when the animal was presented with branches from a mango tree. The branch was sniffed and rubbed vigorously and repeatedly over mouth, chin, and sternal gland. During rubbing, the male produced a copious amount of saliva, which was distributed over the sternal area. To date, this observation has not been replicated, which is not surprising since very little is known about the behaviour of these rare primates. No interpretation of this behaviour has been given by Hill (1944). However, it bears a strong resemblance to the sternal marking of South American woolly monkeys (*Lagothrix lagotricha*), which also rub small objects over mouth and chest, producing copious amounts of saliva, which is spread over the sternal area (Epple and Lorenz 1967; Schifter 1968; Ullrich 1954). Moreover, several other South American primates respond to environmental items which produce strong odours with excitement, salivation, and various behaviour patterns involving their scent glands (cf. Epple and Lorenz 1967). *Callicebus moloch*, for example, rubs chewed sweet gum leaves over its sternal gland and also salivates heavily in response to tobacco smoke. The saliva produced in response to smoke is rubbed into the sternal gland with both hands (personal observation). *Cebus apella* and *C. capucinus*, on the other hand, anoint the fur with a variety of pungent foreign substances such as Eau de Cologne, onion juice, and, in the wild, strong-smelling plants, apparently without salivation (Nolte 1958; Oppenheimer 1977). Perhaps all these behaviours are variations on a common theme, i.e. to mix body odours from various sources together and to modify them by adding environmental scents. As Eisenberg and Kleiman (1972) have suggested, one of the functions of these behaviours may be to establish an optimum odour field and provide a sense of security. An additional function may be that of individual identification.

*Example*—Macaca mulatta

The last example deals with chemical cues in the communication of female receptivity in an Old World monkey, the rhesus macaque. Most of the Cercopithecoidea and Hominoidea do not possess specialized scent glands or scent-marking behaviour. However, in many species, such as several macaques, the chimpanzee, and the gorilla, males intensively sniff the external genitalia of females or insert their fingers into the vagina and sniff the material adhering to them (cf. Keverne 1978). These behaviours suggest that odour cues from the genital tract are used to monitor the female's reproductive condition in many Old World simians.

The study of the role of these cues, however, has been greatly complicated by the fact that the sexual behaviour of these primates is highly complex. Courtship and mating are not only influenced by a variety of sensory signals and an intricate exchange of behavioural interactions, but also by previous experiences and personal 'likes' and 'dislikes' (cf. Goldfoot 1981; Keverne 1978; Michael and Bonsall 1977). Rhesus monkey males, for instance, are perfectly capable of mating successfully after having been made anosmic, although olfactory cues from females may normally influence their sexual behaviour (Goldfoot, Essock-Uitale, Asa, Thornton, and Leshner 1978; Michael and Keverne 1968). This flexibility and complexity in the sociosexual interactions of Old World monkeys is responsible for what appeared to be contradictory results obtained from studies on chemical sex attractants in the rhesus monkey.

These studies have been reviewed in some detail by Goldfoot (1981), Keverne (1978), and Michael and Bonsall (1977) and will only briefly be discussed here. Earlier studies by Michael and his co-workers suggested that the sexual attractiveness of female rhesus monkeys and the sexual arousal of males depended strongly on odour cues of vaginal origin. Ovariectomized females whose perineum had been painted with vaginal secretions collected from ovariectomized donor females injected with oestrogen, were very attractive to males and stimulated male copulatory behaviour in a laboratory situation (Michael and Keverne 1968). Ether extracts of vaginal secretions proved also to be effective in stimulating male sexual behaviour (Keverne and Michael 1971) and the active components were identified as a mixture of short-chain aliphatic acids, produced in the vagina by bacterial action (Curtis, Ballantine, Keverne, Bonsall, and Michael 1971; Bonsall and Michael 1971; Michael, Keverne and Bonsall 1971). These studies, although based on very few animals, suggested a powerful influence of vaginal olfactory stimuli on male sexual arousal. Later studies by Goldfoot, Kravetz, Goy, and Freeman (1976), conducted under conditions somewhat different from those employed by Michael and co-workers, have not supported this notion.

Vaginal lavages from oestrogenized females, as well as synthetic mixtures of aliphatic acids, failed to stimulate significantly sexual behaviour in males when applied to ovariectomized females (Goldfoot *et al.* 1976). The reasons for these contradictory results have been discussed in detail by Goldfoot (1981), Keverne (1978), and Michael and Bonsall (1977). These discussions, as well as some additional studies by Goldfoot (1981) and Bonsall and Michael (1980), all of which cannot be reviewed here in detail, have resulted in a modified view of the role of sexual odours in this species. It now appears that male rhesus monkeys probably use odour cues of vaginal origin during sexual interactions with females. However, there are individual differences regarding the type of male-female interaction influenced by vaginal secretions. In some males, odour cues may influence mounting and ejaculation while in others they may influence grooming of the female or agression toward the female (cf. Keverne 1978). In addition to the aliphatic acids which were identified as the active compounds, other compounds present in vaginal secretions (e.g. phenylpropanoic and para-hydroxyphenylpropanoic acid) seem to be necessary for the full effect (cf. Keverne 1978). However, odour cues are not necessary for fertile mating, since anosmic males show normal reproductive behaviour (Goldfoot *et al.* 1978; Michael and Keverne 1968) nor are odour cues alone capable of rendering the females sexually attractive for the majority of males (cf. Goldfoot 1981). In addition, sexual experience and individual preferences for a particular partner strongly influence a male's response to these odour cues (cf. Goldfoot 1981; Keverne 1978). Furthermore, male-female sexual interactions can even be influenced by perfuming females with arbitrary odours and also depend strongly on the behavioural action of the female (cf. Goldfoot 1981; Michael and Bonsall 1977).

**Summarizing remarks**

The present survey shows that chemical signals are used as sociosexual communicants throughout the order Anthropoidea. However, the relative importance of chemical as compared to visual or vocal communication, and the degree to which the signals elicit strong and predictable responses, varies widely among different primate groups. Thus, the importance of chemical communication parallels to a certain degree the reduction in the complexity of the olfactory and accessory olfactory systems and the increasing degree of neocorticalization as one moves from the prosimians to simian primates (cf. Epple and Moulton 1978; Keverne 1978; Maier 1981).

Chemical communication is highly important to some prosimians. Although the South American primates share a reduction in the complexity of the nasal structures with Old World monkeys and apes, the abundance of scent glands and of behaviours concerned with their use shows that chemical communication is of great importance in this primate group as well. Finally, the relative scarcity of specialized glands and marking behaviours in Old World monkeys and apes suggest that in these primates, chemical communication plays a less powerful role than in the previously mentioned groups. Chemical cues are only one of a great number of stimuli which influence and control the highly complex socio-sexual interactions of these primates.

## References

Andrew, R. J. and Klopmann, R. B. (1974). Urine-washing: comparative notes. In *Prosimian biology* (ed. R. D. Martin, G. A. Doyle, and A. C. Walker) pp. 303–12. Duckworth, London.

Arao, T. and Perkins, E. (1969). Further observations on the Philippine tarsier (*Tarsius syrichta*). *Am J. phys. Anthrop.* 31 93–6.

Aron, C. (1979). Mechanisms of control of the reproductive function by olfactory stimuli in female mammals. *Physiol. Rev.* 59, 229–84.

Baldwin, J. D. (1968). The social behavior of adult male squirrel monkeys (*Saimiri sciureus*) in a seminatural environment. *Folia primat.* 9, 281–314.

— (1970). Reproductive synchronization in squirrel monkeys (*Saimiri*). *Primates* 11, 317–26.

Bearder, S. K. and Doyle, G. A. (1974). Ecology of bushbabies, *Galago senegalensis* and *Galago crassicaudatus*, with some notes on their behaviour in the field. In *Prosimian biology* (ed. R. D. Martin, G. A. Doyle, and A. C. Walker) pp. 109–30. Duckworth, London.

Bonsall, R. W. and Michael, R. P. (1971). Volatile constituents of primate vaginal secretions. *J. Reprod. Fert.* 27, 478–9.

— — (1980). The externalization of vaginal fatty acids by the female rhesus monkey. *J. chem. Ecol.* 6, 499–509.

Box, H. O. and Morris, J. M. (1980). Behavioural observations on captive pairs of wild caught tamarins (*Saguinus mystax*). *Primates* 21, 53–65.

Brinkmann, A. (1909). Über das Vorkommen von Hautdrüsen bei den anthropomorphen Affen. *Anat. Anz.* 34, 513.

— (1923). Nachlese zu meinen Hautdrüsenuntersuchungen. *Bergens Mus. Årb.* 1, 1–34.

Candland, D. K., Blumer, E. S., and Mumford, M. D. (1980). Urine as a communicator in a New World primate *Saimiri sciureus*. *Anim. Learn. Behav.* 8, 468–80.

Castell, R. and Heinrich, B. (1971). Rank order in a captive female squirrel monkey colony. *Folia primat.* 14, 182–9.

— and Maurus, M. (1967). Das sogenannte Urinmarkieren von Totenkopfaffen (*Saimiri sciureus*) in Abhängigkeit von umweltbedingten und emotionalen Faktoren. *Folia primat.* 6, 170–6.

Cave, A. J. E. (1967). Observations on the platyrrhine nasal fossa. *Am. J. phys. Anthrop.* 26, 277–88.

Cheal, M. L. (1975). Social olfaction: a review of the ontogeny of olfactory influences on vertebrate behaviour. *Behav. Biol.* **15**, 1–25.

Christen, A. (1974). Fortpflanzungsbiologie und Verhalten bei *Cebuella pygmaea* und *Tamarin tamarin*. *Z. Tierpsychol.* Suppl. **14**, 1–78.

Clark, A. B. (1978). Olfactory communication, *Galago crassicaudatus* and the social life of prosimians. In *Recent advances in primatology*, Vol. III (ed. D. J. Chivers and K. A. Josey) pp. 109–17. Academic Press, New York.

Curtis, R. F., Ballantine, J. A., Keverne, E. B., Bonsall, R. W., and Michael, R. P. (1971). Identification of primate sexual pheromones and the properties of synthetic attractants. *Nature, Lond.* **232**, 396–8.

Dixson, A. F. (1976). Effects of testosterone on the sternal cutaneous glands and genitalia of the male greater galago (*Galago crassicaudatus crassicaudatus*). *Folia primat.* **26**, 207–13.

Dobroruka, L. J. (1972). Social communication in the brown capuchin *Cebus apella*. *Int. Zoo Yb.* **12**, 43–6.

Eisenberg, J. F. and Kleiman, D. G. (1972). Olfactory communication in mammals. *Ann. Rev. Ecol. Systemat.* **3**, 1–32.

— and Kuehn, R. E. (1966). The behaviour of *Ateles geoffroyi* and related species. *Smithsonian misc. Coll.* **151**, 1–63.

Ellis, R. A. and Montagna, W. (1964). The sweat glands of the Lorisidae. In *Evolutionary and genetic biology of the primates*, Vol. I (ed. J. Buettner-Janusch) pp. 197–228. Academic Press, New York.

Epple, G. (1972). Social communication by olfactory signals in marmosets. *Int. Zoo Yb.* **12**, 36–42.

— (1974*a*). Primate pheromones. In *Pheromones* (ed. M. C. Birch) pp. 366–85. North Holland, Amsterdam.

— (1974*b*). Olfactory communication in South American primates. *Ann. NY Acad. Sci.* **237**, 261–78.

— (1975). The behavior of marmoset monkeys (Callithricidae). In *Primate behaviour*, Vol. 4 (ed. L. A. Rosenblum) pp. 195–239. Academic Press, New York.

— (1976). Chemical communication and reproductive processes in non-human primates. In *Mammalian olfaction, reproductive processes, and behavior*. (ed. R. L. Doty) pp. 257–82. Academic Press, New York.

—Alveario, M. C., and Katz, Y. (1982). The role of chemical communication in aggressive behavior and its gonadol control in the tamarin *Saguinus fuscicollis*. In *Primate communication* (ed. C. T. Snowdon, C. H. Brown, and R. Petersen) pp. 279–302. Cambridge University Press.

— and Lorenz, R. (1967). Vorkommen, Morphologie und Funktion der Sternaldrüse bei den Platyrrhini. *Folia primat.* **7**, 98–126.

— and Moulton, D. G. (1978). Structural organization and communicatory functions of olfaction in nonhuman primates. In *Sensory systems of primates* (ed. C. R. Noback) pp. 1–22. Plenum Press, New York.

Evans, C. S. and Goy, R. W. (1968). Social behaviour and reproductive cycles in captive ring-tailed lemurs (*Lemur catta*). *J. Zool., Lond.* **156**, 171–97.

Fass, B. and Stevens, D. A. (1977). Pheromonal influences on rodent agonistic behavior. In *Chemical signals in vertebrates* (ed. D. Müller-Schwarze and M. Mozell) pp. 185–206. Plenum Press, New York.

French, J. A. and Snowdon, C. T. (1981). Sexual dimorphism in intergroup spacing behaviour in the tamarin, *Saguinus oedipus*. *Anim. Behav.* **29**, 822–9.

Gartlan, J. A. and Brain, C. K. (1968). Ecology and social variability in *Cercopithecus aethiops* and *C. mitis*. In *Primates: studies in adaptation and variability* (ed. P. Jay) pp. 253–92, Holt, Reinhart and Winston, New York.

Glander, K. E. (1980). Reproduction and population growth in free-ranging mantled howling monkeys. *Am. J. phys. Anthrop.* **53**, 25–36.

Goldfoot, D. A. (1981). Olfaction, sexual behavior and the pheromone hypothesis in the rhesus monkey: a critique. *Am. Zool.* **21**, 153–64.

— Essock-Vitale, S. M., Asa, C. S., Thornton, J. E., and Leshner, A. I. (1978). Anosmia in male rhesus monkeys does not alter copulatory activity with cycling females. *Science, NY* **199**, 1095–6.

— Kravetz, M. A., Goy, R. W., and Freeman, S. K. (1976). Lack of effect of vaginal lavages and aliphatic acids on ejaculatory responses in rhesus monkeys: behavioral and chemical analyses. *Horm. Behav.* **7**, 1–27.

Hanson, G. and Montagna, W. (1962). The skin of the owl monkey (*Aotus trivirgatus*). *Am. J. phys. Anthrop.* **20**, 421–30.

Harrington, J. E. (1975). Field observations of social behavior of *Lemur fulvus fulvus*. E. Geoffrey 1812. In *Lemur biology* (ed. I. Tattershall and R. W. Sussman) pp. 259–79, Plenum Press, New York.

Hennessy, M. B., Coe, C. L., Mendoza, S. P., Lowe, E. P., and Levine, S. (1978). Scent marking and olfactory investigatory behavior in the squirrel monkey (*Saimiri sciureus*). *Behav. Biol.* **24**, 57–67.

Hershkovitz, P. (1977). *Living New World monkeys (Platyrrhini)*, Vol. I. University of Chicago Press.

Hill, W. C. O. (1944). An undescribed feature in the drill (*Mandrillus leucophaeus*). *Nature, Lond.* **153**, 199.

— (1951). Epigastric gland of *Tarsius*. *Nature, Lond.* **167**, 994.

— (1954). Sternal glands in the genus *Mandrillus*. *J. Anat.* **88**, 582.

— (1960). *Primates. Comparative anatomy and taxonomy*. IV *Cebidae*. Edinburgh University Press.

— (1970). *Primates. Comparative anatomy and taxonomy*. VIII. *Cyncopithecinae*. Edinburgh University Press.

Hopf, S., Hartman-Weisner, E., Kühlmorgen, B., and Mayer, S. (1974). The behavioral repertoire of the squirrel monkey (*Saimiri*). *Folia primat.* **21**, 225–49.

Jolly, A. (1966). *Lemur behavior*. University of Chicago Press.

— (1967). Breeding synchrony in wild *Lemur catta*. In *Social communication among primates* (ed. S. Altmann) pp. 3–14, University of Chicago Press.

Kaplan, J. N. (1978). Olfactory recognition of mothers by infant squirrel monkeys. In *Recent advances in primatology*. I. *Behaviour* (ed. D. J. Chivers and J. Herbert) pp. 103–5. Academic Press, New York.

Keverne, E. B. (1978). Olfactory cues in mammalian behaviour. In *Biological determinants of sexual behavior*. (ed. J. E. Hutchinson) pp. 727–63. Wiley, Chichester.

— and Michael, R. P. (1971). Sex attractant properties of ether extracts of vaginal secretions from rhesus monkeys. *J. Endocr.* **51**, 313–22.

Kirschofer, R. (1963). Einige bemerkenswerte Verhaltensweisen bei Saimiris im Vergleich zu verwandten Arten. *Z. Morph. Anthrop.* **53**, 77–91.

Klein, L. L. (1971). Observations on copulation and seasonal reproduction of two species of spider monkeys *Ateles belzebuth* and *A. geoffroyi*. *Folia primat.* **15**, 233–48.

— and Klein, D. (1971). Aspects of social behavior in a colony of spider monkeys. *Int. Zoo Yb.* **11**, 175–81.

Knappe, H. (1964). Zur Funktion des Jacobsonschen Organs (Organon vomero-nasale Jacobsoni). *Zool. Gart., Lpz.* **28**, 188–94.

Kneeland-Sisson, J. K. and Fahrenbach, W. H. (1967). Fine structure of steroidogenic cells of a primate cutaneous organ. *Am. J. Anat.* **121**, 337–68.

Latta, J., Hopf, S., and Ploog, D. (1967). Observation on mating behavior and sexual play in the squirrel monkey (*Saimiri sciureus*). *Primates* **8**, 229–46.

Lorenz, R. (1972). Management and reproduction of the Goeldi's monkey *Callimico goeldii* (Thomas, 1904) Callimiconidae, Primates. In *Saving the lion marmoset. Proc. WAPT Golden Lion Marmoset Conference* (ed. D. D. Bridgwater) pp. 92–109. Wild Animal Propagation Trust, Oglebay Park, Wheeling, West Virginia.

Machida, H., Perkins, E., and Hu, F. (1967). The skin of the squirrel monkey (*Saimiri sciurea*). *Am. J. phys. Anthrop.* **26**, 45–54.

— — and Montagna, W. (1964). A comparative study of the skin of the green monkey (*Cercopithecus aethiops*) and Sykes' monkey (*Cercopithecus mitis*). *Am. J. phys. Anthrop.* **22**, 453–66.

Mack, D. S. and Kleiman, D. G. (1978). Distribution of scent marks in different contexts in captive lion tamarins *Leontopithecus rosalia* (Primates). In *Biology and behavior of marmosets* (ed. H. Rothe, H. J. Wolters, and J. P. Hearn) pp. 181–8. Eigenverlag Rothe, Göttingen.

Macrides, F., Bartke, A., and Svare, B. (1977). Interactions of olfactory stimuli and gonadal hormones in the regulation of rodent social behavior. In *Olfaction and taste* VI (ed. J. LeMagnen and P. MacLeod) pp. 143–7. Information Retrieval, London.

Maier, W. (1981). Nasal structures in Old and New World primates. In *Evolutionary biology of the New World monkeys*. (ed. R. L. Liochon and A. B. Chiarelli) pp. 219–41, Plenum Press, New York.

Marler, P. 1975). Communication in monkeys and apes. In *Primate behavior: field studies of monkeys and apes* (ed. I. DeVore) pp. 544–84. Holt, Rinehart and Winston, New York.

Mason, W. A. (1966). Social organization of the South American monkey *Callicebus moloch*: a preliminary report. *Tulane Stud. Zool.* **13**, 23–8.

Mertl, A. (1975). Discrimination of individuals by scent in a primate. *Behav. Biol.* **14**, 505–9.

— (1977). Habituation to territorial scent marks in the field by *Lemur catta*. *Behav. Biol.* **21**, 500–7.

Michael, R. P., and Bonsall, R. W. (1977). Chemical signals and primate behavior. In *Chemical signals in vertebrates* (ed. D. Müller-Schwarze and M. Mozell) pp. 251–71. Plenum, New York.

— and Keverne, E. B. (1968). Pheromones in the communication of sexual status in primates. *Nature, Lond.* **218**, 746–9.

— — and Bonsall, R. W. (1971). Pheromones: isolation of male sex attractants from a female primate. *Science, NY* **172**, 964–6.

Milton, K. (1975). Urine rubbing behavior in the manteled howler (*Alouatta palliata*). *Folia primat.* **23**, 105–12.

Montagna, W. (1962). The skin of lemurs. *Ann. NY Acad. Sci.* **102**, 190–209.

— (1972). The skin of non-human primates. *Am. Zool.* **12**, 109–24.

— Machida, H., and Perkins, E. M. (1966a). The skin of the angwantibo (*Arctocebus calabarensis*). *Am. J. phys. Anthrop.* **25**, 277–90.

— — — (1966b). The skin of the stump-tail macaque (*Macaca speciosa*). *Am. J. phys. Anthrop.* **24**, 71–86.

— and Yun, J. S. (1962*a*). The skin of the Anubis baboon (*Papio doguera*). *Am. J. phys. Anthrop.* **20**, 131–41.

— — (1962*b*). Further observations on *Perodicticus potto*. *Am. J. phys. Anthrop.* **20**, 441–50.

Moynihan, M. (1964). Some behavior patterns of platyrrhine monkeys. I. The night monkey (*Aotus trivirgatus*). *Smithson. misc. Coll.* **146**, 1–81.

— (1966). Communication in the Titi monkey, *Callicebus*. *J. Zool., Lond.* **150**, 77–127.

— (1970). Some behavior patterns of platyrrhine monkeys. II. *Saguinus geoffroyi* and some other tamarins. *Smithson. Contrib. Zool.* **28**, 1–77.

Neville, M. (1972). Social relations within troops of red howler monkeys (*Alouatta seniculus*). *Folia primat.* **18**, 47–77.

Nolte, A. (1958). Beobachtugen über das Instinktverhalten von Kapuzineraffen (*Cebus apella* L.) in der Gefangenschaft. *Behaviour* **12**, 183–207.

Oppenheimer, J. R. (1977). Communication in New World monkeys. In *How animals communicate* (ed. T. E. Sebeok) pp. 851–89. Indiana University Press, Bloomington, In.

Perkins, E. M. (1966). The skin of the black-collared tamarin (*Tamarinus nigricollis*). *Am. J. phys. Anthrop.* **25**, 41–69.

— (1968). The skin of the pygmy marmoset—*Callithrix* (= *Cebuella*) *pygmaea*. *Am. J. phys. Anthrop.* **29**, 349–64.

— (1969*a*). The skin of the cotton top pinché *Saguinus* (= *Oedipomidas*) *oedipus*. *Am. J. phys. Antrop.* **30**, 13–27.

— (1969*b*). The skin of the Goeldi's marmoset (*Callimico goeldii*). *Am. J. phys. Anthrop.* **30**, 231–49.

— (1969*c*). The skin of the silver marmoset—*Callithrix* (= *Mico*) *argentata*. *Am. J. phys. Anthrop.* **30**, 361–87.

— Arao, T., and Uno, H. (1968). The skin of the red uacari (*Cacajao rubicundus*). *Am. J. phys. Anthrop.* **29**, 57–79.

Pocock, R. I. (1925). The external characters of the catarrhine monkeys and apes. *Proc. zool. Soc. Lond.* 1479–579.

Poirier, F. E. (1970). The communication matrix of the Nilgiri langur (*Presbytis johnii*) of South India. *Folia primat.* **13**, 92–136.

— (1972). The St. Kitts green monkey (*Cercopithecus aethiops sabaeus*): ecology, population dynamics, and selected behavioral traits. *Folia primat.* **17**, 20–55.

Rumpler, Y. and Andriamiandra, A. (1971). Étude histologique des glandes de marquage de la face antérieure du cou des Lémuriens malagaches. *C. r. Séanc. Soc. Biol.* **165**, 436–42.

— and Oddou, J. H. (1970). Comportement de marquage et structure histologique des glandes de marquage chez *Lemur macaco*. *C. r. Séanc. Soc. Biol.* **164**, 2686–90.

Schaffer, J. (1940). *Die Hautdrüsenorgane der Säugetiere*. Urban und Schwarzenberg, Berlin.

Schifter, H. (1968). Zucht und Markierverhalten von Wollaffen im Zoologischen Garten Zürich. *Zool. Gart., Lpz.* **36**, 107–32.

Schilling, A. (1979). Olfactory communication in prosimians. In *The study of prosimian behavior* (ed. G. A. Doyle and R. D. Martin) pp. 461–542, Academic Press, New York.

Schultz, A. H. (1921). The occurrence of a sternal gland in the orang-utan. *J. Mammal.* **5**, 194–6.

Schwartz, G. G. and Rosenblum, L. A. (1980). Novelty, arousal and nasal marking in the squirrel monkey. *Behav. neural Biol.* **28**, 116–22.

Schwarz, W. (1937). Die Sternaldrüse bei dem Klammeraffen Ateles. *Morph. Jb.* **79**, 600–33.

Seitz, E. (1969). Die Bedeutung geruchlicher Orientierung beim Plumplori, *Nycticebus coucang* Boddaert 1785 (Prosimii, Lorisidae). *Z. Tierpsychol.* **26**, 73–103.

Sonntag, C. F. (1924). *Morphology and evolution of the apes and man.* John Bale, London.

Starck, D. (1969). Die circumgenitalen Duftdrüsenorgane von *Callithrix* (*Cebuella*) *pygmaea* (Spix 1823). *Zool. Gart., Lpz.* **36**, 312–26.

Sutcliffe, A. G. and Poole, T. B. (1978). Scent marking and associated behaviour in captive common marmosets (*Callithrix jacchus jacchus*) with a description of the histology of scent glands. *J. Zool., Lond.* **185**, 41–56.

Talmage-Riggs, G. and Anschel, S. (1973). Homosexual behavior and dominance hierarchy in a group of captive female squirrel monkeys (*Saimiri sciureus*). *Folia primat.* **19**, 61–72.

Thiessen, D. D. (1976). *The evolution and chemistry of aggression.* Thomas, Springfield, Ill.

Ullrich, W. (1954). Zur Frage des Sichbespuckens bei Säugetieren. *Z. Tierpsychol.* **11**, 150.

Wislocki, G. B. (1930). A study of scent glands in the marmosets, especially *Oedipomidas goeffroyi. J. Mammal.* **11**, 475–82.

— and Schultz, A. H. (1925). On the nature of modifications of the skin in the sternal region of certain primates. *J. Mammal.* **6**, 236–44.

Wojcik, J. F. and Heltne, P. G. (1978). Tail marking in *Callimico goeldii*. In *Recent advances in primatology*, Vol. I (ed. D. J. Chivers and J. Herbert) pp. 507–9. Academic Press, New York.

# 19 The primates II: a case study of the saddle-back tamarin, *Saguinus fuscicollis*

GISELA EPPLE AND AMOS B. SMITH III

## Introduction

Chapter 18 illustrated the abundance of scent glands and other sources of chemical signals in primates. Among the simian primates, the South American monkeys in particular are characterized by the occurrence of specialized scent glands, as well as variable and complex patterns of scent marking, resulting in the application of glandular secretions, urine, saliva, and even environmentally derived odoriferous materials, such as pungent tree sap, to their own bodies, to those of conspecifics and, of course, to the environment. All South American primates are arboreal, and one species, the night monkey, is also nocturnal. They live in a dense environment, where visual and vocal communication may not always be effective. Moreover, compared to most Cercopithecoidea and Hominoidea, the South American primates possess relatively few facial expressions and visual displays. These facts undoubtedly favoured the development of chemical communication in many areas of social and sexual behaviour. They may have been of particular importance in the evolution of chemical communication in the Callitrichidae, the marmosets and tamarins, which are smaller in body size than the Cebidae and possess even fewer visual displays (Moynihan 1976). The occurrence of scent glands and of scent-marking behaviour in all species, and the predominance of sniffing and licking in the investigation of objects and conspecifics, suggest that most callitrichids rely heavily on chemical signals for information on their physical and social environment.

The importance of chemical signals in these primates, their small body size, and the fact that some species are maintained and bred in the laboratory with relative ease (cf. Epple 1978*d*) make the callitrichids ideal subjects for experimental studies of chemical communication. This chapter will discuss various aspects of chemical communication in the Callitrichidae, using our own studies on the saddle-back tamarin as a case in point.

During recent years, our research has been concerned with an analysis of chemical communication in the saddle-back tamarin, *Saguinus fuscicollis*. We have attempted to determine the nature of the information communicated in the scent marks of this species, to discover in which areas of behaviour scent-marking and chemical communication play a role, and to analyse some of the chemical properties of the scent marks. In addition, the role of gonadal hormones in the development of the scent glands and marking behaviour and in the display of aggression has also been studied. This work will be reviewed in some detail below as an example of the complex problems encountered in an attempt to understand the role of chemical communication in a single species.

**Chemical communication in the Callitrichidae—general**

All members of the family Callitrichidae studied so far possess specialized scent glands in the circumgenital area and on the midchest, above the sternum. The morphological differentiation and the size of these glands differ among the various species. In *Saguinus fuscicollis*, for example, scent glands cover the external genitalia and a thick conspicuous suprapubic gland pad which extends across the *symphysis pubis* (Plates 19.1–19.3). Suprapubic pads, very similar in appearance to that of *Saguinus fuscicollis*, are found in *Saguinus oedipus* spp., *Saguinus leucopus*, and *Saguinus midas* but are absent in *Saguinus labiatus* (personal observation). In the common marmoset (*Callithrix jacchus*) the circumgenital gland is predominantly located on the scrotum and the *labia pudendi* (Sutcliffe and Poole 1978). The histology of the glands has been studied in only a few species (see Table 18.1, p. 744). From these studies it appears that circumgenital scent glands are generally composed of sebaceous and apocrine units, while sternal glands are mostly apocrine in nature. The relative predominance of sebaceous and apocrine glands, however, varies among species (Perkins 1966, 1968, 1969a,b,c; Starck 1969; Sutcliffe and Poole 1978). There are also species differences in the extent to which the scent glands are sexually dimorphic. In *Saguinus o. oedipus* and *Saguinus fuscicollis*, for example, the circumgenital-suprapubic gland is larger in females than in males (Epple, Alveario, and Katz 1980; Perkins 1969a). In *Callithrix jacchus* the sternal gland is better developed in males but the circumgenital glands are not sexually dimorphic (Epple and Lorenz 1967; Sutcliffe and Poole 1978).

Sternal as well as circumgenital glands are used to mark objects. The circumgenital gland is also used to mark conspecifics ('partner marking'). Most marmosets and tamarins seem to mark more frequently with the circumgenital than with the sternal gland (Epple 1975a, unpublished; Moynihan 1970; Sutcliffe and Poole 1978). *Leontopithecus rosalia*, the golden lion tamarin, however, marks predominantly with the sternal gland (Epple and Lorenz 1967; Mack and Kleiman 1978). Marking with the circumgenital gland results in a complex deposit of secretions and excretions. Most species not only rub the secretions of the scent gland on to the substrate, but regularly discharge a small amount of urine when scent marking, which is mixed with the glandular secretions. In females, there may be vaginal discharge or vulval secretions, since one can sometimes see the labia deflected and the vaginal orifice exposed during marking. Sternal marking results mostly in the deposit of sternal secretions. However, in addition to these, any contaminants adhering to the chest fur will also be rubbed off. These contaminants may contain urine and secretions from the circumgenital gland. In at least one tamarin species, *Saguinus o. geoffroyi*, the female wipes her looped tail across the genitalia, apparently impregnating it with urine and circumgenital scent-gland secretions (Epple 1976). This material could later be passively transferred to the chest hair as the animal rests with the tail rolled up in front of her and in contact with abdomen and chest (Moynihan 1970). So far, this pattern of tail impregnation has not been described for other callitrichids. In Goeldi's monkey (*Callimico goeldii*), however, it has evolved into an elaborate display during which both sexes mix urine with the secretions of the circumgenital and sternal glands by forcefully wiping the rolled tail over the whole ventral body surface (Lorenz 1972; Wojcik and Heltne 1978).

Although the motor patterns involved in scent marking are rather stereo-typed rubbing movements, they are variable enough to be adapted to the nature and position of the substrate, the parts of the glands involved and the level of arousal of the animal. The most common form of marking is 'sit-rubbing' (Moyni-han 1970). The animal sits, presses the circumgenital (mainly the external genitalia) and the circumanal areas against the substrate and rubs back and forth and/or from side to side. Secretions from the suprapubic part of the gland are applied when the animal assumes a prone position, pressing the lower abdomen and the suprapubic gland to the substrate and either pulling itself forward with its hands, or pushing the body with the feet, or both ('pull-rubbing') (Moynihan 1970). The suprapubic gland may also be rubbed against small proturberances while in a sitting position. Moreover, marking with the suprapubic and sternal glands may be combined into one pattern during which the animal rubs the entire ventral surface over the substrate (Epple 1975a; French and Snowdon 1981; Mack and Kleiman 1978; Moynihan 1970). Sit-rubbing and pull-rubbing and their variations are also employed for partner marking. Plate 19.4 shows the various marking patterns of marmosets and tamarins and Plate 19.5 illustrates the variability in the basic marking pattern of a single *Saguinus fuscicollis* female.

In *Saguinus o. oedipus* sit-rubbing and pull-rubbing appear to serve different functions. Sit-rubbing (= 'anogenital marking') may be used in a sociosexual context, perhaps communicating the stage of the female's reproductive cycle, while pull-rubbing (= 'suprapubic marking') appears to function in intergroup aggression and spacing (French and Snowdon 1981). Similar functional differences may exist between circumgenital marking and sternal marking in *Leontopithecus rosalia* (Mack and Kleiman 1978). In *Saguinus fuscicollis*, on the other hand, sit- and pull-rubbing apparently reflect only differences in the degree of arousal of the marker. Sit-rubbing is performed at low levels of arousal, pull-rubbing at higher levels (Epple, unpublished). Marmosets and tamarins scent-mark frequently; marking, indeed, is one of the most commonly observed behaviours in these pri-mates. It is elicited by a wide variety of stimuli. Among the variables which influence scent-marking activity are age, sex, reproductive condition, social status, aggressive motivation, and environmental novelty (Epple 1970, 1974a,b, 1978a, b, 1981a,b; French and Snowdon 1981; Kleiman 1978; Sutcliffe and Poole 1978). Since marking activity is so strongly influenced by such a variety of vari-ables, any meaningful analysis involving comparisons of the marking activities of individuals has to be performed under carefully controlled social and environ-mental conditions.

Marmoset and tamarins frequently sniff and sometimes lick their own scent marks and those of conspecifics. Moreover, the first contact these primates make with any object, as well as with conspecifics, involves sniffing (Epple 1967; Sutcliffe and Poole 1978). This indicates that chemical signals from the social and physical environment provide important information. Most of these signals are probably odours, perceived via the main olfactory system. The fact, however, that a vomeronasal organ and accessory olfactory bulbs are present in these primates—although the vomeronasal organ is not as prominent as in some Cebidae such as the squirrel monkey (Maier 1981)—suggests that the accessory olfactory system may play an important role in monitoring chemical stimuli. This has recently been demonstrated for several rodents (Beauchamp, Wellington, Wysocki, Brand, Kubie, and Smith 1980; Powers and Winians 1975; Reynolds and Keverne 1979).

Observational as well as some experimental evidence suggest that scent signals play a role in a variety of sociosexual interactions. Pair-bonding and precopulatory behaviour (Christen 1974; Epple 1976; French and Snowdon 1981; Kleiman 1978), parent–infant interactions and infant attachment (Cebul, Alveario, and Epple 1978), and spacing and aggressive behaviour (Epple 1970, 1974a, 1978a, b; French and Snowdon 1981; Mack and Kleiman 1978; Sutcliffe and Poole 1978) are among the behaviours influenced by chemical signals.

## Chemical communication in *Saguinus fuscicollis*

### The scent glands and their hormonal control

The apocrine sternal gland and the mixed sebaceous and apocrine circumgenital-suprapubic scent pad of *Saguinus fuscicollis* (Perkins 1966) develop as macroscopically recognizable structures in males and females when the animals are 7–10 months old and grow until they reach adult size around the age of two years. In adults, the sternal gland is often associated with a small brush of stiff, coarse hair (Epple and Lorenz 1967). It is surrounded by the dense, silky chest hair and only detectable upon close inspection. The extent to which the sternal gland is differentiated is not only dependent on the age of the animal but also appears to be influenced by social rank. While adult, low-ranking males and females often do not possess a macroscopically differentiated sternal gland, dominant individuals usually have well-developed glands. Similar observations were made in *Callithrix jacchus*, where the sternal glands are well developed, both macroscopically and microscopically, only in dominant adult males (Epple and Lorenz 1967; Sutcliffe and Poole 1978).

In contrast to the sternal gland, the circumgenital-suprapubic gland is a very conspicuous structure, which is always present in adults, although its size possibly is also sensitive to social rank (personal observations). It involves the skin of the external genitalia (scrotum and labia majora) and extends rostrally across the symphysis pubis as a thick, largely hairless, and pigmented glandular pad.

Although there is no dramatic sexual dimorphism in gland size, female glands tend to be larger than male glands (Epple *et al.* 1982). The existence of a true sexual dimorphism, however, awaits histological confirmation.

In many species of mammals, the ontogenetic development and adult maintenance of scent glands is under the control of gonadal hormones, particularly testosterone (cf. Thiessen 1976). Among prosimians, the brachial and antebrachial glands of *Lemur catta* and the sternal gland of *Galago crassicaudatus* are testosterone dependent, showing a certain degree of atrophy after castration (Andriamiandra and Rumpler 1968; Dixson 1976). In *Saguinus fuscicollis*, we have found that the gonads are necessary for the full morphogenetic development of the circumgenital–suprapubic scent gland in both sexes. In an experiment comparing the development of the scent glands and of marking behaviour (see below) in castrated and control subjects, five males were castrated, four sham-castrated when 5½ months old, i.e. prior to the macroscopically evident development of the circumgenital-suprapubic scent gland (Epple 1981*a*).* Prepubertal castration inhibited the development of the scrotum, which developed normally in sham-castrated control males. Moreover, it permanently retarded the development of the suprapubic part of the scent gland but did not inhibit its growth completely. At one year of age, the scent gland of sham castrates had developed normally, although it had not yet reached its adult size. The suprapubic part of the pad showed an average length of 21.5 ± 1.8 mm and the scrotum an average width of 16.1 ± 0.1 mm (Plate 19.1). One-year-old castrates, on the other hand, showed an unpigmented, flat patch of glandular skin of an average length of 4.6 ± 0.7 mm in the area above the *symphysis pubis* (Plate 19.1). No glandular differentiation was noticeable on the underdeveloped scrotum. By the time the subjects were two years old, the scrotum (average width 20.3 ± 0.9 mm) and suprapubic scent glands of the control males had reached adult size. Their suprapubic gland (average length of 30 ± 2.9 mm) was fully pigmented and raised well above the surrounding skin. The suprapubic scent pad of the castrates had also increased in size. It now formed a largely unpigmented, only slightly raised, narrow strip of glandular skin with an average length of 17.6 ± 1.4 mm. The scrotum also showed some glandular differentiation. The difference between the glands of castrated and control males persisted to their present age of four years. Similar results were obtained on females ovariectomized when six months old. The development of the scent gland was permanently retarded but not completely inhibited (Epple 1982).

In contrast to these findings, gonadectomy of adult males and females did not cause any macroscopically recognizable changes in the suprapubic glands (Epple 1980). In adult males, gonadectomy resulted in atrophy of the scrotum, which forms part of the gland, thereby decreasing its surface area. However, the appearance of the suprapubic part of the gland was little changed and its average length of 26.9 ± 2.3 mm was in the range of that of intact males (27.8 ± 0.9 mm; Plate

* The sternal gland was not considered in this study.

19.2). The morphology of female scent glands also does not seem to be strongly affected by ovariectomy in adulthood. The average length of the glands of three ovariectomized females (33.7 ± 3.3 mm) was also within the range of that of intact females (33.6 ± 1.4; Plate 19.3).

Preliminary results of a histological investigation of biopsies taken from the scent glands of animals gonadectomized as adults indicate that the sebaceous glands do show a certain degree of atrophy, while the apocrine glands do not. This finding will have to be substantiated as further material is studied.

The experiments reviewed above show that gonadal hormones are necessary for the full morphogenetic development of the circumgenital-suprapubic scent glands do show a certain degree of atrophy, while the apocrine glands do not. This ment completely. As in intact males, the glands of prepubertal castrates grew larger as the animals aged beyond one year. Gonadectomy of adult tamarins, on the other hand, had little effect on the morphological appearance of the scent glands of males and females, although it may result in some atrophy of the sebaceous component. It appears therefore that gonadal hormones are essential for the full development of the glands. However, once these structures have developed fully, at least the apocrine units may become largely hormone independent or may be maintained by extragonadal hormones. Adrenal androgens, which provide a source of extragonadal steroids, may possibly support the glands after gonadectomy in adulthood and also allow a certain amount of development of the glands in prepubertal castrates.

*Scent-marking behaviour and its communicatory functions*

Male and female tamarins use the sternal as well as the circumgenital-suprapubic gland during scent marking. Sternal marking, however, is a relatively infrequent behaviour, which is mostly shown in situations of high social arousal. Since it is such a rare pattern, all quantitative data on scent marking reported below relate to the use of the circumgenital–suprapubic gland. The use of this gland is very frequent. Scent marking is regularly shown by social dominant adults. Similar to *Saguinus o. oedipus* and in contrast to *Callithrix jacchus*, females mark more frequently than males (Epple 1980; French and Snowdon 1978; Sutcliffe and Poole 1978). Items in the environment, and occasionally conspecifics, are marked with these glands, applying gland secretions, urine, and perhaps genital discharge. Circumgenital–suprapubic marking is easily evoked by a wide variety of stimuli, particularly environmental and social novelty. It seems to be one of the species' responses to general arousal in the absence of strong, fear-inducing stimuli (animals who are thoroughly frightened by a conspecific or a predator do not mark). In all the various situations during which marking is shown, the animal appears to express 'self confidence' by applying its odour in an act of 'self-advertisement', a term proposed by Jolly (1966) for the complex marking behaviour of *Lemur catta*.

The chemical signals deposited by scent markings are of considerable importance as a means of social communication. In a series of experiments published previously, we studied the communicatory contents of the circumgenital-suprapubic marks and of urine. All of these experiments employed choice tests in which the subjects were simultaneously presented with two different scent samples. During each choice test, which lasted either five or 15 minutes, the subjects were given two stimulus objects. Fresh pieces of white pine or clean aluminium plates which had been marked by donor monkeys, or to which a pre-determined amount of stimulus fluid had been applied, served as stimulus objects.

The tamarins were tested in one unit of a double unit home-cage while other members of their group remained in the other half of the cage within sight and hearing. Both stimulus objects were simultaneously placed into the test cage, parallel to each other, about 30 cm apart. Their left–right position was counter-balanced across the subjects of each experiment, and across all tests on each subject. Each test was divided into 5-s intervals. For every interval, the subjects received a score of 1 for each of the following behaviours, if this behaviour was shown within the interval: contacting, sniffing, and marking either stimulus with the circumgenital glands. Mean scores for all behaviours were computed for each subject and analysed by non-parametric statistics following Siegel (1956).

Table 19.1 summarizes the results obtained during all of these experiments. As the table shows, tamarins prefer homospecific scent marks and urine over marks and urine of other marmoset species. Moreover, the odour of *Saguinus f. fuscicollis* is preferred over that of *Saguinus f. illigeri*, showing that the scent is also subspecies specific (Epple, Golob, and Smith 1979). Male conspecific marks and urine are preferred over female conspecific odours when tested under a variety of conditions, and male marks are even recognized in a mixture of male and female scents (Epple 1973, 1974b, 1978c). Intact male marks are preferred over castrated male marks, and the scent marks of dominant males are preferred over those of submissive males (Epple 1973, 1979). Discrimination between individual marks, but not urine samples, is shown when the monkeys are allowed to interact aggressively with one of two familiar individuals of the same sex, both members of another group. When tested hours or days after such aggressive encounters, a preference is displayed for the scent marks of the recent opponent (Epple 1973, 1978c). In another experiment on individual discrimination, the animals were allowed to investigate two plates marked by the same individual for 5 min. When after this period they were given a choice between another plate marked by the previous donor and a plate marked by a novel donor of the same sex, they preferred the novel donor (Epple *et al.* 1979). This demonstrates their ability to discriminate between individual odours in a non-aggressive context.

In summary, the complex circumgenital-suprapubic marks of *Saguinus fusci-collis* communicate the identity of the species, subspecies, and individual; gender; hormonal condition; and social rank. Additional signals, particularly those relating to the stage of the female's reproductive cycle, may also be present (Katz and Epple, unpublished).

Table 19.1. Mean scores for scent-marking, sniffing, and contacting wooden or metal stimulus plates carrying different scent samples.

| | Choice of stimuli A | B | Scent-marking scores on: A | | B | Sniffing scores on: A | | B | Contacting scores on: A | | B | Reference |
|---|---|---|---|---|---|---|---|---|---|---|---|---|
| Conspecific v. heterospecific scent marks | *Saguinus fuscicollis* v. | *Callithrix jaccus* | 3.4 | * | 1.2 | 11.5 | * | 7.0 | 11.7 | * | 7.2 | 1 |
| | *Saguinus fuscicollis* v. | *Saguinus labiatus* | 2.9 | | 2.3 | 6.5 | * | 4.8 | 10.8 | * | 7.5 | 1 |
| | *Saguinus f. fuscicollis* v. | *Saguinus f. illigeri* | 5.0 | * | 3.4 | 10.7 | * | 8.5 | 12.5 | * | 10.0 | 1 |
| Conspecific scent marks | Males v. | Females | 4.7 | * | 3.4 | 5.6 | * | 4.5 | 14.8 | * | 12.9 | 2 |
| | Male/female mixtures v. | Females | 2.9 | * | 1.9 | 5.7 | | 6.5 | 9.3 | | 9.7 | 3 |
| | Intact male v. | Castrated male | 6.0 | * | 2.6 | 10.8 | * | 6.5 | 12.3 | * | 5.9 | 4 |
| | Dominant male v. | Submissive male | 5.3 | * | 3.6 | 10.0 | * | 7.4 | 16.5 | * | 14.2 | 5 |
| | Individual males opponent v. | neutral | 8.0 | * | 5.1 | 14.5 | * | 9.0 | — | | | 5 |
| | Individual females opponent v. | neutral | 5.2 | * | 3.3 | 9.3 | * | 6.2 | — | | | 5 |
| | Individuals (♂♂ + ♀♀) novel v. | preadapted | 5.0 | | 3.9 | 14.0 | * | 11.1 | 14.6 | * | 11.4 | 1 |
| Urine samples (0.4 ml per sample) | *Saguinus fuscicollis* v. | *Callithrix jacchus* | 1.2 | | 0.8 | 4.2 | * | 2.3 | 6.6 | * | 4.8 | 6 |
| | Conspecific males v. | Conspecific females | 4.0 | * | 1.4 | 8.2 | | 5.6 | — | | | 5 |

A signifies the sample to which a higher response was given ('preferred scent'). * indicates that the difference between both scores was statistically significant. References: 1—Epple *et al.* (1979). 2—Epple (1974b). 3—Epple (1978c). 4—Epple (1979). 5—Epple (1973). 6—Epple (unpublished data).

## The longevity of the scent marks

One of the advantages of chemical communication is the fact that scent marks deposited into the environment retain their message for some time, even in the absence of the signalling animal (Regnier and Goodwin 1977). In their natural environment saddle-back tamarins probably scent mark branches as they travel through their territories (Dawson 1978) and one would expect these scent marks to retain some of their information for extended periods of time. In an attempt to establish the period of time for which the scent marks of *Saguinus fuscicollis* attract conspecifics and communicate the gender of the marker, we tested the tamarins' responses to aluminium plates carrying scent marks of various ages (Epple, Alveario, Golob, and Smith 1980). Their sniffing and scent-marking activities on clean aluminium plates and on plates carrying conspecific scent marks of various ages or synthetic odorants were tested in four experiments. In each experiment, the tamarins received several series of 5-min tests during which they were allowed to investigate either a single plate carrying one type of scent or were given a choice between two plates, each carrying a different type of scent.

In the first experiment, the relative intensities of the monkeys' responses to clean, dry aluminium plates, to plates carrying a novel arbitrary odour (exalto-lide) which is not a component of conspecific scent marks, to synthetic mixtures of components which form part of the natural scent marks (see below, p. 782) and to natural marks of various ages were assessed. Plates introduced into the monkeys' cages represent relatively novel objects which usually elicit investigatory behaviour and scent marking in the tamarins. It was therefore necessary to assess the monkeys' interest in variously aged conspecific scent marks relative to their interest in clean plates and in plates carrying arbitrary odours. Figure 19.1 shows the results of the first experiment. Clean, dry plates elicited sniffing and scent marking, but the responses to these plates were relatively low. Exaltolide, an odour to which the monkeys had not been exposed before, did not elicit significantly more sniffing or marking than unscented plates did. Plates carrying mixtures of synthetic butyrates and squalene, compounds which constitute part of the natural scent mark (see below, p. 782) were also sniffed and marked with approximately the same frequency as clean plates. Plates with fresh scent marks were sniffed and marked significantly more frequently than clean plates. One-day-old scent marks were still quite attractive. The subjects scent marked them as frequently as plates carrying fresh marks and more than clean plates. However, they tended to sniff day-old marks slightly less frequently than fresh marks.

Sniffing and marking activities gradually decreased with increasing age of the marks. By the time the scent was three days old, the plates were sniffed significantly less frequently than freshly marked ones. Scent marking on plates carrying two- and three-day-old marks did not differ significantly from marking on plates carrying fresh scent but by four and five days the plates were marked significantly less frequently than freshly scented ones.

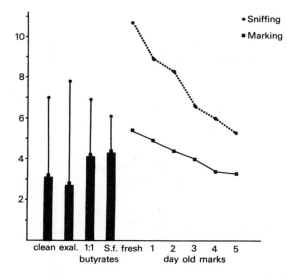

Fig. 19.1. Mean scores for sniffing and scent-marking clean and differentially scented single plates. The following scents were offered: Exaltolide (exal.); synthetic mixtures of the butyrates and squalene representing major compounds of the natural mark in a 1:1 ratio (1:1 butyrates) and in the same ratio found in marks (S.f. butyrates); fresh and 1–5-day-old natural scent marks. (From Epple *et al.* (1980).)

In the second experiment, we gave the tamarins a choice between a clean, dry aluminium plate and a plate carrying either fresh male scent marks or marks applied to the plate by the same donor one, two, three, four, five, seven, nine, or 11 days prior to testing. As in the first experiment, the attractiveness of the scent marks decreased with increasing age. While plates with fresh marks and one-day-old marks were significantly more frequently sniffed and marked than clean ones, plates carrying two- or more day-old scent marks were not preferred to clean plates (Fig. 19.2).

The third experiment tested the responses of the tamarins when given a choice between a plate marked by a male donor immediately before the test and a plate marked by the same donor one, two, three, or four days before the test. The results of these tests, pictured in Fig. 19.3, show that by one day of age, scent marks appeared to be less attractive than fresh marks from the same donor. Although one-day-old marks were sniffed approximately as frequently as fresh marks, there was significantly more scent marking on the plates carrying fresh scent. The attractiveness of the old scent as compared with the fresh marks decreased further with increasing age of the marks.

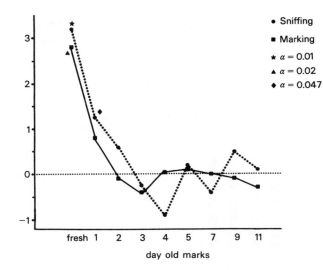

Fig. 19.2. Mean preference scores for sniffing and scent-marking plates carrying differently aged natural marks, offered as an alternative to clean plates. The preference scores were computed by deducting the scores on the clean plate from those on the marked plate. (From Epple *et al.* (1980).)

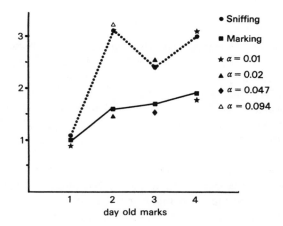

Fig. 19.3. Mean preference scores for sniffing and scent-marking plates carrying differently aged marks, offered as an alternative to freshly marked plates. The preference scores were computed by deducting the scores on the plate carrying aged scent from those on the plate carrying fresh scent. (From Epple *et al.* (1980).)

In the last experiment, we attempted to determine for which period of time the scent marks retain those properties which motivate the tamarins to prefer male over female marks (see Table 19.1), i.e. how long the marks identify gender. In this experiment, the subjects were given a choice between either two clean, dry plates or between a plate carrying male scent marks and a plate carrying female scent marks. The plates carrying male and female odours had either been marked by the donors immediately before, or one, two, three, and four days before the test. Each choice test offered male and female marks of the same age. Figure 19.4 illustrates the results of this experiment. When two clean plates were presented the monkeys did not prefer either one of two plates. When they were given scented plates they preferentially marked the plate carrying fresh male scent while sniffing male and female odours with about equal frequency. This marking preference for male scent gradually decreased with increasing age of the scent marks. Plates with one-day-old male odours were still significantly more frequently marked than plates with one-day-old female odours. The same was true for two-day-old scents. No discriminatory responses were shown to three- and four-day-old marks but the response levels to these scents were still above those to clean plates.

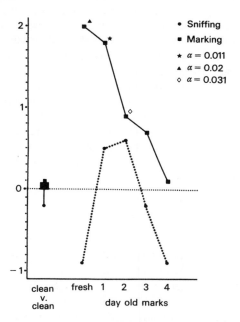

Fig. 19.4. Mean preference scores for sniffing and scent-marking plates carrying differently aged male marks, offered as an alternative to a plate carrying female marks of the same age. The preference scores were computed by deducting the scores on the female marked plate from those on the male-marked plate. (From Epple *et al.* (1980).)

The experiments reviewed above show that fresh scent marks present complex and specific stimuli of high sniffing and marking activity. With the laboratory conditions under which they were aged (windowless room, 21 °C, 25 per cent relative humidity) they lost some of their attractiveness within 24 h. It further decreased and by three to four days of age the marks did not elicit more interest than unscented objects or arbitrary odours and no longer appeared to identify sex. The rate at which behaviourally active signals are released from complex mammalian scent marks is influenced by the physiochemical characteristics of all compounds in the mixture as well as by a number of environmental factors such as the nature of the substrate, ambient temperature, relative humidity, and air flow (Regnier and Goodwin 1977). Therefore, the longevity of the marks under natural conditions may differ somewhat from that established in the laboratory and probably fluctuates depending on substrate and weather conditions. The differences under natural and laboratory conditions, however, should not be dramatic (i.e. in the range of weeks). Thus, it appears that the behaviourally active components of the scent marks are of medium volatility when compared to other mammalian chemical communicants whose longevity has been established. The flank marks of male and the vaginal marks of female golden hamsters, for example, appear to retain attractiveness for at least 25 days when aged under laboratory conditions (Johnston and Lee 1976) and dried guinea-pig urine communicates sex after 40 days of storage (Wellington, Beauchamp, and Smith 1981). Our experiments on the longevity of tamarin marks, of course, give no information on how long signals such as species identity are retained. It is conceivable that this particular signal survives for a longer time than that identifying sex. However, it may become so weak that the scent marks are no longer very attractive, at least under the testing conditions employed in our experiments. Under natural conditions tamarins may obtain information about the approximate age of a scent mark by assessing its attractiveness and/or signal content and may use this information in a variety of social situations such as spacing and territorial defence.

### The chemical properties of the scent marks

Concurrently with our behavioural studies, we conducted work on the chemical analysis of the major volatile constituents of the scent marks. Details of the procedures involved in collecting and analysing the material have been reported elsewhere (Smith, Yarger, and Epple 1976; Yarger, Smith, Preti, and Epple 1977) and will not be repeated here. In summary, combined micropreparative gas chromatographic and mass spectrometric analysis of male and female scent marks, which were collected on frosted glass plates and taken up into hexane, revealed that 96 per cent (by weight) of the volatile material of male and female marks consist of 15 esters of *n*-butyric acid and of squalene (Fig. 19.5). Squalene was the only previously known constituent. Examination of the structure of the

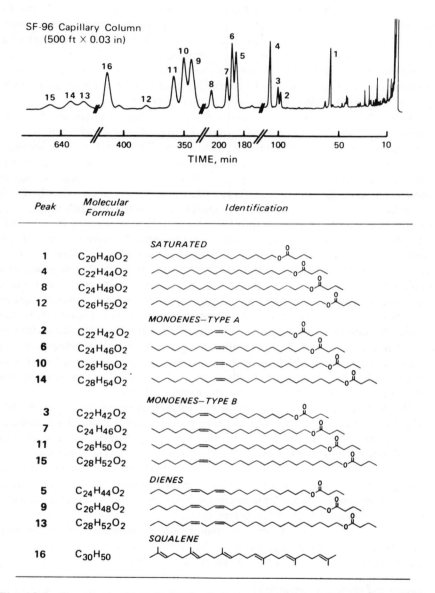

Fig. 19.5   Gas chromatogram (top) of the scent marks of *Saguinus fuscicollis* and their major volatile constituents.

butyrates revealed the presence of four major classes, namely the saturated butyrates, the monounsaturated butyrates of two structural types (A and B) dependent on the location of the olefinic linkage, and the diunsaturated butyrates. As occurs with other mammalian fatty-acid systems, each member of a class differs from the next by a two-carbon unit. Golob, Yarger, and Smith (1979) have performed the stereospecific total synthesis of each of the 15 major compounds. In each case the synthetic material was identical in all respects (i.r., 220 MHz n.m.r. and m.s.) with the natural material.

Although the mixture of squalene and the butyrates comprise about 96 per cent (by weight) of the volatile material of a scent mark, the remaining 4 per cent may have considerable behavioural significance. We have therefore recently undertaken studies to complete the chemical analysis of the scent material. Interestingly, most of the remaining material was found to be comprised of highly volatile compounds; that is, those that elute prior to butyrate 1 (see Fig. 19.5) upon gas chromatography. While this analytical programme is still in progress, our preliminary results are illustrated in Fig. 19.6. Here we present two chromatograms of the highly volatile components, one derived from a male and the other from a female, along with a list of those components identified to date as common to both sexes. Several of these components, in particular the aldehydes, could conceivably derive from the appropriate butyrate esters in the scent mark via microbial degradation. More specifically, cleavage of the olefinic linkages of the respective butyrates via an oxidative process (i.e. formal ozonolysis) would afford the observed aldehydes. This hypothesis is supported by observation of diminished amounts of highly volatile components as well as by an unusual butyrate ester pattern obtained upon gas chromatography of glandular material obtained when secretions from the gland are expressed by gentle manual pressure after cleaning the surface of the skin with 40 per cent ethanol.

The level of complexity of the highly volatile components in conjunction with the 15 butyrate esters and squalene provides a rich resource of chemical constituents for formulating a communicatory code. Indeed, the question of the relationships between chemical composition and behavioural messages is of paramount importance. Even so, and given the rather extensive chemical analysis both past and ongoing, we are a long way from establishing which compounds and in which relative concentrations are involved in encoding the specific messages that our behavioural research has shown to exist.

In the following paragraphs we shall report some additional preliminary findings on the relationships between the relative concentrations of squalene and the butyrates and some of the behavioural messages.

*Species and subspecies specificity*

The results of our behavioural studies suggest that the scent marks are species specific (Table 19.1). This result is supported by our gas chromatographic studies. Figure 19.7 displays the gas chromatographic traces of three tamarin species and of the common marmoset. Although retention time data alone are not sufficient to establish the identity of a compound, similar retention times are certainly suggestive that the other species have, to varying extents, components similar to those already identified. Notable differences also exist. Studies to confirm similarities as well as to identify new components are in progress.

The saddle-back tamarins in our colony belong to two subspecies, *Saguinus f. fuscicollis* and *Saguinus f. illigeri*, who interbreed and produce fertile hybrids. Behavioural testing has shown that the animals can discriminate between the scent marks of both pure-bred subspecies. *Saguinus f. fuscicollis* significantly preferred the odour of their own subspecies. *Saguinus f. illigeri* scent marked preferentially on top of the odour of the opposite subspecies but did not show differences in sniffing and contacting while hybrids did not discriminate (Epple *et al.* 1979).

Since the pure-bred tamarins were able to discriminate between subspecific odours, it was of interest to investigate possible differences in the chromatographic profiles of both subspecies and their hybrids. Male scent marks were collected and analysed by gas chromatography. Both pools of marks from several males of the same subspecies as well as marks of individual donors were analysed as described by Epple *et al.* (1979). In order to facilitate comparisons, an internal standard was added and consistent gas chromatographic conditions were

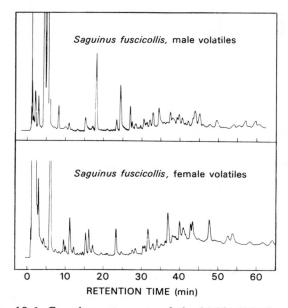

Ethanol
Butanol
Hexanal
Undecane
Heptanal
2–Pentyl furan
Nonanal
Methyl heptanoate
Hexadecane
Isopropyl cyclopentene
Geranyl acetone
Phenethyl alcohol
Phenol
*m*-Anisaldehyde
*f*-Cresol
Nonanoic acid
Eugenol
Methyl palmitate

Fig. 19.6. Gas chromatograms of the highly volatile constituents of the scent marks of *Saguinus fuscicollis* and list of compounds identified. Note that these volatiles elude prior to peak 1 on Fig. 19.5.

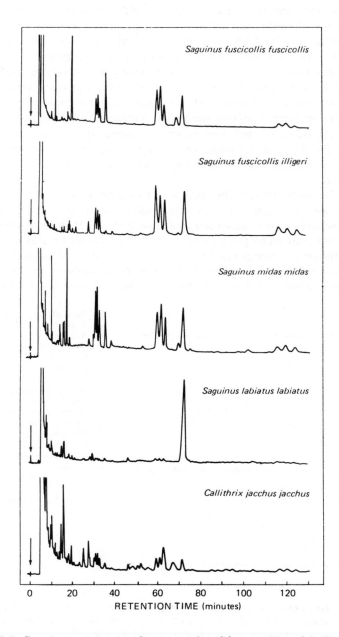

Fig. 19.7. Gas chromatograms of scent marks of four species of Callitrichids.

established and maintained throughout the study. The chromatographic profiles shown in Figs. 19.8 and 19.9 are graphical representations in which the relative peak area (i.e. peak area of each component divided by the total area of the 16 components) is plotted for each of the 16 components at their respective retention times. Most profiles are averages of several daily profiles; the height of the bar for each component represents the mean for that component averaged over the number of days examined, and the small 'T' attached to the bar represents the standard deviation.

Figure 19.8 shows that there is a considerable difference between the profiles of the subspecies. Thirteen of the 16 components occur in different relative ratios, while only three are present in the same amounts. Members of the same subspecies, on the other hand, share similar profiles. In both subspecies the pattern of an individual animal closely resembles that obtained from the pooled marks of its subspecies (Fig. 19.8). Hybrid animals show pronounced differences from both subspecies. Moreover, the profiles of hybrid individuals differ from each other to a larger degree than those of pure-bred animals. Individual pure-bred males and hybrids also maintain very consistent profiles over time (see below). If an individual's profile is consistent over long periods of time and individuals of the same subspecies share quite similar profiles, then it is possible that the ratios of the compounds present in the scent mark may participate in encoding the individual's subspecies. The availability of the synthesized esters and commercial squalene provided us with the opportunity to investigate this hypothesis by testing the tamarin's responses to synthetic subspecies-specific mixtures of esters and squalene (Epple *et al.* 1979).

Synthetic mixtures of squalene and butyrates yielding profiles very similar to those of the authentic marks of individual *Saguinus f. fuscollis* and *Saguinus f. illigeri* males were obtained as described by Epple *et al.* (1979). The synthetic formulations were then bioassayed using choice tests. The quantity of synthetic material presented for each test was 1.00 mg (after evaporation of the solvent), an amount equivalent to approximately three natural marks. In these choice tests, the subjects obtained a slightly, but significantly higher average marking score on top of the formulation mimicking *Saguinus f. fuscicollis* marks, a result which replicates their responses to natural material. The difference in the marking scores, however, was extremely small and their sniffing and contacting scores on both samples did not differ. Although this result suggests that the subjects discriminated between both samples, their average scores for investigating the synthetic material were much lower than those obtained for investigating natural marks (Epple *et al.* 1979). This obvious decrease in interest in the synthetic marks as compared to the natural material suggests a large loss of activity of the formulations. Therefore, our present working hypothesis is that the identity of the subspecies may, in part, be encoded by the specific ratio of esters present in the marks but that further synergists, probably present in the highly volatile material, are necessary to mimic the biological activity of the natural scent mark.

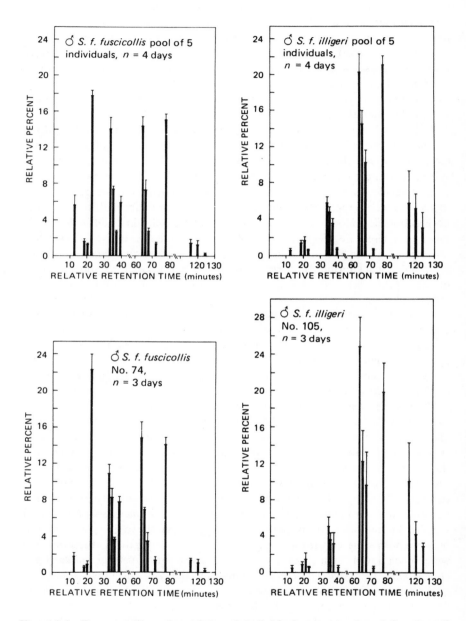

Fig. 19.8. Ester profiles of pooled and individual scent marks of *Saguinus f. fuscicollis* and *Saguinus f. illigeri*. (From Epple *et al.* (1979).)

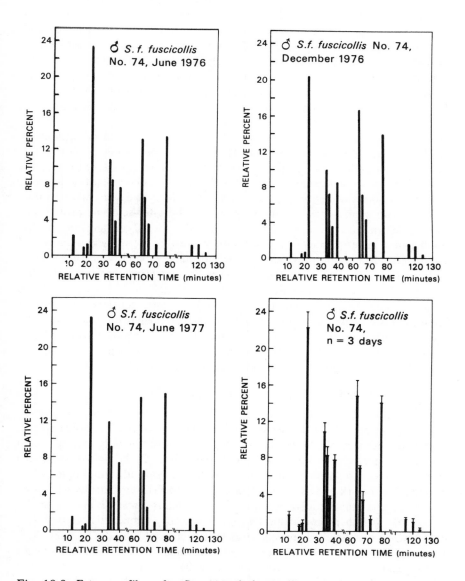

Fig. 19.9. Ester profiles of a *Saguinus f. fuscicollis* male. Note the continuity in the profiles over a period of one year.

*Individual specificity*

The scent marks of *Saguinus fuscicollis* identify individuals of the same sub-species (Table 19.1). Because of these findings, it was of interest to monitor chemically the individual variability of the known components of the scent marks. Scent marks were collected from three adult males at regular intervals of time and gas chromatographic profiles were obtained under standard experi-mental conditions. The results of this study demonstrate that profiles for pure-bred individual males not only closely resemble those of the appropriate subspecies but are also extraordinarily constant over extended periods of time. The relative ratios of each of the 16 components tend to fluctuate less than 20 per cent.

Figure 19.9 shows three profiles of the scent mark of a male *Saguinus f. fuscicollis* taken at six-month intervals. The pattern is remarkably consistent over the 12-month period. The profile in the lower-right box represents the mean of three days and the standard deviation of each peak is small. Similar findings were obtained from a *Saguinus f. illigeri* male and a male hybrid who was monitored for over two years. These results suggest that the ester–squalene profile may form fingerprint-like patterns which may be useful in identifying individual males. Studies on a larger number of pure-bred individuals are currently in progress and should provide more information on interindividual variability of ester profiles within subspecies. Behavioural experiments demonstrated that the esters and squalene *alone* do not carry the information on individual identity. In choice tests, we presented the tamarins with synthetic ester–squalene preparations mimicking two individual donor males after adapting them to one of the synthetic formulations as described on p. 787. During these experiments the subjects investigated both samples at random and with little obvious interest, giving no indication that they could discriminate between them (Epple *et al.* 1979).

*Social functions and hormonal control of scent-marking*

For any species, the sociosexual functions of signalling behaviour and of communi-catory signals have to be discussed with consideration of the social organization of the species. *Saguinus fuscicollis* is one of the most frequently used species in biomedical research. However, little is known about its behaviour and social structure in its natural environment. Our own laboratory studies (cf. Epple 1975*a*) as well as field work by Castro and Soini (1978) and Terborg (personal communi-cation) suggest that the social organization of *Saguinus fuscicollis* may be similar to that of two other tamarin species (*Saguinus oedipus* spp., and *Saguinus nigri-collis*) which were studied in some detail in their natural habitats (Dawson 1978; Izawa 1978; Neyman 1978). Both species form small groups, each containing a stable nucleus consisting of a breeding female, her dependent young, and a male with whom she probably has a monogamous pair-bond. Although these groups may contain several adults of both sexes, there appears to be only one breeding female in each group. Maturing offspring and non-breeding adults show a tendency to leave their families and immigrate into other conspecific groups with which they stay associated for various lengths of time. These transient animals may change group affiliations repeatedly (Dawson 1978; Izawa 1978; Neyman 1978).

Although tamarins may inhabit overlapping home-ranges and occasionally form temporary feeding aggregations consisting of more than one group (Castro and Soini 1978; Izawa 1978; Moynihan 1970; Thorington 1968), the occupation and defence of territories in the strictest sense has also been reported. Which kind of spacing behaviour is adopted appears to depend on a number of ecological factors (Dawson 1978; Moynihan 1970; Neyman 1978; Thorington 1968). At a study site in Peru, *Saguinus fuscicollis* is strictly territorial and, besides a number of displays, aggression forms part of their territorial defence (Terborg, personal communication).

In the laboratory, *Saguinus fuscicollis*, like other marmosets and tamarins, establishes monogamous pair-bonds and is characterized by high levels of aggression directed at non-related adult conspecifics (Epple 1978a; Kleiman 1977). In laboratory groups consisting of a bonded pair and its progeny, as well as those containing several non-related males and females, only one female ever bears offspring (Epple 1975a). In families, the mother remains the only reproductively active female and there is little sexual activity among her adult offspring (Epple, unpublished). In groups containing several non-related adults, one female usually establishes dominance over all other females, pair-bonds to a dominant male, and appears to inhibit pair-bonding in other females, although these may engage in sexual activities (Epple 1978a). Although only one female bears young, all group members usually participate in the care of her offspring, which are carried sometimes predominantly by animals other than their mother (cf. Epple 1975b).

In the laboratory all members of tamarin groups scent mark, but the frequency of marking is influenced by their age, sex, social and hormonal status, group composition, and short-term motivational state. The study of the effect of some of these variables on the marking activity of individuals under controlled conditions provides one of the means by which the role of marking in sociosexual behaviour can be analysed. Such studies have indicated that one area in which scent-marking is important is the demonstration of aggressive motivation and the establishment and maintenance of social dominance. Aggressive interactions among members of the same group are usually accompanied by a strong increase in the marking activity of the dominant animal and a decrease in that of the submissive animal. Similarly, aggressive interactions with conspecifics belonging to different groups are associated with an increase in the marking activity of dominant adults as compared to their marking activity in undisturbed situations, but may result in a complete suppression of marking in submissive adults introduced into established groups (Epple 1978a). These data suggest that one of the functions of marking is the communication of aggressive motivation. In this context, the scent may actually serve as an aggressive threat display. The high marking activity of a dominant animal results in a large amount of this individual's odour in the environment. Since the tamarins sniff conspecific marks frequently, a subdominant animal must be aware of the identity and amount of the dominant's odour. This awareness, combined with the experience of previous defeat or with other aggressive signals received from a social partner, may effectively control the recipient's behaviour. The following two experiments demonstrate the relationships between aggression and scent-marking and investigate the role of the gonads in the control of both behaviours (Epple 1978b, 1980, 1981a).

*The role of scent-marking in aggression and its hormonal control*

Scent-marking activity and injurious aggression are associated in many mammalian species, including some primates (cf. Schilling 1979; Ralls 1971). With a few exceptions, aggression, the activity of the scent glands, and marking behaviour are dependent on gonadal hormones in most non-primate species (Thiessen 1976). However, the aggressive behaviour of simian primates is often more strongly influenced by social experience than by hormones (e.g. Dixson 1980; Rose, Bernstein, Gordon, and Catlin 1974). Therefore, we studied the relative importance of the gonads and a number of social factors in influencing these characteristics in the tamarins. In two experiments, we studied the social interactions of established, permanently cohabiting pairs with adult, strange conspecifics in situations analogous to territorial defence in the wild.

In the first experiment, the effects of aggressive motivation on scent-marking activity and the effects of gonadectomy in adulthood on aggression and marking behaviour of 14 adult male–female pairs were studied (Epple 1978b, 1980). The behaviour of these pairs was studied, as described below, prior to gonadectomy. Following these tests, the males of seven pairs and the females of two pairs were surgically gonadectomized while their partners remained intact. The males of the remaining five pairs underwent sham-castration.

The behaviour of each pair was tested in the following manner before and after surgery. Each pair lived by itself in a small test-room, their 'home-territory'. The scent-marking activity of both partners under undisturbed trial free conditions was scored for a series of 20 min observation sessions during which each partner was observed for 10 min. The observation period was divided into intervals of 15 s and the subject obtained a score of 1 for each interval in which it scent-marked. The animals were confined to a home-cage within the test-room during the trial-free observation. However, as established previously, their scent-marking activity during trial-free conditions was not affected by confinement to the home-cage (Epple 1980).

In addition to observations under trial-free conditions, the interactions of the subjects (not confined to their cage) with strange conspecifics 'intruding' into their home-territory were studied. Every pair encountered 10 different adult males and 10 females, belonging to other groups, once before and once after surgery. For each encounter, the stranger was wheeled into the test-room in a

2 X 3 X 4-foot wire-mesh cage and remained confined to the cage during the entire encounter, allowing subjects and intruder to interact through the ¼-inch wire mesh. Each 10-min encounter was divided into 40 intervals of 15 s. A score of 1 per interval was given for performing a number of behaviour patterns, described in detail by Epple (1978*b*), regardless of their actual frequencies of occurrence. In the present review, only the scores for scent-marking and patterns of injurious aggression are presented. An accumulative score for injurious aggression was computed by summing the scores per interval of attacking the intruder, fighting with it through the wire mesh of the transport cage, and chasing it along the cage wall. Average scores for each individual before and after castration were computed and all data was analysed by non-parametric statistics following Siegel (1956).

The second experiment examined the effect of castration prior to puberty on the development of the scent glands, scent-marking behaviour, and aggression of males (Epple 1981*a*). Nine males, all born in the laboratory and living with their families until they were 5½ months old, served as subjects. At the age of 5½ months, five males were castrated and four sham-castrated. After surgery, the males remained with the families for two weeks. At the age of six months, each male was removed from his family and paired with an adult, sexually experienced female with whom he cohabited throughout the experiment.*

When the males were between 10 and 12 months old, their social interactions with strange conspecifics and their scent-marking behaviour were tested. Each pair was observed under trial-free conditions in its home-cage and also received seven social encounters with seven different adult, intact female intruders and seven encounters with seven intact, adult male intruders. The encounters were conducted and evaluated in the same way as described above for the first experiment. Following completion of all encounter tests, each subject pair was housed in a home-cage located in a colony room until the males were two years old. At that time, all castrates and three control males† received five social encounters with five strange males and five social encounters with five strange females, using the same strangers as before. Their marking behaviour under trial-free conditions was not tested at that time. The methods of conducting the encounters were identical to those described above, with one exception. Since a sufficient number of test rooms to house all subject pairs was no longer available, the following procedure was adopted. Each subject pair lived in a home-cage in a room housing several groups. Prior to every encounter the subject pair, inside its home-cage, was wheeled into the test room, released from its home-cage and allowed to run free in the room for two hours. This habituation period was followed by a 10-min social encounter test. Fifteen minutes after the encounter the pair was placed into its home-cage and was transferred back into the colony room.

---

* The partner of one male died and was replaced by another female when the male was eight months old.
† One of the control males had died.

Figure 19.10 shows the mean behaviour scores obtained by males and females gonadectomized as adults. As the figure demonstrates, aggressive interactions with strange conspecifics resulted in a strong increase in the scent-marking activity of both sexes. Under the test conditions used here, the territorial owners were almost always dominant over the intruder and directed much more aggression at the stranger than they received from it (Epple 1978*a,b*). An increase in the scent-marking activity of dominant males and females but not in that of submissive

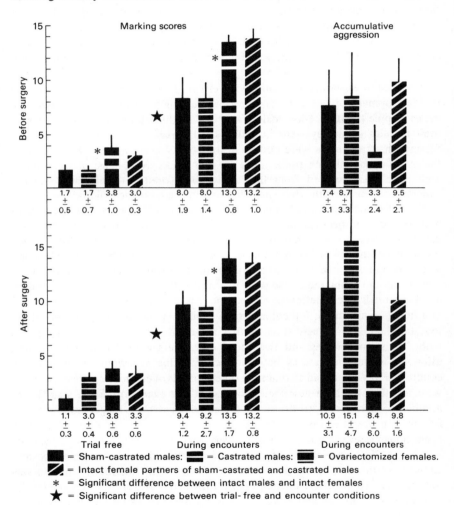

Fig. 19.10. Mean scent-marking scores under trial-free conditions and during social encounters and mean aggression scores before and after surgery.

individuals during aggressive interactions has been found in several other experiments conducted in our laboratory (unpublished data) and also occurs in *Callithrix jacchus* (Epple 1970). As Fig. 19.10 shows, this response to aggressive arousal was not influenced by gonadectomy.

When scent-marking and aggression scores obtained before surgery are compared with the postsurgical scores, it becomes obvious that gonadectomy and sham-castration of adult, socially experienced subjects had no statistically significant effect on their display of injurious aggression and on scent-marking. Although the postsurgical aggression scores and the marking scores of the adult males tended to be higher than their presurgical scores, particularly in the castrates, this increase was not statistically significant. The two ovariectomized females also showed an increase in their aggression scores. However, no statistical analysis was done on only two subjects.

The scent-marking scores of females were higher than those of males, both under trial-free conditions as well as during encounters. However, the tendency of the castrates to mark more frequently following surgery resulted in a disappearance of this sexual dimorphism in the postsurgical marking scores obtained under trial-free conditions.

Figure 19.11 shows the behaviour scores of prepubertally castrated and sham-castrated males. At one year of age, these males showed less aggression than adult males showed before and after castration. The sham-castrated juveniles tended to have higher aggression scores than the castrates. However, the average scores of the subject groups were not statistically different. By the time the subjects were two years old, the aggression scores of both groups were in the range of adults, those of the sham-castrates being very high. The large standard errors reflect the fact that there was no difference in the scores of experimental and control males. The scent-marking activity of the sham-castrates, on the other hand, was significantly higher than that of the castrates, both at one and two years of age. The scent-marking activity under trial-free conditions was only recorded in the one-year-old subjects. Under these conditions, there was no difference between experimental and control males. Aggressive encounters resulted in an increase in marking in the sham-castrates but in a decrease in the castrates. These results indicate that gonadal hormones influence the development of aggression and scent-marking. However, other factors also play a role in their development. Once fully developed in the socially experienced adult, these behaviours seem to be independent of the gonads.

The fact that year-old castrates and sham-castrates marked with approximately the same frequency under trial-free conditions suggested that the decreased marking of castrates during aggressive encounters was specific to this social situation. This, and other casual observations on the influence of social stimuli on scent-marking, prompted us to test the scent-marking activity of the subjects gonadectomized as adults and prior to puberty under a condition in which they were not confronted with conspecifics. Since the tamarins readily investigate and

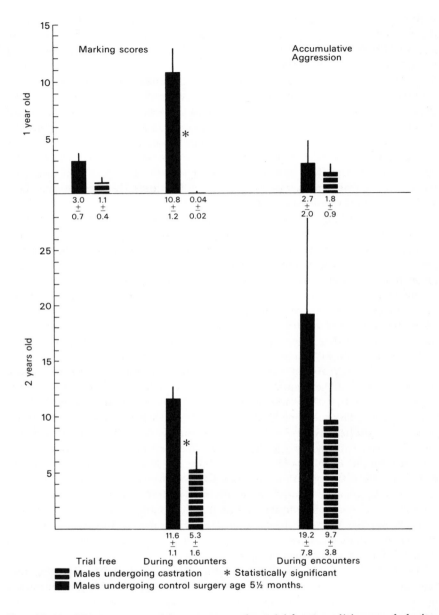

Fig. 19.11. Mean scent-marking scores under trial-free conditions and during social encounters and mean accumulative aggression scores. Scores obtained at one and two years of age.

scent-mark novel objects introduced into their home-cages, we tested the scent-marking activity of the subjects on novel objects in series of 5-min tests. During each test, the subject was allowed to investigate a single aluminium plate which had either been scent-marked by a conspecific immediately before the test or was clean. The 5-min tests were divided into 60 intervals of 5 s each. For each interval the subject received a score of 1 if it scent-marked the stimulus plate or any other place in the cage. In all tests the plate was placed into the animal's home-cage, while the subject was confined to one half of its two compartment home-cage and its mate remained in the adjacent compartment.

The males castrated in adulthood and their controls only received plates which had been marked by donor males. These animals were tested with male-scented bars before and after surgery as detailed by Epple and Cerny (1979). After surgery, the castrated males were tested while living under three different social conditions. For several months after surgery they continued to cohabit with their original female partners. Between 13 and 25 months after surgery, four of the castrates were removed from their females and each was placed into an experimental trio consisting of an intact male, the castrated male and a female, all non-related adults. At 26, 38, and 51 months after surgery, three of the trio living castrates were again placed with a permanent female partner alone. These changes in the social setting provided an opportunity to assess the relative importance of gonadal hormones and social conditions in influencing scent-marking behaviour.

The scent-marking activity of the prepubertal castrates and sham-castrates on novel objects was evaluated while all subjects cohabited with their original females (Epple 1981*a*). The marking responses to clean aluminium plates, to plates scent-marked by an adult, intact male immediately before the test and to plates scent-marked by an adult, intact female immediately before the tests were studied under conditions identical to those used for the males castrated as adults. Figure 19.12 shows the results of these experiments. The studies of the marking behaviour of the males castrated as adults, and their controls, on male-scented plates showed no changes in marking activity attributable to gonadectomy. Changes in the social conditions under which the castrates lived, however, had quite a dramatic effect on their marking behaviour. While the marking activity of the castrates during cohabitation with their original females was in the range of that of the control males, each of the four castrates studied when living in a trio with an intact male and an intact female strongly decreased his scent-marking behaviour. The decline in the marking activity of the castrates during life in trios was completely reversible when their social situation changed after removal from the presence of an intact male. The average score of these three males who had been castrated 2, 4, and 4½ years ago was now well above the overall average for all castrates.

The marking activity of prepubertally castrated and sham-castrated one- and two-year-old males on scented plates is also shown in Fig. 19.12. The marking

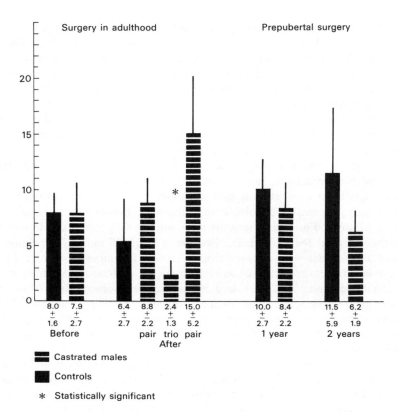

Fig. 19.12. Mean scent-marking scores on novel objects. The scores of the males castrated as adults when living with their original females, when living in trios and when living with a single female again were given. The scores of the prepubertal castrates and control males at one and two years of age are presented separately.

activity of castrates and sham-castrates did not differ under these conditions and some of the scores received by these males tended to be slightly but not significantly higher than the overall scores of the adult males.

### Summarizing remarks

Summarizing our work on chemical communication in *Saguinus fuscicollis*, we conclude that these primates make use of chemically highly complex scent marks to communicate a number of basic messages such as the identity of the species, subspecies and individual, and its sex. Although many of the chemical compounds constituting the scent marks have been identified, the manner in which they encode communicatory messages remains largely unknown.

The meaning of these messages to the animal which perceives them, and their roles in controlling behavioural events, are probably influenced by a number of variables. Marmosets and tamarins, like most simian primates, possess multimodular communication systems. They have rich vocal repertoires and make use of facial expressions and other visual displays (Epple 1975*a*). Furthermore, social experiences and learning influence their behaviour a great deal. Therefore, chemical signals are placed into a complex contextual network and the information they contain may take on a large number of meanings. The meaning attributed to the message by its recipient, in turn, will only be one of a number of internal and external stimuli which influence the recipient's behaviour. For example, chemical information on the sex of an individual might function as an attractant for an animal of the opposite sex, but as a repellent or aggression-eliciting signal for an animal of the same sex, particularly when combined with information on reproductive state. Similarly, individual odours may serve as social or sexual attractants to recipients who have experienced friendly interactions with the sender, while they may be regarded as threat signals by recipients who have experienced agonistic interactions with the sender. Thus, scent marks of very similar chemical composition are adaptable to a number of different functional roles. Behavioural processes such as group cohesion, spacing and territoriality, parent–infant interactions, recognition of and attraction to a sexual partner at the right state of reproductive activity, sexual arousal, and pair-bonding, may, in part, be controlled by chemical messages.

One area in which scent-marking and the messages deposited in this way appear to be of particular importance is aggressive behaviour. Here, scent-marking may serve as an aggressive threat display in intergroup as well as intragroup conflicts and in the maintenance of social dominance. Furthermore, overt aggression and aggressive marking may play a role in competition for a mate and enforcement of the pair-bond as discussed in more detail by Epple (1978*a*,*b*). Considering the fact that social dominance is a prerequisite for reproductive success of females, it is not surprising that in most laboratory groups the dominant female shows the highest marking activity. A high scent-marking activity may demonstrate her high rank and help her to maintain it. In addition, it is conceivable that the marks of the dominant female contain compounds which are involved in inhibiting reproduction in subdominant females. We are presently investigating this hypothesis.

As our studies on the hormonal control of the scent glands, marking and aggression have shown, the ontogenetic development of these characteristics is, to a certain degree, influenced by gonadal hormones, at least in males. Prepubertal castration influenced the social interactions of the males with strange conspecifics. However, these effects were not permanent and have to be evaluated in the context of social experience. Our results suggest that, in addition to gonadal hormones, social learning is an important factor which influences the display of aggression and scent-marking behaviour. The role of social learning is further documented by an additional study in which six-month-old intact males and females were removed from their families, permanently paired with either a sexual partner of their own age or with an adult partner and studied during encounters with strange conspecifics when one year old (Epple 1981*b*). Under these conditions young males and females who had cohabited with adult mates from the age of six months on were more aggressive and scent-marked more frequently than youngsters who had lived with a partner of their own age. The results obtained with males and females gonadectomized as socially and sexually experienced adults further underline the importance of social learning in influencing scent-marking, suggesting that, at this stage in life, these behaviours have become more or less independent of the gonads but that their display is highly sensitive to social stimuli.

## Acknowledgements

The long-term support of this research programme by the National Science Foundation, is gratefully acknowledged. Our current studies are supported by NSF grants BNS 78 06172 and BNS 78 19875.

## References

Andriamiandra, A. and Rumpler, Y. (1968). Rôle de la testostérone sur le déterminisme des glandes brachiales et antebrachiales chez le *Lemur catta*. *C. r. Séanc. Soc. Biol.* **162**, 1651–8.

Beauchamp, G. K., Wellington, J. L., Wysocki, C. J., Brand, J. G., Kubie, J. L., and Smith, A. B. III (1980). Chemical communication in the guinea pig: urinary components of low volatility and their access to the vomeronasal organ. In *Chemical signals: vertebrates and aquatic invertebrates* (ed. D. Müller-Schwarze and R. M. Silverstein) pp. 327–39. Plenum Press, New York.

Castro, R. and Soini, P. (1978). Field studies on *Saguinus mystax* and other callitrichids in Amazonian Peru. In *The biology and conservation of the Callitrichidae* (ed. D. G. Kleiman) pp. 73–8. Smithsonian Institution Press, Washington, DC.

Cebul, M. S., Alveario, M. C., and Epple, G. (1978). Odor recognition and attachment in infant marmosets. In *Biology and behaviour of marmosets* (ed. H. Rothe, H. J. Wolters and J. P. Hearn) pp. 141–6. Eigenverlag Rothe, Göttingen.

Christen, A. (1974). Fortpflanzungsbiologie und Verhalten bei *Cebuella pygmaea* and *Tamarinus tamarin* (Primates, Platyrrhina, Callithricidae). *Z. Tierpsychol.* Suppl. 14.

Dawson, F. (1978). Composition and stability of social groups of the tarmarin *Saguinus oedipus geoffroyi* in Panama: ecological and behavioral implications. In *The biology and conservation of the Callitrichidae* (ed. D. G. Kleiman) pp. 23–37. Smithsonian Institution Press, Washington, DC.

Dixson, A. F. (1976). Effects of testosterone on the sternal cutaneous glands and genitalia of the male greater galago (*Galago crassicaudatus crassicaudatus*). *Folio primat.* **26**, 207–13.

—— (1980). Androgens and aggressive behavior in primates: a review. *Aggress. Behav.* **6**, 37–67.

Epple, G. (1967). Vergleichende Untersuchungen über Sexual- und Sozialverhalten der Krallenaffen (*Hapalidae*). *Folia primat.* **7**, 37–65.

— (1970). Quantitative studies on scent marking in the marmoset (*Callithrix jacchus*). *Folia primat.* **13**, 48–62.

— (1973). The role of pheromones in the social communication of marmoset monkeys. *J. Reprod. Fert.* Suppl. **19**, 445–52.

— (1974a). Olfactory communication in South American primates. *Ann. NY Acad. Sci.* **237**, 261–78.

— (1974b). Pheromones in primate reproduction and social behavior. In *Reproductive behavior* (ed. W. Montagna and W. A. Sadler) pp. 131–55. Plenum Press, New York.

— (1975a). The behavior of marmoset monkeys (Callithriciade). In *Primate behavior*, Vol. 4 (ed. L. A. Rosenblum) pp. 195–239. Academic Press, New York.

— (1975b). Parental behavior in *Saguinus fuscicollis* spp. (Callitrichidae). *Folia primat.* **24**, 221–38.

— (1976). Chemical communication and reproductive processes in nonhuman primates. In *Mammalian olfaction, reproductive processes, and behavior* (ed. R. L. Doty) pp. 252–82. Academic Press, New York.

— (1978a). Notes on the establishment and maintenance of the pair bond in *Saguinus fuscicollis*. In *The biology and conservation of the Callitrichidae* (ed. D. G. Kleiman) pp. 231–7. Smithsonian Institution Press, Washington, DC.

— (1978b). Lack of effects of castration on scent marking, displays and aggression in a South American primate (*Saguinus fuscicollis*). *Horm. Behav.* **11**, 139–50.

— (1978c). Studies on the nature of chemical signals in scent marks and urine of *Saguinus fuscicollis* (Callitrichidae, Primates). *J. chem. Ecol.* **4**, 383–94.

— (1978d). Reproductive and social behavior of marmosets with special reference to captive breeding. In *Marmosets in experimental medicine* (ed. N. Gengozian and F. Deinhardt) pp. 50–62. Karger, Basel.

— (1979). Gonadal control of male scent in the tamarin *Saguinus fuscicollis* (Callitrichidae, Primates). *Chem. Senses Flavour* **4**, 15–20.

— (1980). Relationships between aggression, scent marking and gonadal state in a primate, the tamarin *Saguinus fuscicollis*. In *Chemical signals: vertebrates and aquatic invertebrates* (ed. D. Müller-Schwarze and R. M. Silverstein) pp. 87–105. Plenum Press, New York.

— (1981a). Effects of prepubertal castration on the development of the scent glands, scent marking and aggression in the saddle back tamarin (*Saguinus fuscicollis*, Callitrichidae, Primates). *Horm. Behav.* **15**, 54–67.

— (1981b). Effect of pair bonding with adults on the ontogenetic manifestation of aggressive behavior in a primate, *Saguinus fuscicollis*. *Behav. Ecol. Sociobiol.* **8**, 117–23.

— (1982). Effects of prepubertal overiectomy on the development of scent glands, scent marking, and aggressive behaviours of female tamarin monkeys (*Saguinus fuscicollis*). *Horm. Behav.* **16**, 330–42.

— Alveario, M. C., Golob, N. F., and Smith, A. B. III (1980). Stability and attractiveness related to age of scent marks of saddle back tamarins (*Saguinus fuscicollis*). *J. chem. Ecol.* **6**, 735–48.

— — and Katz, Y. (1982). The role of chemical communication in aggressive behavior and its gonadal control in the tamarin *Saguinus fuscicollis*. In *Primate communication* (ed. C. T. Snowdon, C. H. Brown, and M. P. Petersen) pp. 279–302. Cambridge University Press.

— and Cerny, V. A. (1979). Effects of castration and social change on scent marking behavior of *Saguinus fuscicollis* (Callitrichidae). *Folia primat.* **32**, 252–62.

— Golob, N. F., and Smith, A. B. III (1979). Odour communication in the tamarin *Saguinus fuscicollis* (Callitrichidae): behavioral and chemical studies. In *Chemical ecology: odour communication in animals* (ed. F. J.Ritter) pp. 117–30. Elsevier/North Holland, Amsterdam.

— and Lorenz, R. (1967). Vorkommen, Morphologie und Funktion der Sternaldrüse bei den Platyrrhini. *Folia primat.* **7**, 98–126.

French, J. A. and Snowdon, C. T. (1981). Sexual dimorphism in intergroup spacing behavior in the tamarin, *Saguinus oedipus*. *Anim. Behav.* **29**, 822–9.

Golob, N. F., Yarger, R. G., and Smith, A. B. III (1979). Primate chemical communication. Part III. Synthesis of the major volatile constituents of the marmoset (*Saguinus fuscicollis*) scent mark. *J. chem. Ecol.* **5**, 543–55.

Izawa, K. (1978). A field study of the ecology and behavior of the black-mantle tamarin (*Saguinus nigricollis*). *Primates* **19**, 241–74.

Johnston, R. E. and Lee, N. A. (1976). Persistence of the odour deposited by two functionally distinct scent marking behaviours of golden hamsters. *Behav. Biol.* **16**, 199–210.

Jolly, A. (1966). *Lemur behavior: a Madagascar field study*. University of Chicago Press.

Kleiman, D. G. (1977). Monogamy in mammals. *Q. Rev. Biol.* **52**, 39–69.

— (1978). The development of pair preferences in the lion tamarin (*Leontopithecus rosalia*) male competition or female choice? In *Biology and behaviour of marmosets* (ed. H. Rothe, H. J. Wolters, and J. P. Hearn) pp. 203–7. Eigenverlag Rothe, Göttingen.

Lorenz, R. (1972). Management and reproduction of the Goeldi's monkey *Callimico goeldii* (Thomas, 1904) Callimiconidae, Primates. In *Saving the lion marmoset. Proc. Wild Animal Propagation Trust, Golden Lion Marmoset Conference* (ed. D. D. Bridgewater) pp. 92–109. Wild Animal Propagation Trust, Oglebay Park, West Virginia.

Mack, D. S. and Kleiman, D. G. (1978). Distribution of scent marks in different contexts in captive lion tamarins *Leontopithecus rosalia* (Primates). In *Biology and behaviour of marmosets* (ed. H. Rothe, H. J. Wolters, and J. P. Hearn) pp. 181–8. Eigenverlag Rothe, Göttingen.

Maier, W. (1981). Nasal structures in Old and New World primates. In *Evolutionary biology of the New World monkeys and continental drift* (ed. R. L. Ciochon and A. B. Chiarelli) pp. 219–41. Plenum Press, London.

Moynihan, M. (1970). Some behavior patterns of platyrrhine monkeys. II. *Saguinus geoffroyi* and some other tamarins. *Smithson. Contrib. Zool.* **28**, 1–77.

— (1976). *The New World primates*. Princeton University Press.

Neyman, P. F. (1978). Aspects of the ecology and social organization of free ranging cotton-top tamarins (*Saguinus oedipus*) and the conservation status of the species. In *The biology and conservation of the Callitrichidae* (ed. D. G. Kleiman) pp. 39–71. Smithsonian Institution Press, Washington, DC.

Perkins, E. M. (1966). The skin of the black-collared tamarin (*Tamarinus nigricollis*). *Am. J. phys. Anthrop.* **25**, 41–69.

— (1968). The skin of the pygmy marmoset—*Callithrix* (= *Cebuella*) *pygmaea*. *Am. J. phys. Anthrop.* **29**, 349–64.

— (1969a). The skin of the cotton-top pinché *Saguinus* (= *Oedipomidas*) *oedipus*. *Am. J. phys. Anthrop.* **30**, 13-27.

— (1969*b*). The skin of Goeldi's marmoset (*Callimico goeldii*). *Am. J. phys. Anthrop.* **30**, 231–49.

— (1969*c*). The skin of the silver marmoset—*Callithrix* (= *Mico*) *argentata*. *Am. J. phys. Anthrop.* **30**, 361–87.

Powers, J. B. and Winans, S. S. (1975). Vomeronasal organ: critical role in mediating sexual behavior of the male hamster. *Science, NY* **187**, 961–3.

Ralls, K. (1971). Mammalian scent marking. *Science, NY* **171**, 443–9.

Regnier, F. E. and Goodwin, M. (1977). On the chemical and environmental modulation of pheromone release from vertebrate scent marks. In *Chemical signals in vertebrates* (ed. D. Müller-Schwarze and M. M. Mozell) pp. 115–33. Plenum Press, New York.

Reynolds, J. and Keverne, E. B. (1979). The accessory olfactory system and its role in the pheromonally mediated suppression of oestrus in grouped mice. *J. Reprod. Fert.* **57**, 31–5.

Rose, R. M., Bernstein, I. S., Gordon, T. P., and Catlin, S. F. (1974). Androgens and aggression: a review and recent findings in primates. In *Primate aggression, territoriality and xenophobia* (ed. R. L. Holloway) pp. 275–304. Academic Press, New York.

Schilling, A. (1979). Olfactory communication in prosimians. In *The study of prosimian behavior* (ed. G. A. Doyle and R. D. Martin) pp. 461–542. Academic Press, New York.

Siegel, S. (1956). *Nonparametric statistics for the behavioral sciences.* McGraw-Hill, New York.

Smith, A. B. III, Yarger, R. G., and Epple, G. (1976). The major volatile constituents of the marmoset (*Saguinus fuscicollis*) scent mark. *Tetrahedron Lett.* 983–6.

Starck, D. (1969). Die circumgenitalen Duftdrüsenorgane von *Callithrix* (*Cebuella*) *pygmaea* (Spix 1823). *Zool. Gart., Lpz.* **36**, 312–26.

Sutcliffe, A. G. and Poole, T. B. (1978). Scent marking and associated behaviour in captive common marmosets (*Callithrix jacchus jacchus*) with a description of the histology of scent glands. *J. Zool., Lond.* **185**, 41–56.

Thiessen, D. D. (1976). *The evolution and chemistry of aggression.* Thomas, Springfield, Ill.

Thorington, R. W. Jr (1968). Observations of the tarmarin *Saguinus midas*. *Folia primat.* **9**, 95–8.

Wellington, J. L., Beauchamp, G. K., and Smith, A. B. III (1981). Stability of chemical communicants of gender in guinea pig urine. *Behav. & neural Biol.* **32**, 364–75.

Wojcik, J. F. and Heltne, P. G. (1978). Tail marking in *Callimico goeldii*. In *Recent advances in primatology*, Vol. I (ed. D. J. Chivers and J. Herbert) pp. 507–9. Academic Press, New York.

Yarger, R. G., Smith, A. B. III, Preti, G., and Epple, G. (1977). The major volatile constituents of the scent mark of a South American primate *Saguinus fuscicollis*, Callithricidae. *J. chem. Ecol.* **3**, 45–56.

# 20 The primates III: humans

RICHARD L. DOTY

## Introduction

Although human beings, since the dawn of civilization, have used scents to add pleasantness to their environment and to change or mask their own body odours, little is known about their abilities to communicate biological information via natural body secretions. It is commonly assumed that the human sense of smell is somewhat vestigial and not capable of mediating a number of communicative functions subserved by olfaction in many other mammals. Despite this assumption, psychophysical studies indicate that man is keenly sensitive to a number of odorants and a large anecdotal literature suggests that odours play an important role in the rituals and social interactions of a number of societies (cf. Block 1934; Doty 1976; Ellis 1936; Kiell 1976). Indeed, olfaction can aid in the diagnosis of many diseases. As noted by Mace, Goodman, Centerwall, and Chinnock (1976) regarding metabolic diseases of infancy, 'By intelligent use of smell, an astute physician may make a presumptive diagnosis of a rare disorder of metabolism and institute life-saving therapy while awaiting laboratory confirmation. There are a group of disorders in metabolism which lead to unusual odours of the body or urine which individually are rare but collectively make up a sizable portion of acute life-threatening illnesses of infancy.' Examples of such disorders are presented in Table 20.1. Although a number of other diseases can be diagnosed on the basis of associated odours (e.g. gout, yellow fever, pellegra, scrofula, cirrhosis of the liver, uraemia, typhoid, diptheria, scurvy, rubella, and some respiratory and gastrointestinal problems), the use of smell for diagnostic purposes is fast becoming a lost art (for reviews, see Hayden 1980; Mace *et al.* 1976; Pope 1928; Liddell 1976).

In light of the widespread current interest in the chemical communication of mammals, it is of both academic and practical interest to establish if man, just as most other mammals, can determine information related to gender, reproductive state, emotion, and subspecies or race from body odours. Furthermore, it is of interest to determine if odours are used by infants in recognizing their own mothers, or by mothers in recognizing their infants, as occurs in many other mammals. The present review examines available studies from the English literature on these interesting but largely unexplored topics.

## Sources of odorous volatiles in humans

Literally thousands of volatiles, many with odours, are excreted or secreted from the human body. These chemicals reflect, in varying degrees, (i) environmental

Table 20.1. Metabolic disorders of infancy reportedly associated with unusual odours. (Modified from Mace et al. (1976).)

| Disease | Odour | Enzyme defect | Clinical features |
|---|---|---|---|
| Diabetes mellitus | Acetone breath | Lack of insulin or insulin activity | Polyphagia, polyuria, weight loss, acidosis, coma, polydipsia |
| Fish odour syndrome | Like dead fish | Unknown | Stigmata of Turner's syndrome, neutropaenia, recurrent infections, anaemia, splenomagaly |
| Maple syrup urine disease | Maple syrup | Branched-chain decarboxylase | Marked acidosis, seizures, coma leading to death or mental sub-normality without acidosis or intermittent acidosis without mental retardation |
| Oasthouse urine disease | Yeast-like, dried-celery like | Defective transport of methionine, branched-chain amino acids, tyrosine, and phenyl-alanine | Mental retardation, spasticity, hyperpnoea, fever, oedema |
| Cat's odour syndrome | Like cat's urine | Beta-methyl-crotonyl-CoA carboxylase | Neurological disorder resembling Werdnig–Hoffman's disease. Ketoacidosis, failure to thrive |
| Odour of rancid butter syndrome | Rancid butter | Unknown | Poor feeding, irritability, progressive neurological deterioration with seizures and death. Hepatic dysfunction. Possibly same as acute tyrosinosis |
| Odour of sweaty feet, syndrome I | Sweaty feet | Isovaleryl CoA dehydrogenase | Recurrent acidosis bouts, vomiting, dehydration, coma, aversion to protein foods |
| Odour of sweaty feet, syndrome II* | Sweaty feet | Green acyl dehydro-genase | Onset of symptoms in first week of life with acidosis, dehydration, seizures and death |
| Phenylketonuria | Musty, 'mousey', 'horsey', | Phenylalanine hydroxylase | Progressive mental retardation, eczema, decreased pigmentation, seizures, spasticity |

*There is evidence suggesting that persons reported with this disorder actually had isovalericacidaemia.

factors (e.g. chemicals in the air, drinking water, diet, drugs, and personal hygiene products) and (ii) organismal factors (e.g. the individual's gender, reproductive state, race, age, health, exercise schedule, hygiene, and emotional state). Such volatiles are exuded from a number of sources, including sweat, urine, faeces, breath, saliva, breast milk, sexual secretions, and oily secretions of the skin (cf. Sastry, Buck, Janák, Dressler, and Preti 1980). Because little scientific attention has been paid to the potential role of odours from many of these sources in serving communicative functions, the present chapter focuses only on odour sources for which studies of odour discrimination, preference, or communication have been performed (i.e. the oral cavity, the hands, the breasts, and the axillary and genital regions).

### Studies of body odour arising from skin secretions

In humans, chemicals are secreted to the surface of the skin from three primary sources: (i) eccrine sweat glands, (ii) apocrine sweat glands, and (iii) sebaceous glands. Major features of these glands are pictured in an artist's rendering in Fig. 20.1.

The eccrine glands, which are commonly identified as 'the' sweat glands in humans, have coiled tubules, which communicate with the skin surface independently of hair follicles. These glands are distributed over nearly all areas of the body and secrete an aqueous solution of inorganic salts and amino acids. In humans, the primary function of eccrine glands is thermoregulatory, and their secretions are generally considered non-odourous. However, dietary factors, such as garlic, can make the sweat odorous, and certain metabolic diseases clearly influence the character of these secretions.

The sebaceous glands, many of which are commonly associated with acne and other skin problems, are also located in numerous body areas, although their density varies from place to place, being greatest on the forehead, face, and scalp. For example, as many as 900 glands/cm$^2$ are present on the scalp or face and fewer than 50 glands/cm$^2$ on the forearm. No sebaceous glands are present on the palms of the hands or soles of the feet (Nicolaides 1974). Specialized sebaceous glands occur at all points of entry into the body, including the eyelids (Meibomian glands), the ear canal (Menous glands), the nostrils, the lip, the buccal mucusa (Fordyce's glands), the nipple (Montgomery's glands), the prepuce, and the anogenital region. Sebaceous glands secrete most of the lipids of the skin (e.g. triacylglycerols), and function to keep the skin supple and waterproof. In addition, they produce products which pose metabolic problems to potentially pathogenic microorganisms. As reviewed in an interesting article by Nicolaides (1974), skin produces lipids quite different from those produced by internal tissues, making only odd instead of only even chains, branched instead of only straight chains, and free instead of only esterified acids. Skin places double bonds in unusual positions in the fatty chains, produces extremely long chains, and accumulates a number of intermediary products as the result of synthesizing biologically important compounds, such as cholesterol. Although human sebaceous secretions reportedly have a weak and pleasant odour, to my knowledge no psychophysical experiments have been performed to document this suggestion.

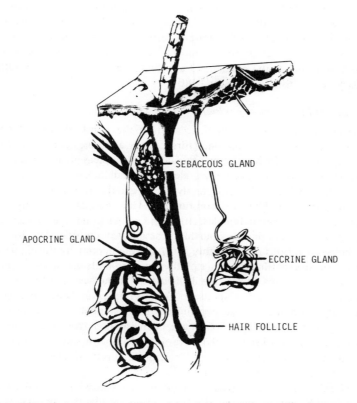

Fig. 20.1. Schematic representation of eccrine, apocrine, and sebaceous glands' relationships to the skin surface and hair follicle. (Modified from Champion (1970).)

The apocrine sweat glands are the major source of 'body odour' in humans. Although these glands serve thermoregulatory functions in a number of species, such functions are not apparent in man. Unlike eccrine glands, the distribution of apocrine glands is limited to skin areas paralleling the evolutionary regression of terminal hair. Thus, they are found primarily in the axillae, the sternal region, the anogenital region, the mammary areola, the cheek region, the eyelid, the ear canal, and regions of the scalp. These glands are usually part of an 'apo-pilosebaceous unit', emptying into the shaft of the hair follicle and being derived embryologically from the primary epithelial germ (Hurley and Shelley 1960). Most of these glands are located in the axillae, where bacteria operate upon their viscous secretions to produce the characteristic body odour. The two major axillary bacteria are the diphtheroids (lipophilic and large colony) and the micrococci. The latter produce isovaleric acid, which has an acrid, sweat-like odour. The former also produce isovaleric acid, and, in addition, other odorous chemicals which are more similar to typical body odour (Labows 1979). The isolation of the specific combination of chemicals responsible for axillary odour has not yet been made, although androstenone and androstenol, two odorous steroids, have been detected in the axillae and appear to be important contributions to the axillary odour (Bird and Gower 1980).

*Behavioural studies of axillary odour*

Apocrine gland secretions are of interest from the point of view of chemical communication for several reasons. First, since apocrine glands produce the typical body odour which is frequently judged as unpleasant when experienced in situations such as crowded buses, elevators, and locker rooms, they clearly influence our social behaviour. Secondly, these glands are functional only after the time of puberty and become non-functional at menopause. Thirdly, they are much larger in males than in females, although females sometimes possess a larger number of them than males. Fourthly, they differ considerably in size and number between various races. Fifthly, they excrete maximally during periods of excitement or stress. Finally, some of their components are steroids similar or identical to compounds which are believed to influence, in other mammals, behavioural or endocrine responses. For example, androsterone sulphate, androst-4-en-3-17-dione, dehydroepiandrosterone sulphate, 5-$\alpha$-androst-16-en-3-$\alpha$ol, and 5-$\alpha$-androst-16-en-3-one have been isolated from secretions of the axillary region (cf. Brooksbank 1970; Brooksbank, Brown, and Gustafsson 1974; Claus and Alsing 1976; Gower 1972; Labows, Preti, Hoelzle, Leyden, and Kligman 1979).

Several behavioural studies have sought to establish whether axillary odours can provide information to humans regarding sexual identity and whether such odours are differentially preferred by males and females. Russell (1976) had 13 women and 16 men wear T-shirts for 25 h without bathing or using deodorants. The armpit regions of these shirts were then presented to subjects in a triangle odour test where (a) the subject's own T-shirt, (b) a strange male's T-shirt, and (c) a strange female's T-shirt were used as stimuli. Each participant was asked to first identify his or her own odour, and then to report which of the two remaining odours came from a male. Nine of the 13 females and 13 of the 16 males performed both of these tasks correctly, suggesting to Russell that 'at least the rudimentary communications of sexual discrimination and individual identification can be made on the basis of olfactory cues'.

It is conceivable that the sex discrimination noted in this novel study was due to the tendency of subjects to associate stronger (and less pleasant) odours with male donors and weaker (and less unpleasant) odours with female donors, since women have smaller apocrine glands than men (apocrine gland size reportedly correlates positively with body odour intensity) and most women in our culture shave their axillae (a procedure which limits the surface area for diffusion). To examine this possibility, a series of psychophysical studies was performed in our laboratory, three of which are described in this section (cf. Doty 1977; Doty, Kligman, Leyden, and Orndorff 1978).

In the first of these studies, a set of axillary odours from nine males (collected on gauze pads that had been taped in the armpits for approximately 18 h) and a blank control pad were presented in sniff bottles to 10 male and 10 female judges who had received these instructions:

You will be presented with a series of sniff bottles containing human sweat odor. We wish you to tell us which sex each of the odor samples comes from. The set of odors may include samples from both men and women, or from only men or from only women. Thus, some may be from females, some from males, or, alternatively, all may be from females or all from males. Therefore, don't allow yourself to assume that some predetermined number of one or the other sex is represented. If you have any questions, please feel free to ask them.

Following this task, the subjects gave magnitude estimates of the relative intensity and pleasantness of the stimuli. In the second study, a set of nine female odours and a bland control were presented under identical conditions. In the third study a set of five male odours, four female odours, and a blank control were similarly presented.

The results of these studies confirmed our initial hypothesis, namely that the stronger (and more unpleasant) the axillary odour, the more likely it was to be assigned to a male gender category, regardless of the true sex of its donor. In addition, these studies demonstrated that the intensity and pleasantness of axillary odours are strongly and inversely related ($r_s \approx -0.95$).

The data from the study where five male odours, four female odours, and a blank control were presented are depicted in Fig. 20.2. It is clear that a positive relation between the perceived odour intensity and the assignment of the odour to the male category was present in this study. One female odour was called male by more than half of the judges and one male odour was called female by all but one of the judges. Figure 20.3 shows that this positive relation between perceived intensity and assignment to the male category still held when only female odours were presented. Thus, over 40 per cent of the samples (i.e. those with the most intense odours) were incorrectly assigned to the male category by the majority of judges. These general findings were also observed in an *in vivo* experiment where blindfolded subjects sniffed the axillae directly, indicating that these results are not idiosyncratic to an *in vitro* presentation of the stimuli (Doty, unpublished).

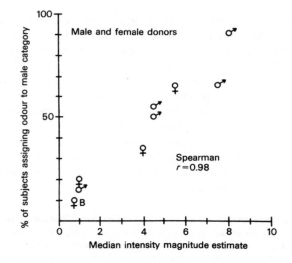

Fig. 20.2. Relation between perceived intensity, sex of odour donor, and percentage of subjects assigning each axillary odour to the male gender category. Male symbols indicate axillary odours from male donors, female symbols indicate axillary odours from female donors, and B indicates a blank control.

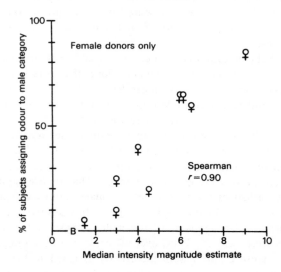

Fig. 20.3. Relation between perceived intensity and percentage of subjects assigning each axillary odour to the male gender category. Female symbols indicate axillary odours from the female donors and B indicates a blank odour control.

These results suggest that the ability of humans to 'detect' gender from axillary odours may depend upon quantitative, rather than qualitative, aspects of the odours, and that strong odours are judged as having come from males and weak odours as having come from females, regardless of the true sex of the odour donor. Thus, it is likely that correct sexual assignments from axillary odours are probabilistic, analogous to the estimation of gender from parameters such as height or weight. Schleidt (1980) has recently noted that axillary hygiene decreases the ability of subjects to assign correct gender to axillary odours, a finding in accord with this hypothesis.

If axillary odours serve as sexual attractants or repellants (without extensive conditioning between sexual partners), one might hypothesize that women would rate male axillary odours as more pleasant than do men, or that men might rate female axillary odours as more pleasant than do women. Such general relations have not been found in our experiments. Data from two identical studies using different judges and donors are presented in Table 20.2. In both of these studies, male odours were perceived as more intense and less pleasant than female odours by both sexes. High relative agreement between the sexes is indicated by correlation coeficients computed across their pleasantness and intensity magnitude estimates ($r_s > 0.90$).

Table 20.2. Median intensity and pleasantness magnitude estimates given to axillary odours from male and female donors

| | | Median intensity magnitude estimates | | | |
|---|---|---|---|---|---|
| | | Male odours | | Female odours | |
| **Experiment 1** | | | | | |
| Male subjects ($n = 10$) | ns* | 4.50 | $p < 0.01$[†] | 3.00 | ns* |
| Female subjects ($n = 10$) | | 4.00 | $p < 0.001$ | 3.00 | |
| **Experiment 2** | | | | | |
| Male subjects ($n = 10$) | ns* | 3.50 | $p < 0.01$[†] | 2.82 | ns* |
| Female subjects ($n = 10$) | | 6.50 | $p < 0.01$ | 3.00 | |

| | | Median pleasantness magnitude estimates[‡] | | | |
|---|---|---|---|---|---|
| | | Male odours | | Female odours | |
| **Experiment 1** | | | | | |
| Male subjects ($n = 10$) | ns* | −2.00 | $p < 0.01$[†] | 0.00 | ns* |
| Female subjects ($n = 10$) | | −1.00 | ns | 0.00 | |
| **Experiment 2** | | | | | |
| Male subjects ($n = 10$) | ns* | −2.00 | $p < 0.05$[†] | 0.00 | ns* |
| Female subjects ($n = 10$) | | −3.00 | $p < 0.05$ | 0.00 | |

*Mann-Whitney U-test; ns = not significant

[†]Wilcoxin matched pairs signed ranks test.

[‡]In the pleasantness magnitude estimation task, subjects assigned negative numbers to unpleasant odours, positive numbers to pleasant odours, and 0 to neutral odours. The numbers were assigned in relative proportion to the degree of the pleasantness or unpleasantness.

It is conceivable that the aforementioned findings might depend upon cultural factors, and that different results might be found in cultures other than North American or Northern European. In response to this problem, Schleidt, Hold, and Attili (1981) examined whether German, Japanese, and Italian subjects differed in their abilities to distinguish between (i) their own axillary odours, (ii) the axillary odours of opposite-sex cohabiting partners, and (iii) the axillary odours of strangers, as well as in their abilities to assign such odours to correct gender categories. In addition, measures of hedonic responses to the odours were also obtained. The primary expectation from this work was that differences might be observed between 'contact cultures' (e.g. Italy) and 'non-contact cultures' (e.g. Germany and Japan), the latter being characterized by relatively less sensory involvement, eye contact and touching, and relatively greater interpersonal distance, including a less direct body orientation during social interactions (see Hall 1966; Altmann and Vinsel 1977; Watson 1970). Twenty-four German and 25 Italian couples took part in the study, as did seven Japanese couples and 15 male and 15 female singles. Each participant wore a cotton shirt for seven consecutive nights and used no perfume or deodorants during the study. Within each cultural group, the subjects were presented with 10 shirts and the following questions: (i) Which shirt has your own smell? (ii) Which shirt has your partner's smell? (iii) Which shirts smell male? (iv) Which shirts smell female? and (v) Which shirts smell pleasant, which indifferent, and which unpleasant? This task was repeated three times.

The results of this study indicated that one-quarter to slightly more than one-third of the subjects correctly reported their own odour on two of the three trials (Table 20.3). From 20 to 64 per cent of the subjects (males and females combined) distinguished male from female odours (Table 20.4), with women of all cultures being more accurate than men in this regard. A larger proportion of

Table 20.3. Recognition of individual odour. (Modified from Schleidt *et al.* (1981).)

|  | German sample (*n* = 45) | Italian sample (*n* = 50) | Total Japanese sample (*n* = 44) | Japanese couples (*n* = 14) |
|---|---|---|---|---|
| % of subjects who recognized their own odour | 31 | 38 | 25 | — |
| % of subjects who recognized their partner's odour | 33 | 30 | — | 21 |
| % of subjects who recognized their own or their partner's odour | 49 | 52 | — | 29 |

the Japanese subjects discriminated correctly between the sexes, although it was not clear if this was due to sex differences in odour production, reception, or some combination of both. The hedonic responses to male and female odours were the same in the three cultures: male odour was judged more unpleasant or less pleasant than female odour; men judged their own odour as more unpleasant than women did their own odour. Unlike the German women, the Japanese and Italian women classified their partners' odours more often as unpleasant than as pleasant. The odour of the supposed partner (i.e. the odour noted correctly or incorrectly as the partner's) was classed as predominantly pleasant by the Italian and German, but not by the Japanese women. No marked cultural variation was noted in the responses of the males, who generally judged their partners' odours as pleasant. Schleidt *et al.* suggest that the finding that the supposed partner's odour was generally classed as pleasant may reflect an underlying attitude that 'my partner smells good'. Overall, no strong support of differences in axillary odour perception between 'contact' and 'non-contact' cultures was found in these studies.

In summary, the following conclusions appear warranted from the few studies of human axillary odour which relate to social communication. First, given a heterogeneous collection of male and female axillary odours, subjects can correctly assign such odours to male and female gender categories at a frequency unlikely to be due to chance. Secondly, such assignments are probably based, at least in part, on intensity and/or pleasantness cues, since stronger and less pleasant odours are assigned to the male category and weaker and less unpleasant odours are assigned to the female category, regardless of the true gender of the odour donors. Thirdly, at least in American subjects, men and women exhibit relatively similar intensity and pleasantness responses to axillary odours. Fourthly, odours believed to come from familiar sexual partners are rated, on the average, as more pleasant than odours believed to come from strangers, regardless of whether or not they truly come from familiar sexual partners. Fifthly, although subtle differences between cultures may be present in regards to the perceived hedonicity of axillary odours, considerable similarity in such ratings is the rule. When differences between cultures have been found, the possibility that they reflect cultural biases in responding to a somewhat taboo topic cannot be ruled out.

Table 20.4. Percentage of subjects distinguishing correctly between male and female odours.* (Modified from Schleidt *et al.* (1981).)

| | German sample ($n = 50$) | Italian sample ($n = 50$) | Total Japanese sample ($n = 44$) |
|---|---|---|---|
| Male subjects | 20 | 16 | 50 |
| Female subjects | 44 | 24 | 82 |
| Male + female subjects | 32 | 20 | 64 |

*Percentages based on subjects who performed discriminations beyond $p < 0.01$.

In addition to the aforementioned experiments on the potential role of axillary odours in human social communication, a recent experiment has been published which claims to support the notion that the time of menstrual onset can be modified by human axillary odours (Russell, Switz, and Thompson 1980). Menstrual synchrony, which was first documented scientifically by McClintock (1971) and subsequently demonstrated by Graham and McGrew (1980) and Quadangno, Shubeita, Deck, and Francoeur (1981), is the phenomenon where close friends or room-mates exhibit temporally closer onsets of menses as a function of the duration of living together. To date, the factor(s) involved in producing this effect have not been established. In the experiment of Russell *et al*. (1980), axillary secretions were collected on gauze pads taped under the arm of a woman who had a history of regular 28-day menstrual cycles and 'a previous experience of "driving" another woman's menstrual cycle on three separate occasions, over three consecutive years; i.e. a friend had become synchronous with her when they roomed together in summer and dissynchronous when they moved apart in the fall'. These pads were then removed, cut into four squares, combined with four drops of 70 per cent alcohol, and frozen in dry ice. Following thawing, this material was rubbed on the upper lips of five women, three times a week for four months. A control group of six women were similarly rubbed with gauze pads that had received only an alcohol treatment. A mean pretreatment difference of 9.3 days was observed between the day of onset of the donor's menses and that of the experimental subjects. This average difference decreased to 3.4 days after the fourth month of treatments, suggesting a trend toward menstrual synchrony. Pretreatment/posttreatment differences for the control group were 8.0 and 9.2 days, respectively, suggesting a lack of a trend towards menstrual synchrony. The authors concluded that 'The data indicate that odors from one woman may influence the menstrual cycle of another and that these odors can be collected from the underarm area, stored as frozen samples, for at least short periods, and placed on another woman. Further, the experiment supports the theory that odor is a communicative element in human menstrual synchrony, and that at least a rudimentary form of olfactory control of the hormonal system is occurring in humans in a similar fashion to that found in other mammals.' The authors acknowledge, however, that the induction of synchrony may have been due to non-odorous properties of the stimuli.

The conclusion of these authors, in my opinion, must be viewed with caution in light of methodological limitations in their experiment. For example, this study was not performed in either a single- or a double-blind manner. Thus, the woman who donated the samples (the second author of the paper) also acted as one of two female experimenters who rubbed the stimuli on the subjects (Russell, personal communication). In addition, 'The purpose of the experiment was explained to each subject' (p. 737). Such considerations are important, since social stimuli are well-known Zeitgebers for a number of human biological rhythms (including rhythms of certain hormones) and have the potential to override even the dominant mammalian Zeitgeber, light (Aschoff, Fatranska, and Giedke 1971; Rusak and Zucker 1975). The importance of demand character-istics on human experiments is well documented (cf. Orne 1962; Rosenthal and Rosnow 1969) and cannot be excluded from consideration in this context. If the major finding of this work is supported by more rigorous experiments employing double-blind procedures and additional control groups (e.g. ones using axillary odours from women with no history of 'driving' menstrual cycles), then an exciting field of olfactory/endocrine interactions in humans will be born. For the time being, it is probably prudent to consider the issue of the role of axillary stimuli in the production of menstrual synchrony as still open.

*Behavioural studies of components of axillary secretions*

Several behavioural studies have focused on compounds, present in human axillae, which are found in some animal secretions, where they play a role in reproductive function. Two compounds that have received attention in this regard are the steroids androstenone ($5\alpha$-androst-16-en-3-one) and its related alcohol (androstenol), both of which are responsible, in part, for the character-istic 'tainted breath' of boars, and both of which, when inhaled by sows, increase the probability of lordosis.

In the first human experiment using such substances, Cowley, Johnson, and Brooksbank (1977) had college students evaluate hypothetical applicants for a campus job while wearing surgical-type masks impregnated with either a mixture of fatty acids or with androstenol. Other subjects wore control masks not impregnated with either of these substances. The students were told that the purpose of the mask was to prevent their observing each other's facial expressions. The results suggested that the odours influenced the female's assessments of male applicants (the fatty acid mixture and androstenol had opposite effects) and that the odorants interacted with the applicant's characeristics in affecting the evaluations. Unfortunately, 'neutral' control odours were not included in this study, so that the potential uniqueness of these substances in producing these complex and difficult-to-evaluate findings cannot be assessed.

In another experiment, Kirk-Smith, Booth, Carroll, and Davis (1978) asked male and female students to rate photographs of men, women, animals, and buildings while wearing, on two separate occasions counterbalanced in terms of order, masks either containing or not containing the androstenol odour. On three of 15 psychological rating scales (sex, attractive, good) the males, when wearing the androstenol-impregnated masks, rated the photographed women more positively than when wearing the control masks. Androstenol did not significantly influence the ratings for most of the 15 scales, although animals were rated as being 'rasher' and buildings 'less sensitive' under this condition. Given the large numbers of ratings performed and the lack of 'neutral' control odours, this study is similar to the preceding one in being difficult to interpret meaningfully.

In a third, quite different, experiment, Kirk-Smith and Booth (1980) sprayed, on different occasions, three concentrations of androstenone (3.2, 16 or 32µg) on a seat in a dentist's waiting room. The seating patterns of both men and women were subsequently recorded over a series of days to see if they were influenced by the introduction of the odour. The authors report that more women used the odourized seat when it had been sprayed with 3.2 or 32µg of androstenone and fewer men when it had been sprayed with 32µg (relative to an initial four-day control period, when no women sat in the seat in question). Several factors suggest that these findings should be cautiously interpreted. For example, since repeated returns to a dentist's office over a series of days or weeks are quite common, the observation of this study are unlikely to be independent, invalidating the authors' use of the Chi square statistic in analysing their results.

Overall, the results and interpretations provided by the authors of the three studies of this section are not convincing and, in my opinion, must be viewed conservatively in light of methodological considerations. Before the conclusions reached in most of these studies can be given much credence, similar results from more carefully analysed and designed studies should be obtained. None the less, the procedures used in these experiments are quite novel and, aside from providing information on 'human pheromones', could be modified for the evaluation of a number of odours on subtle psychological processes not previously measured in olfactory studies.

*Behavioural studies of hand odour*

Aside from the apocrine sweat studies mentioned above, the only other published report of which I am aware that has examined gender discrimination in humans using odours from skin secretions is a study of hand odours by Wallace (1977). Eight men and eight women served as subjects in her initial experiment. Each subject was blindfolded and placed his or her head inside a glass 'olfactorium', which had an opening at the side to allow placement of hands within a half-inch of the nose. Air was circulated throughout the equipment by a fan. On the evening before the experiment, the odour donors (several men and women 18–21 years of age) showered and then rinsed their left hand in a sequence of distilled water, ethanol, and acetone. Each fifteen minutes prior to testing on the next day, as well as between test trials, each donor wore a disposable plastic glove that induced perspiration. The subjects were asked to discriminate, in this initial phase of the study, between the hands of two men, two women, and a man and a woman. Feedback as to the correctness of the responses was given on each trial. More than half of the subjects made these discriminations on 21 or more of 30 trials ($p < 0.05$), with the highest accuracy occurring in the male v. female condition. Just as in the axillary experiments, the women were generally more accurate than the men in performing these discriminations.

In the second experiment, nine women were tested for their ability to distinguish between (i) two unrelated females who had been on the same diet for three previous days and (ii) identical twins who were also on the same diet. In a series of subsequent tests, ten women were tested for their ability to discriminate between (i) two female siblings on the same diet, (ii) identical twin females on the same diet, and (iii) identical twin females on different diets. In the latter case, one of the twins ate only bland foods for three days, whereas the other ate foods containing garlic, onion, and spices.

The results of these experiments indicated that the unrelated females on the same diet were more easy to discriminate between than either of the two identical twin pairs or the female siblings. Twins on different diets were more easy to discriminate between than twins on the same diet. Wallace interpreted these findings as indicating that humans can discriminate between individuals on the basis of hand odour, and that females perform this task more efficiently than males. Although one might argue that the male v. female discrimination could be due to the larger surface area of the male hands, the discriminations between identical twins (whose hands, presumably, were similar in size) is difficult to explain on this basis.

### Studies of human breath odours

The oral cavity is a well-known source of both endogenous (e.g. lung) and exogenous (e.g. oral bacterial) odours. Although anecdotal reports of breath odours (e.g. Ellis 1936) and scientific studies of oral volatiles (e.g. Sastry *et al.* 1980; Tonzetich, Preti, and Huggins 1978) suggest that changes may occur in the smell of breath during the menstrual cycle, no psychophysical data are available on this point. However, a determination of whether males and females exhibit different oral odours was recently made in our laboratory, in conjunction with the Clinical Research Center of the University of Pennsylvania School of Dental Medicine (Doty, Green, Ram, and Yankell 1982*a*).

In this study, five male and five female judges evaluated the likely gender, as well as the perceived intensity and pleasantness, of breath odours from 19 females and 14 males on a daily basis for five consecutive days. The breath donors had above-average oral health, the majority being dental students at the University. The donors used no oral hygiene during the five days of the study. The odour-evaluation sessions were held between 7 a.m. and 9 a.m. in a large well-ventilated room, and the donors were separated visually from the judges by a series of interconnected plywood barriers (Fig. 20.4).

The results of this study are summarized in Figs. 20.5 and 20.6. Fig. 20.5. depicts the proportion of the judges assigning the correct sexual identity to the male and female odours. On average, both the male and female odours were judged correctly by the majority of the judges. As in the case of axillary and hand odours, the female judges were more accurate than the male judges in making the correct sexual assignments. Fig. 20.6 depicts the mean intensity

Fig. 20.4. Human odour panel shown assessing breath odours, which are blown through glass ports. The panelists are separated from the odour donors by plywood partitions and curtains at the top and bottom of the partitions. (From Doty, Green, Ram, and Yankell (1982*a*).)

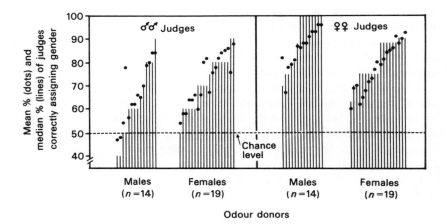

Fig. 20.5. Mean and median percentages of judges correctly assigning breath odours to male and female gender categories during two test occasions on each of five test days. Dots signify means and lines medians. Values ordered arbitrarily in order of increasing medians. (From Doty, Green, Ram, and Yankell (1982*a*).)

and pleasantness ratings given to the male and female odours by the male and female judges on each of the test days. On average, (i) the male odours were judged as more intense and less pleasant than the female odours, (ii) females tended to judge the odours as more intense and less pleasant than did males, and (iii) the intensity of the odours increased, and the pleasantness decreased, across the test days. As in the axillary odour work, an inverse relation between perceived odour intensity and pleasantness was present. Furthermore, the stronger (and more unpleasant) the odour was rated, the more likely it was to be assigned to the male category (Table 20.5).

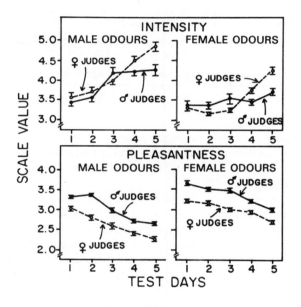

Fig. 20.6. Rating scale values (possible range = 1–7) of perceived intensity and pleasantness of breath odours given by male and female judges on five consecutive test days. Values signify arithmetic means ± 1 SEM. (From Doty, Green, Ram, and Yankell (1982*a*).)

### Studies of human vaginal secretion odours

The odours of human vaginal secretions and exudate, particularly during or around the time of menses, have profoundly influenced the behaviour and social customs of a number of societies (cf. Delaney, Lupton, and Toth 1977; Himes 1970). For example, vegetable juices with smells similar to those of seminal fluid or vaginal secretions are believed to have aphrodisiacal powers by some primative societies, and men in one southwest Pacific society use fish odours with smells

Table 20.5. Pearson correlation coefficients between (a) rated intensity, (b) rated pleasantness, and (c) percentage assignment to the male gender category of breath odours from 14 female and 19 male odour donors. Data collapsed across five test days before calculation of correlations. Significance values indicated in parentheses. (From Doty *et al.* (1981)).

| Donor sex | Judge sex | r, % male category rated intensity | r, % male category rated pleasantness | r, % rated intensity, rated pleasantness |
|---|---|---|---|---|
| Male | Male | 0.931 (0.0001) | −0.908 (0.0001) | −0.918 (0.0001) |
| Male | Female | 0.656 (0.0108) | −0.734 (0.0028) | −0.947 (0.0001) |
| Female | Male | 0.523 (0.0214) | −0.640 (0.0032) | −0.429 (0.0666)NS* |
| Female | Female | 0.179 (0.4637)NS* | −0.477 (0.0391) | −0.645 (0.0028) |
| Male | Male + female | 0.646 (0.0001) | −0.701 (0.0001) | −0.785 (0.0001) |
| Female | Male + female | 0.472 (0.0055) | −0.563 (0.0007) | −0.880 (0.0001) |
| Male + female | Male + female | 0.511 (0.0009) | −0.629 (0.0001) | −0.860 (0.0001) |

NS = not significant at $p < 0.005$ level.

similar to vaginal secretions as magic potents to attract women (Davenport 1965). The association of foul smells with menstruation is believed to be associated with the development of sexual taboos related to this time of the cycle (cf. Delaney *et al.* 1977).

The volatiles produced in the human vagina arise from several sources: (i) vulvar secretions from sebaceous, sweat, Bartholin's, and Skene's glands; (ii) exfoliated cellular debris and leukocytes; (iii) cervical mucus; (iv) endometrial and tubal fluid; (v) transudate exuded through the vaginal epithelium; and (vi) male semen from intercourse. It is well documented that much of the odour associated with vaginal secretion arises from the microfloral bacteria in the region, and numerous remedies for malodour problems prescribe douches which alter the pH of the region to decrease the number of such organisms. As with sweat, however, a portion of the odour may stem from non-bacterial sources, including diet.

A relatively large number of compounds, both volatile and non-volatile, have been chemically isolated from vaginal secretions, and recent work has focused on the characterization of low-molecular-weight organic compounds from this region and their changes during the cycle and during sexual arousal (e.g. Huggins and Preti 1976; Michael, Bonsall, and Kutner 1975; Michael, Bonsall, and Warner 1974; Preti and Huggins 1975; Preti, Huggins, and Silverberg 1979). Gas chromatographic and gas chromatrographic/mass spectrometry analyses of vaginal secretions indicate that there are two types of women in regard to the production of $C_2$–$C_5$ acids in their vaginas—those who produce large amounts of these acids and those who produce little or none of these compounds. Amongst those who produce these acids, the patterning of changes during the cycle are controversial, some laboratories suggesting increases near the time of ovulation and others reporting increases only during the luteal phase (Preti and Huggins 1978; Huggins and Preti 1976). Although sexual arousal produces a quantitative change in the amount of volatile organic constituents of human vaginal secretions, no consistent qualitative changes have been observed (Preti *et al.* 1979).

Based upon observations that vaginal secretions appear to be a major source of materials associated with arousing sexual interest in a number of mammals, including primates (see, for example, Doty and Dunbar (1974) and Michael and Keverne (1968)), we sought to establish whether or not vaginal odours change markedly during the menstrual cycle in a way that would facilitate the detection of the time of optimal fertility (Doty, Ford, Preti, and Huggins 1975). In this experiment, pleasantness and intensity magnitude estimates of the odours of human vaginal secretions sampled across 15 ovulatory menstrual cycles of four women were obtained. Both male and female judges rated the odours, which were presented in an *in vitro* situation. On average, secretions from preovulatory and ovulatory phases of the cycle were judged as weaker and less unpleasant in odour than those from menstrual, early luteal, and late luteal phases (Fig. 20.7).

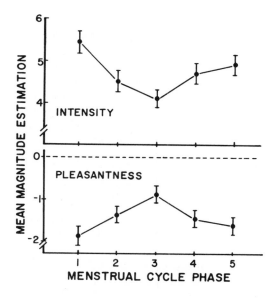

Fig. 20.7. Mean magnitude estimates of perceived intensity and pleasantness of human vaginal secretion odours sampled from consecutive phases of the menstrual cycle. Vertical bars indicate ± 1 SEM. Each data point represents an average of 256 cases. Phases are designated as follows: 1, menstrual; 2, preovulatory; 3, ovulatory; 4, early luteal; 5, late luteal. Ovulatory phase based on basal body temperature data. (Adapted from Doty, Ford, Preti, and Huggins (1975).)

However, considerable variation was present across the cycle days of the same donor, as well as across cycles·from different donors (Fig. 20.8). These data suggest that the odours of human vaginal secretions change in both pleasantness and intensity during the menstrual cycle, but do not support the notion that such odours are markedly attractive in such an out-of-context *in vitro* situation. The perception of such odours in *in vivo* situations presumably differs from the results of this study, at least in some cases, depending upon the partners involved, their sexual proclivities, ages, and numerous situational variables. It should be noted that the average pleasantness ratings did not extend far from the neutral zero point in our experiment, however (Fig. 20.7), and that in some cases the odours were judged as pleasant (Fig. 20.8). The marked cycle-

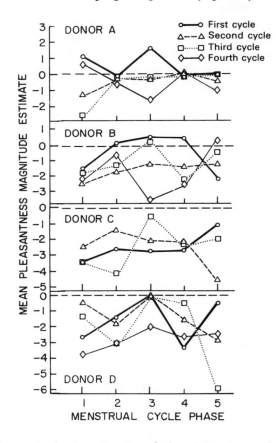

Fig. 20.8. Mean magnitude estimates of the pleasantness of human vaginal secretion odours sampled from consecutive phases of 15 menstrual cycles of four donors. Phase designations the same as in Fig. 20.7. (Adapted from Doty, Ford, Preti, and Huggins (1975).)

to-cycle and day-to-day variability in the intensity and unpleasantness magnitude estimates suggests that it is unlikely that humans can reliably determine the time of ovulation on the basis of such cues.

The question of whether odorous volatiles from vaginal secretions can influence human sexual behaviour was addressed by Morris and Udry (1978), in the light of early and now somewhat controversial reports of Michael and his associates that a series of aliphatic acids (present in both human and rhesus monkey vaginal secretions) act as 'pheromones' in eliciting male rhesus copulatory behaviour (e.g., Michael and Keverne, 1968; Curtis, Ballatine, Keverne, Bonsall, and Michael 1971). Morris and Udry (1978) sought to establish whether such a mixture of aliphatic acids, when applied to the chest region of women at bedtime influence the self-reported frequency of intercourse, desire for sex, occurrence of sex play, or male/female orgasm of 63 married couples tested over three complete menstrual cycles. No influences of the synthetic secretion on any of these measures were found.

### Studies of infant discrimination of breast odours

There is an emerging body of evidence, as indicated in other chapters of this volume, that many infant mammals recognize their mothers by means of odours, although the relative importance of such cues probably varies from species to species. Kittens recognize the odours of individual teats soon after birth (Rosenblatt 1972), and maternal odours play a predominant role in nursing, nipple selection, and nest orientation of many forms of rodents that have been examined during early infancy (e.g. Alberts 1981; Leon and Moltz 1972; Teicher and Blass 1976). Whether human newborns can recognize their mothers by odours (e.g. from the breast region or from breast milk) has been the subject of several experiments.

MacFarlane (1975) performed two studies designed to establish if infants can detect the odours of their mothers' breasts. In the first study, a clean breast-pad and a breast-pad previously worn by the lactating mother were lowered, side-by-side, into a test crib where the child was lying flat on its back. The breast-pads were hung $\approx 1$ cm apart from an adjustable arm such that they touched the baby's cheeks. Nasal contact was possible with either pad by turning the head about $20°$ from the midline. Total time orienting towards the breast-pads was determined from videotapes of two, one minute-long test sessions of each infant. Seventeen of the 20 babies spent more time orienting towards the mother's breast-pad than towards the clean breast-pad (binomial test, $p < 0.001$). Seventy one per cent of the total time oriented towards either breast-pad was directed towards the mother's pad, and over 75 per cent of the test period was spent orienting towards this pad. MacFarlane concluded that 'a neonate is able to distinguish between the smell of a clean breast pad and a breast pad that has been in contact with mother's breast' (p. 108).

In the second experiment, 32 neonates were similarly tested on postpartum days 2 and 6 for their orienting responses to breast-pads previously worn by their mothers versus breast-pads previously worn by other lactating women. In addition, a separate group of babies was tested on days 8–10 of age. The total time spent turned towards *both* breast-pads increased from 72.3 per cent on day 2 to 82.6 per cent on day 6 to 89.4 per cent on days 8–10. The percentage time spent with the head turned towards the breast-pad of their own mother (as a percentage of the total time spent with both breast-pads) for the three age groups was 51.8, 60.3, and 68.2 per cent, respectively. At two days of age, 17 of the 32 babies turned more towards their own mother's pad (n.s., binomial test); at six days of age, 22 of the 32 did so ($p < 0.01$); and by 8–10 days of age, 25 of the 32 did so ($p < 0.001$), suggesting a gradual increase in preference for the mother's pad as a function of age or familiarity with the test situation.

Using a somewhat less-structured presentation procedure, Russell (1976) tested 14 healthy infants at two days of age, at two weeks of age, and at six weeks of age for responses to odours from three types of breast-pads: (i) a clean moist pad, (ii) a pad previously worn by its own lactating mother, and (iii) a pad previously worn by another lactating mother. The pads were held by the experimenter between the thumb and forefinger and placed near the baby's nose for about 30 seconds. The infants 'were tested while sleeping whenever possible and not tested when distressed or crying' (p. 521). On day 2, only one infant responded to any of these stimuli, making a suckling response to both the strange and familiar breast odours. At two weeks of age, seven of the 10 babies made responses to the strange mother and eight to the odour of their own mother. Three responded solely to the odour of their own mother. At six weeks of age, six babies responded only to their own mother's odour, and one responded to both the odour of its own mother and to that of the unfamiliar mother. In this case, the responses to the strange mother were negative, consisting of a head jerk and a cry. Russell concluded that 'The existence of olfactory maternal attraction suggests that humans have a pheromonal system and that it operates at a very early age' (p. 521).

Although supporting data are needed from experiments using well-defined behavioural measures and double-blind scoring techniques, these pioneering experiments strongly suggest that human infants can distinguish the odour of their own mothers' breasts from those of strange mothers' breasts, and that this preference increases with age and/or testing experience. Such results suggest that humans, like most other mammals which have been tested, can discriminate odours in the suckling situation. The extent to which such perception influences behaviours in later life is not known, although in rodents species preferences may develop at this time (Doty 1974). Whether odour 'imprinting' is possible in humans is currently a point of conjecture open to experimental test.

### Studies of sex differences and menstrual-cycle influences on odour perception and detection, and implications for theories of human odour communication

In popular accounts of human odour communication (or potential odour communication), much is made of studies which report that women are more sensitive than men to certain compounds (e.g. the synthetic musk of exaltolide) and that cyclical changes occur in olfactory sensitivity to such compounds during the menstrual cycle. To quote from Comfort's (1971) speculative article on the likelihood of human pheromones, 'It is known that women have the greater olfactory sensitivity to most mammalian odours, and that this is oestrogen-dependent and, in the case of exaltolide, cyclical. Thus women detect, and react to, boar taint in pork far more readily than men, the substance detected being apparently 5α-androst-16-en-3-one. A similar material occurs in human male urine, and in female urine during the luteal phase . . .' (p. 432-3).

If such sex differences or fluctuations were specific to 'mammalian odours' (presumably imparting biological significance), circumstantial support for the notion of an endocrine-related heterosexual-odour communication system in humans might be at hand. Unfortunately, such specificity has not been borne out by subsequent experimentation, and control odorants with physiochemical properties similar to musks and steroids (but without presumed 'biological significance') have yet to be tested. For example, recent experiments have demonstrated that sex differences and menstrual-cycle changes in olfactory sensitivity occur for a number of compounds in addition to musks (Koelega and Köster 1974; Doty, Snyder, Huggins, and Lowry 1981). Furthermore, data from our laboratory clearly demonstrate that menstrual-cycle-related fluctuations in olfactory sensitivity also occur in women taking oral contraceptives, suggesting they are probably not caused by the waxing and waning of circulation levels of ovarian steroids (Doty *et al.* 1981). Although more compounds need to be examined, these findings, which are briefly summarized in the next few paragraphs, suggest that the simple notion of biological meaningfulness of musks and certain steroids in the context of a human heterosexual odour communication system is questionable.

As indicated by a number of studies reviewed in this chapter, women consistently perform better than men in sexual identification or detection tasks based upon the evaluation of axillary odours (Doty 1977; Schleidt *et al.* 1981), hand odours (Wallace 1977), and breath odours (Doty *et al.* 1982*a*). However, it is important to note, in the present context, that their superior performance is not restricted to body odours. For example, women consistently perform better than men in identifying odours which, using the apparent criteria of authors such as Comfort (1971), presumably play no potential role in biological odour communication (Cain 1980). Furthermore, women excel in taste tests where various concentrations of quinine, sugar, acids, sodium chloride, and sodium bicarbonate are sorted into perceptual categories such as sweet, bitter, sour, salty, acid, and alkaline, and exhibit lower thresholds to both electrical and chemical stimulation of the tongue (see review by Doty 1978). Superior female performance has also been noted in auditory tasks requiring an ability to distinguish between changes in stimulus intensity (e.g. Shuter 1968), as well as in a number of visual tasks (cf. McGuinness 1976). The possibility that some of these differences have a clear-cut biological basis is suggested by findings such as those of Jerger and Hall (1980), which indicate that women consistently exhibit shorter latency and larger amplitude responses than men to the auditory-evoked potential brainstem response.

As briefly mentioned earlier in this section, we now have data which throw into question the widely-held belief that changes in olfactory sensitivity during the menstrual cycle are dependent upon ovarian hormones, in particular, oestrogens. For example, in a recent study (Doty *et al.* 1981) we monitored olfactory sensitivity to furfural, as well as blood pressure, heart rate, body temperature, nasal airflow, respiration rate, and circulating levels of luteinizing hormone (LH), follicle-stimulating hormone (FSH), oestrone ($E_1$), oestradiol ($E_2$), progesterone (P), and testosterone (T), during 17 menstrual cycles of women not taking oral contraceptives, six menstrual cycles of women taking oral contraceptives, and six equivalent time periods of three men. The major finding of this study—that both women taking and women not taking oral contraceptives exhibit cyclical changes in olfactory sensitivity across the cycle—is illustrated in Fig. 20.9. As can be seen in this figure, the fluctuations in olfactory sensitivity ($d'$), although roughly correlated with the changes in serum oestrogen levels in women not taking oral contraceptives, are not related to oestrogen levels in women taking oral contraceptives. We have repeated this finding using another odorant, the rose-like smelling substance phenyl ethyl alcohol. Furthermore, Dr James Hall and I have recently shown in a subject taking oral contraceptives that several measures of auditory function follow a rhythm similar to that observed in olfactory sensitivity. These results suggest that the fluctuation in olfactory sensitivity may reflect general changes in several modalities, which possibly depend upon cyclical changes within the central nervous system (Fig. 20.10).

### Summary and conclusions

Despite the relatively limited amount of information on odour communication in humans, the studies reviewed in this chapter suggest that humans, like many other mammals, have the potential for communicating basic biological information via the smell medium. Anatomically, humans possess a variety of secretory and excretory systems which potentially provide a rich substrate for olfactory communication, although odours from only a few of these sources have received much attention. Behaviourally, humans can establish gender, at least probabilistically, from breath, axillary, and hand odours, although the degree to which such determinations depend upon quantitative factors is not known. Human infants appear to be able to detect odours from their mothers' breasts, and exhibit a preference for odours from their own mother to those from a strange mother. Similar in general respects to results from some animal studies, fluctuations in the intensity and pleasantness of vaginal odours have been noted during the menstrual cycle, as have fluctuations in perceptual sensitivity to several types of odorants. Interestingly, recent studies demonstrate that the cyclical changes in smell sensitivity during the menstrual cycle are not, as has been popularly believed, primarily dependent upon ovarian hormones, such as the oestrogens.

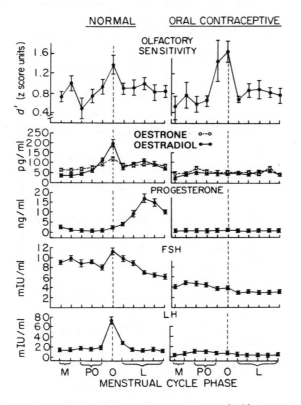

Fig. 20.9. Fluctuations in signal detection measures of olfactory sensitivity ($d'$) and five major reproductive hormones across the phases of 17 menstrual cycles from normally-cycling women and six menstrual cycles from women taking an oral contraceptive. Values indicate arithmetic means ± 1 SEM. Data normalized and collapsed into phase designations using the procedure of Doty (1979). M = menstrual phases, 1 and 2; PO = preovulatory phases 1, 2, and 3; O = ovulatory phase (designated as day of LH surge in normally-cycling group and as day 14 in oral contraceptive group; L = luteal phases 1–5. (Data from Doty, Snyder, Huggins, and Lowry (1981).)

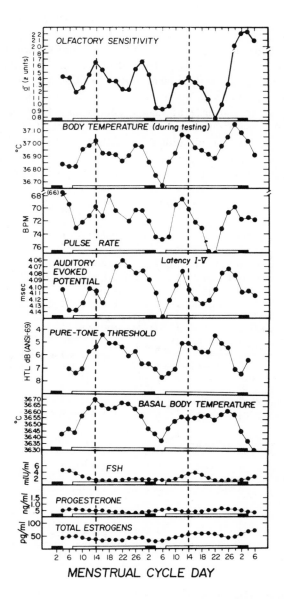

Fig. 20.10. Changes in nine variables across two consecutive menstrual cycles of a subject taking oral contraceptive medication. To diminish noise, a moving average with equal weights attached to three adjacent time points was applied to each series. Dark rectangles on the abscissae signify periods of menstrual bleeding, open rectangles days during which the oral contraceptive medication was taken. Testing took place from 9.30 a.m. to noon on each test day. Pure-tone thresholds are averaged across a wide range of frequencies. (From Doty *et al.* (1982*b*), with permission.)

## Acknowledgements

This work was supported by the National Institutes of Health Grant PO1 NS 16365-01. I thank Jeff Alberts, Jay Hall, John Labows, George Preti, M. Schleidt, and David Stevens for their helpful comments and suggestions on a previous version of the manuscript. The present review appeared as an article in the David Moulton Memorial volume of the *Chemical Senses* **6**, 351-76 (1981) and is dedicated to his memory.

## References

Alberts, J. R. (1981). Ontogeny of olfaction: reciprocal roles of sensation and behavior in the development of perception. In *The development of perception: psychobiological perspectives*, Vol. 1 (ed. R. N. Aslin, J. R. Alberts, and M. R. Petersen) pp. 322-57. Academic Press, New York.

Altmann, I. and Vinsel, A. M. (1977). Personal space: an analysis of E. T. Hall's proxemic framework. In *Human behavior and environment* (ed. I. Altmann and J. F. Wohlwill) Plenum Press, New York.

Aschoff, J., Fatranska, M., and Giedke, H. (1971). Human circadian rhythms in continuous darkness: entrainment by social cues. *Science, NY* **171**, 213-15.

Bird, S. and Gower, D. B. (1980). Measurement of 5-alpha-androst-16-en-3-one in human axillary secretions by radioimmunoassay. *J. Endocr.* **85**, 8P-9P.

Block, I. (1934). *Odoratus sexualis*. Panurge Press, New York.

Brooksbank, B. (1970). Labelling of steroids in axillary sweat after administration of $^3$H-$\Delta^5$-pregnenoline and $^{14}$C-progesterone to a healthy man. *Experientia* **26**, 1012-14.

Brooksbank, B. W. L., Brown, R., and Gustafsson, J. A. (1974). The detection of 5$\alpha$-androst-16-en-3$\alpha$-ol in human male axillary sweat. *Experientia* **30**, 864-5.

Cain, W. S. (1980). Chemosensation and cognition. In H. van der Starre (Editor), *Proc. 7th Int. Symp. Olfaction and Taste and 4th Congr. European Chemoreception Research Organization.* (ed. H. van der Starre) pp. 347-58. IRL Press, London.

Champion, R. H. (1970). Sweat glands. In *An introduction to the biology of the skin.* (ed. R. H. Champion, T. Gillman, A. J. Rook, and R. T. Sims) pp. 175-83. Davis, Philadelphia.

Claus, R. and Asling, W. (1976). Occurence of 5$\alpha$-androst-16-en-3-one, a boar pheromone, in man and its relationship to testosterone. *J. Endocr.* **68**, 483-4.

Comfort, A. (1971). Likelihood of human pheromones. *Nature, Lond.* **230**, 432-4.

Cowley, J. J., Johnson, A. L., and Brooksbank, B. W. L. (1977). The effect of two odorous compounds on performance in an assessment-of-people test. *Psychoneuroendocrinology* **2**, 159-72.

Curtis, R. F., Ballantine, J. A., Keverne, E. B., Bonsall, R. W., and Michael, R. P. (1971). Identification of primate sex pheromones and the properties of synthetic attractants. *Nature, Lond.* **232**, 396-8.

Davenport, W. (1965). Sexual patterns and their regulation in a society of the southwest Pacific. In *Sex and behavior* (ed. F. A. Beach) pp. 164-207. Wiley, New York.

Delaney, J., Lupton, M. J. and Toth, E. (1977). *The curse: a cultural history of menstruation.* New American Library, New York.

Doty, R. L. (1974). A cry for the liberation of the female rodent: courtship and copulation in Rodentia. *Psychol. Bull.* **81**, 159–72.

— (ed.) (1976). *Mammalian olfaction, reproductive processes, and behavior.* Academic Press, New York.

— (1977). A review of recent psychophysical studies examining the possibility of chemical communication of sex and reproductive state in humans. In *Chemical signals in vertebrates* (ed. D. Müller-Schwarze and M. M. Mozell) pp. 273–86. Plenum Press, New York.

— (1978). Gender and reproductive state correlates of taste perception in humans. In *Sex and behavior: status and prospectus* (ed. T. McGill, D. A. Dewsbury, and B. Sachs) pp. 337–62. Plenum Press, New York.

— (1979). A procedure for combining menstrual cycle data. *J. clin. Endocr. Metab.* **48**, 912–18.

— and Dunbar, I. A. (1974). Attraction of beagles to conspecific urine, vaginal and anal sac secretion odors. *Physiol. Behav.* **35**, 729–31.

— Ford, M., Preti, G., and Huggins, G. (1975). Human vaginal odors change in pleasantness and intensity during the menstrual cycle. *Science, NY* **190**, 1316–18.

— Green, P. A., Ram, C., and Yankell, S. L. (1982*a*). Communication of gender from human breath odors: relationship to perceived intensity and pleasantness. *Horm. Behav.* **16**, 13–22.

— Hall, J. W., Flickinger, G. L., and Sondheimer, S. J. (1982*b*). Cyclical changes in olfactory and auditory sensitivity during the menstrual cycle: no attenuation by oral contraceptive medication. In *Olfaction and endocrine regulation* (ed. W. Breipohl) pp. 35–42. IRL Press, London.

— Kligman, A., Leyden, J., and Orndorff, M. M. (1978). Communication of gender from human axillary odors: relationship to perceived intensity and hedonicity. *Behav. Biol.* **23**, 373–80.

— Snyder, P., Huggins, G., and Lowry, L. D. (1981). Endocrine, cardiovascular and psychological correlates of olfactory sensitivity changes during the human menstrual cycle. *J. comp. physiol. Psychol.* **95**, 45–60.

Ellis, H. (1936). *Studies in the psychology of sex.* Random House, New York.

Gower, D. B. (1972). 16-Unsaturated $C_{19}$ steroids—a review of their chemistry, biochemistry and possible physiological role. *J. steroid Biochem.* **3**, 45–103.

Graham, C. A. and McGrew, W. C. (1980). Menstrual synchrony in female undergraduates living on a coeducational campus. *Psychoneuroendocrinology* **5**, 245–52.

Hall, E. T. (1966). *The hidden dimension.* Doubleday, New York.

Hayden, G. F. (1980). Olfactory diagnosis in medicine. *Postgrad-Med.* **67**, 110–16.

Himes, N. E. (1970). *Medical history of contraception.* Schocken, New York.

Huggins, G. R. and Preti, G. (1976). Volatile constituents of human vaginal secretions. *Am. J. Obstet. Gynec.* **126**, 129–36.

Hurley, H. J. and Shelley, W. B. (1960). *The human apocrine sweat gland in health and disease.* Thomas, Springfield, Ill.

Jerger, J. and Hall, J. L. (1980). Effects of age and sex on auditory brainstem response. *Arch. Otolar.* **106**, 387–91.

Kiell, N. (1976). *Varieties of sexual experience.* International Universities Press, New York.

Kirk-Smith, M. D. and Booth, D. A. (1980). Effect of androstenone on choice of location in other's presence. In *Olfaction and taste VII* (ed. H. van der Starre) pp. 397–400. IRL Press, London.

—— Carroll, D., and Davis, P. (1978). Human social attitudes affected by androstenol. *Res. Commun. Psychol. Psychiat. Behav.* 3, 379–84.

Koelega, H. S. and Köster, E. P. (1974). Some experiments on sex differences in odor perception. *Ann. NY Acad. Sci.* 237, 234–46.

Labows, J. N. Jr (1979). Human odors—what they can tell us? *Perfum. Flav.* 4, 12–17.

— Preti, G., Hoelzle, E., Leyden, J., and Kligman, A. (1979). Steroid analysis of human apocrine secretion. *Steroids* 34, 249–58.

Leon, M. and Moltz, H. (1972). The development of the pheromonal bond in the albino rat. *Physiol. Behav.* 8, 683–6.

Liddell, K. (1976). Smell as a diagnostic marker. *Postgrad. med. J.* 52, 136–8.

McClintock, M. K. (1971). Menstrual synchrony and suppression. *Nature, Lond.* 229, 244–5.

MacFarlane, A. (1975). Olfaction in the development of social preferences in the human neonate. In Ciba Foundation Symposium 33: *The human neonate in parent-infant interaction*, pp. 103–77. Elsevier, Amsterdam.

McGuinness, D. (1976). Sex differences in the organization of perception and cognition. In *Exploring sex differences* (ed. B. Lloyd and J. Archer) pp. 123–56. Academic Press, London.

Mace, J. W., Goodman, S. I., Centerwall, W. R., and Chinnock, R. F. (1976). The child with an unusual odor. *Clin. Pediat.* 15, 57–62.

Michael, R. P., Bonsall, R. W., and Kutner, M. (1975). Volatile fatty acids 'copulins', in human vaginal secretions. *Psychoneuroendocrinology* 1, 153–63.

—— and Warner, P. (1974). Human vaginal secretions: volatile fatty acid content. *Science, NY* 186, 1217–19.

— and Keverne, E. B. (1968). Pheromones in the communication of sex status in primates. *Nature, Lond.* 218, 746–49.

Morris, N. and Udry, J. R. (1978). Pheromonal influences on human sexual behavior: an experimental search. *J. biosoc. Sci.* 10, 147–57.

Nicolaides, N. (1974). Skin lipids: their biochemical uniqueness. *Science, NY* 186, 19–26.

Orne, M. T. (1962). On the social psychology of the psychological experiment: with particular reference to demand characteristics and their implications. *Am. Psychol.* 17, 776–83.

Pope, C. (1928). The diagnostic art of smelling. *Am. Med.* 34, 651–3.

Preti, G. and Huggins, G. R. (1975). Cyclical changes in volatile acidic metabolites of human vaginal secretions and their relation to ovulation. *J. chem. Ecol.* 1, 361–76.

—— (1978). Organic constituents of vaginal secretions. In *The human vagina* (ed. E. S. E. Hafez and T. N. Evans) pp. 151–66. Elsevier/North-Holland, Amsterdam.

—— and Silverberg, G. D. (1979). Alterations in the organic compounds of vaginal secretions caused by sexual arousal. *Fert. Steril.* 32, 47–54.

Quadagno, D. M., Shubeita, H. E., Deck, J., and Francoeur, D. (1981). Influence of male social contacts, exercise and all-female living conditions on the menstrual cycle. *Psychoneuroendocrinology* 6, 239–44.

Rosenblatt, J. S. (1972). Learning in newborn kittens. *Scient. Am.* 227, 18–25.

Rosenthal, R. and Rosnow, R. L. (eds) (1969). *Artifact in behavioral research*. Academic Press, New York.

Rusak, B. and Zucker, I. (1975). Biological rhythms and animal behavior. *A. Rev. Psychol.* **26**, 137–71.

Russell, M. J. (1976). Human olfactory communication. *Nature, Lond.* **260**, 520–2.

— Switz, G. M., and Thompson, K. (1980). Olfactory influences on the human menstrual cycle. *Pharmacol. Biochem. Behav.* **13**, 737–8.

Sastry, S. D., Buck, K. T., Janák, J., Dressler, M., and Preti, G. (1980). Volatiles emitted by humans. In *Biochemical applications of mass spectrometry, first supplementary volume* (ed. G. R. Waller and O. C. Dermer) Wiley, New York.

Schleidt, M. (1980). Personal odor and nonverbal communication. *Ethol. Sociobiol.* **1**, 225–31.

— Hold, B., and Attili, G. (1981). A cross-cultural study on the attitude towards personal odours. *J. chem. Ecol.* **7**, 19–31.

Shuter, R. (1968). *The psychology of music*. Methuen, London.

Teicher, M. H. and Blass, E. M. (1976). Suckling in newborn rats: eliminated by nipple lavage, reinstated by pup saliva. *Science, NY* **193**, 422–4.

Tonzetich, J., Preti, G., and Huggins, G. R. (1978). Changes in concentration of volatile sulfur compounds of mouth air during the menstrual cycle. *J. int. med. Res.* **6**, 245–54.

Wallace, P. (1977). Individual discrimination of humans by odor. *Physiol. Behav.* **19**, 577–9.

Watson, O. M. (1970). *Proxemic behaviour*. Mouton, The Hague.

# Author index

# Index of common names

Page numbers in **bold type** refer to Volume 1, those in roman type to Volume 2.

# Index of Linnean names

Page numbers in bold type refer to Volume 1, those in roman type refer to Volume 2.

# Index of odour sources